MICROW

An Introduction to Microwave Theory and Techniques

THIRD EDITION

Pergamon Titles of Related Interest

CHINESE SOCIETY FOR MEASURING
Electromagnetic Metrology

DEBENHAM
Microprocessors

DUMMER
Electronic Inventions & Discoveries

GANDHI
Microwave Engineering and Application, 2nd edition

HAMMOND
Electromagnetism for Engineers

HINDMARSH
Electrical Machines & Their Applications

INFOTECH
Microelectronics

MIYA
Electromagnetics in Advanced Electromagnetic Materials & Solids

MURPHY & TURNBULL
Power Electronic Control of AC Motors

Pergamon Related Journals (*free specimen copy gladly sent on request*)

Electrochimica Acta

Microelectronics and Reliability

Neural Networks

Progress in Quantum Electronics

Solid State Communications

Solid State Electronics

MICROWAVES

An Introduction to Microwave Theory and Techniques

THIRD EDITION

A. J. BADEN FULLER

University of Leicester

PERGAMON PRESS

Member of Maxwell Macmillan Pergamon Publishing Corporation

OXFORD · NEW YORK · BEIJING · FRANKFURT

SÃO PAULO · SYDNEY · TOKYO · TORONTO

U.K.	Pergamon Press plc, Headington Hill Hall, Oxford OX3 0BW, England
U.S.A.	Pergamon Press, Inc., Maxwell House, Fairview Park, Elmsford, New York 10523, U.S.A.
PEOPLE'S REPUBLIC OF CHINA	Pergamon Press, Room 4037, Qianmen Hotel, Beijing, People's Republic of China
FEDERAL REPUBLIC OF GERMANY	Pergamon Press GmbH, Hammerweg 6, D-6242 Kronberg, Federal Republic of Germany
BRAZIL	Pergamon Editora Ltda, Rua Eça de Queiros, 346, CEP 04011, Paraiso, Sao Paulo, Brazil
AUSTRALIA	Pergamon Press (Australia) Pty Ltd, PO Box 544, Potts Point, NSW 2011, Australia
JAPAN	Pergamon Press, 5th Floor, Matsuoka Central Building, 1-7-1 Nishishinjuku, Shinjuku-ku, Tokyo 160, Japan
CANADA	Pergamon Press Canada Ltd, Suite No 271, 253 College Street, Toronto, Ontario, Canada M5T 1R5

Copyright © 1990 A J Baden Fuller

All Rights Reserved. No part of this publication may be reproduced, stored in a retrieval system or transmitted in any form or by any means: electronic, electrostatic, magnetic tape, mechanical, photocopying, recording or otherwise, without permission in writing from the publisher

First edition 1969

Reprinted 1978

Second edition 1979

Reprinted 1985, 1988

Third edition 1990

Library of Congress Cataloging in Publication Data

Baden Fuller, A. J.
Microwaves : an introduction to microwave theory and techniques A. J. Baden Fuller. — 3rd ed.
p. cm.
1. Microwaves. 2. Microwaves. Devices. I. Title.
TK7876.B3 1989 621.381'3—dc20 89-26630

British Library Cataloguing in Publication Data

Baden-Fuller, A. J. (Arthur John)
Microwaves. — 3rd ed.
1. Microwaves
I. Title
621.381'3

ISBN 0-08-040494-4 Hardcover
ISBN 0-08-040493-6 Flexicover

Printed in Great Britain by B.P.C.C. Wheatons Ltd, Exeter

Contents

Preface to the Third Edition

IT is now over twenty years since the first edition of *Microwaves* was prepared. The shape of microwave equipment has changed and microwaves has become a mature science. Then it was common to see microwave laboratories full of waveguide and microwave systems used waveguide as a major transmission medium. Now semiconductor devices and circuits operate successfully in the microwave frequency range and miniaturization may be applied to many systems. Where twenty years ago, low power microwave systems would be constructed of waveguide, they are now made in microstrip or in integrated circuit form. The use of waveguide tends to be confined to high power systems or where the low loss properties of waveguide are needed. However, any microcircuit is enclosed in a box, so that a microwave engineer without a good working knowledge of waveguide theory or waveguide modes is at a disadvantage.

In the second edition, that part of the chapter on Oscillators and Amplifiers dealing with semiconductor devices was rewritten and a completely new chapter on Stripline was added. In this revision, the opportunity has been taken to completely restructure parts of the book so that microwave circuit theory and microstrip devices become an integrated part of the whole. At the same time, waveguide theory is still an important part of microwaves and no waveguide theory has been omitted. Because of the growing importance of microstrip and microwave integrated circuit design, a new chapter about Microwave Circuits has been added to the book which includes an introduction into *s*-parameter theory. Other improvements have been made throughout the text. The order of the material in the descriptive chapters on components and devices has been reorganized and includes both microstrip and waveguide devices. At the same time, a description of all the ferrite nonreciprocal and control devices has been gathered together into one chapter. Consequently, the first part of the book giving a theoretical development of microwave theory now consists of Chapters 2 to 9 and the descriptive part about microwave components and measurements consists of Chapters 10 to 14. The main aims and structure of the book as outlined in the original preface remain unchanged.

Preface to the First Edition

TWENTY years ago the science of microwave propagation was just emerging from the secrecy of war-time research and use in radar systems. Twenty years has seen the growth of a sizeable microwave industry. Today the general use of radar systems for navigation, together with the extension of radio communication links and satellite communications into the microwave frequency range, has meant a wider need for engineers and physicists with a knowledge of those aspects of electromagnetic theory that are applicable at microwave frequencies. Most electrical engineering degree courses include the study of microwave theory. This book seeks to provide an introduction to this subject, suitable for use as a textbook by undergraduate or senior technical college students. The contents fall into two main parts.

Initially, there is an introduction followed by one chapter on general transmission line equations. The first main part, Chapters 2 to 8, consists of a theoretical development of electromagnetic propagation of guided waves starting from Maxwell's equations and the material properties. Attention has been focused on those properties of wave propagation which are dependent on microwave scale: i.e. the wavelength is of the same order as the dimensions of the body handling the electromagnetic wave. No attempt has been made to discuss those topics where the scale of operation is such that the principles of optics are applicable or conversely using microcircuits, high-frequency circuit techniques are applicable. The treatment has been confined to those topics where the application of field theory is appropriate.

The second main part, Chapters 9 to 12, consists of a descriptive treatment of microwave components and measurements. A student, faced with the need to operate microwave equipment in advance of learning the theory, may well dip into these final chapters before studying the earlier part of the book, and he will find that they are largely self-supporting. Alternatively, for the student following a course in microwave theory, these final chapters provide a useful background to the theory.

The book assumes a knowledge of vector analysis equations. Where appropriate, i.e. vector analysis and Bessel functions, the mathematical terms are defined and all the necessary properties are quoted but not derived. Each chapter concludes with a summary and problems are given at the end of the first eight chapters. No worked examples are given in the main text of the book because electromagnetic theory does not lend itself to such treatment, but some worked solutions to selected problems are given at the end.

It is difficult to thank all the many people from whom I have learned and I hope that lack of acknowledgement will not be taken to imply lack of gratitude. I should like to thank Professor P. Hammond for his encouragement which inspired me to write this book and to Dr. A. G. Bailey for reading the manuscript and making helpful suggestions. My thanks are due to Professor G. D. S. MacLellan, Head of the Engineering Department at the University of Leicester, for making the facilities of the department available to me in the preparation of the manuscript.

Introduction

MICROWAVES is the name given to the electromagnetic waves arising as radiation from electrical disturbances at high frequencies. At low frequencies, the radiation aspects of electromagnetic power distribution are negligible and it is only necessary to consider electric charges, stored or flowing as currents, and potential difference. As the frequency of operation increases, however, radiation becomes of more importance. The relative importance of radiation depends on the size of the circuit or system under consideration. Any system of electric charges gives rise to electric and magnetic fields in the surrounding space. At low frequencies with circuit theory, the effects of these fields are generally ignored. At higher frequencies, however, the effect of these fields becomes more pronounced. This is seen initially by the introduction of "stray" capacitance into circuit theory. At high frequencies, even a short length of wire acts as a radiation element, dissipating its electrical signals into surrounding space. The electromagnetic fields become the dominant factor in a study of electrical theory at high frequencies.

As will be shown, the characteristic wavelength of radiation is related to the frequency of the electrical signals by $\lambda = c/f$, where f is the frequency and c is the speed of light. It is only necessary to consider the radiation contribution to the understanding of an electrical system if the dimensions of the system are of the same order or larger than the characteristic wavelength of the electrical signals handled by the system. This means that for medium wave radio broadcasting with a wavelength of a few hundred metres, where the electrical signal goes from the transmitter to the receiver in the form of electromagnetic radiation, the radio receiver can be designed entirely by consideration of electrical currents flowing in wires and ignoring any electromagnetic radiation which is generated by the circuits.

Microwave techniques may be considered to cover those applications of electrical technology where the characteristic wavelength is smaller than the dimensions of the system or circuit and yet where it is not so small that only ray optical techniques need be considered. Microwaves are normally considered to embrace the frequency range 10^9 to 10^{12} Hz or a characteristic wavelength range of 30 cm to 0.3 mm. At these wavelengths, the components of conventional electronic circuits tend to behave like individual antenna, dissipating their electrical signals as radiation. New techniques are necessary to handle electrical signals in the microwave frequency range, leading to new techniques of analysis. Although the microwave frequency range given above is accepted by common usage as the region where these specialized techniques are used most frequently, the relationship of size to characteristic wavelength is the true guideline determining when microwave techniques or analysis are applicable to any particular system. For example, the power system designer has a radiation problem with a.c. transmission lines a few thousand miles long, conversely, microcircuits are designed using conventional circuit techniques to operate at microwave frequencies.

The study of electromagnetic radiation is an exact science because it can be represented exactly by mathematical expressions. The ease with which the mathematical analysis can be performed depends on the complexity of the electromagnetic fields which in turn is determined by the shape of the constraining boundary. The elementary mathematical theory contained in this book is confined to a study of the fields due to simple boundary shapes. Although these simple shapes may not always arise in practice, many practical situations approximate to simple shapes and the simple theory gives good results for many applications.

The elementary mathematical theory is given in Chapters 1 to 9. The behaviour of electromagnetic radiation under various conditions is determined by mathematical analysis. To complement the theory of the first nine chapters, Chapters 10 to 14 contain a brief non-mathematical outline of practical microwaves. They contain a descriptive explanation of microwave components and measurements.

The frequency bands in the microwave region have been given letter codes. The band designations for both radio and microwave frequencies are given in the table.

STANDARD FREQUENCY BAND DESIGNATIONS

Band	Frequency GHz	Characteristic wavelength
h.f.	0.003–0.03	10–100 m
v.h.f.	0.03–0.3	1–10 m
u.h.f.	0.3–1	0.3–1 m
L	1–2	150–300 mm
S	2–4	75–150 mm
C	4–8	37.5–75 mm
X	8–12	25–37.5 mm
Ku	12–18	17.5–25 mm
K	18–27	11–17.5 mm
Ka	27–40	7.5–11 mm
Millimetre	40–300	1–7.5 mm

A new rationalized system of letter codes for the entire broadcast frequency range has been proposed:

New Standard Frequency Band Designations

Band	Frequency GHz	Characteristic wavelength
A	0–0.25	1.2 m–∞
B	0.25–0.5	0.6–1.2 m
C	0.5–1	0.3–0.6 m
D	1–2	150–300 mm
E	2–3	100–150 mm
F	3–4	75–100 mm
G	4–6	50–75 mm
H	6–8	37–50 mm
I	8–10	30–37 mm
J	10–20	15–30 mm
K	20–40	7.5–15 mm
L	40–60	5–7.5 mm
M	60–100	3–5 mm

However, it is expected that the old system will still be in use for a long time.

Microwaves possess certain useful characteristics, one of the most important being that microwave wavelengths are the same size as any structure used to guide or enclose them. Microwave pulses can be very short so that they can be used for distance or time measurement. Also this makes them compatible with high speed computers. The high frequency of microwaves means that very large bandwidths are available for communication links. Microwave radiation penetrates fog and clouds, travels in straight lines and gives distinct shadows and reflections enabling it to be used for distance and direction measurement and in radar systems. Microwaves are necessary for communication with satellites because they can pass through the ionosphere which reflects lower frequency radio waves. Microwave power is absorbed by water or any material containing water so that microwaves can be used for heating and drying. Many atomic and molecular resonances occur at microwave frequencies so that they are a necessary part of some scientific measurements. Certain resonances can be used to make stable atomic clocks. All these properties mean that microwaves are becoming more and more widely used. Some applications are given in the following paragraphs.

Broadcasting. Originally radio broadcasting and television used frequencies below the microwave range. However, increasing congestion of the radio spectrum made reception difficult for some listeners. There are no frequencies available for any large increase in broadcasting at radio frequencies, so that any further increase must occur at higher frequencies, which will be in the microwave region. 12 GHz is being used either for local television stations or for satellite television broadcasting. The domestic consumer has a microwave receiver on the roof as part of a small aerial and a radio frequency signal is transmitted along the aerial cable to the television set. Frequencies in the range 28–45 GHz are also being investigated for local television broadcasting.

Communication. Increased bandwidth for communication channels requires higher carrier frequencies. Line of sight radio relay systems have been operating for a number of years. The microwave system consists of tower mounted directional aerials which receive signals, amplify them and transmit them on to the next tower in the chain. Post Office towers have enabled such microwave relay systems to enter into the centre of many big cities. In many developed countries, all the frequencies available for such links have been fully utilized. However, such a system is ideal for underdeveloped areas or difficult terrain since the relay towers and equipment can be positioned by helicopter and powered by solar cells, small generators or batteries. Laying landline communications is difficult across mountainous or similarly inhospitable country and it is simpler to set up a microwave relay system. Because the ionosphere is opaque to lower frequencies, microwave frequencies have to be used for satellite communications and for communications with satellites. The microwave communication channel has a very large bandwidth and can accommodate thousands of telephone conversations or dozens of television channels at once.

Radar is the traditional use of microwaves. It started at about the beginning of the Second World War. The name is derived from the initial letters of RAdio Detection And Ranging. The simplest form of radar is the pulse radar giving a plan position indication (ppi); it measures the time for an echo to return, operates by echo sounding with a narrow beam like a searchlight, and is used for navigation. The CW (carrier wave) or doppler radar gives a velocity indication; it is used in military applications because it is more difficult for an enemy to jam. The doppler radar also has many industrial and consumer uses; it is used in industrial controls for flow or velocity measurement. It also is used for motion detection. As an intruder alarm, it is difficult to eliminate false alarms, such as those from a cat or from curtains moving in a breeze, but it is very suitable for other applications such as controlling a door opener, or operating temporary traffic signals. It is used for the police speed radar and it is hoped to develop it into an anticollision device for vehicles. A form of radar can be used to detect hidden objects; it is much more sophisticated in application than the simple metal detector since it can locate non-metallic objects such as plastic gas or water pipes. Microwave radiometry, which uses microwave radiation in the same way that photography uses light, can give useful information about the object being observed such as the moisture content of soils and vegetation.

Microwave heating. The rate of microwave power absorption in most materials is proportional to its water content. This property can be used to provide microwave heating. Because the microwave signal penetrates most non-conductors, microwave power provides a most efficient means of applying heat uniformly throughout a body. Because the heat does not have to be conducted through but is generated inside the body, microwave heating reduces the time needed for heating a body to a uniform temperature. The rate of heating usually depends on the water content, so that microwave heating is a most efficient method of drying. Microwave ovens are in use in many homes and catering establishments and microwave heating is used in many process industries for heating, drying, curing or sterilizing.

Moisture measurement. Microwave absorption by water also means that moisture content measurement by microwaves is possible. The attenuation of a microwave signal in passing through the specimen is measured.

Microwave power transmission has been advocated for electric power distribution since it can be used directly for heating and for exciting fluorescent lights. The possibility is being actively investigated of satellite power generation with microwave transmission to earth. The satellite is powered with solar cells and the microwave power generating valves operate in high vacuum without any glass envelope. The microwave power is beamed to earth where it is collected and rectified. The system will be expensive in capital cost but uses a free non-expendable energy supply.

Computers. As computers work at faster rates, high frequency circuits are required so as not to degrade the pulse shape. Application of transmission-line and microwave techniques in the design of computer modules will become necessary.

Clocks. Microwave clocks measure the frequency of some particular atomic transitions and have an accuracy of about one second in a million years. The second is defined as: the duration of 9 192 631 770 periods of the radiation corresponding to the transition between the two hyperfine levels of the ground state of the caesium 133 atom. This corresponds to a frequency of 9192.63177 MHz—right in the middle of the microwave range.

Biological hazards. Microwaves are potentially hazardous because of their heating effect. The effect may not be felt until damage has already been done because the heating may be internal whereas our body is designed more to warn us about externally applied heat. Such heating is especially dangerous where the excess heat is not dissipated easily as in the case of the lens of the eye, and there the most likely effect of excessive microwave exposure is the formation of cataracts. From heat balance considerations of standard man in standard conditions, $100\,\mathrm{W/m^2}$ ($10\,\mathrm{mW/cm^2}$) is considered to be the safe upper limit even during infinite exposure because thermoregulatory systems compensate for any power absorption. A power level of $10\,\mathrm{W/m^2}$ can be considered to give no heating effect even under adverse conditions of ambient temperature and humidity. However there is also some evidence of a non-thermal effect through the nervous system, although the effect is harder to prove and controversy still surrounds it. It is claimed that exposure over a period of years to power levels greater than $2\,\mathrm{W/m^2}$ can lead to nervous system disturbances, although occupational exposure of healthy adults to this power level seems to have no adverse effects. However the probable safe level for continuous exposure of the general population ought to be even lower. In the U.S.A. and many western countries, the only recommended limit is 100 W/m^2 for a safe working environment, based on the proved heating effect, but it is probably safer to aim at a limit lower than this. Poland has a general population limit of 1 W/m^2 for intermittent exposure and 0.1 W/m^2 for continuous exposure. In the U.S.S.R. the general population limit is 0.01 W/m^2 but this is probably too restrictive on the use of microwaves. Another problem is that low power levels have caused other effects in some people such as a ringing in the ears at the pulse repetition frequency of a nearby radar set, although even this effect has been traced to heating in the mechanism of the ear. The difficulty is that evidence for non-thermal effects is disputed and decisions will have to wait until more is known.

1

Transmission Lines

1.1. Transmission Line

Although this book is mainly concerned with the properties of electromagnetic transmission systems for use at such high frequencies that the electrical signals can only be handled in the form of electromagnetic radiation, it is still necessary to understand the general properties of transmission lines for use at any frequency before proceeding to a consideration of microwave transmission lines in particular. This chapter contains a summary of transmission line theory that is relevant to microwave theory. At low frequencies, the properties of an electrical circuit may be specified in terms of currents and potential differences. The consequential electromagnetic fields are often so small that they may be ignored. At microwave frequencies, the field quantities become dominant and the current and potential difference are difficult to measure.

In this chapter the properties of transmission lines are derived in terms of relatively low-frequency currents and potential differences. It is found that there are properties of the transmission line waves, other than currents and potential differences, that may be measured and these properties are then applied to microwave transmission lines. A transmission line consists of any system of conductors that can be used to transmit electrical energy between two or more points. When a voltage generator is connected to the input of a long transmission line, the potential difference on the line cannot rise instantaneously to that of the generator. Time is needed for the transfer of energy corresponding to the potential difference between the lines. An instantaneous change of potential difference along the whole length of line is also deemed impossible by the special theory of relativity. No signal can be transmitted at a speed greater than that of light. Hence time is taken for the charge to travel along a transmission line. It also takes time for any information, usually in the form of electrical signals, to travel along a transmission line. For an a.c. signal, there appears to be a continual flow of energy into a transmission line under steady-state conditions and the signal on the line at any distance from the source is out of phase with that of the source. We start by considering the simplest line which is the two-conductor transmission line.

1.2. Two-conductor Line

A two-wire transmission line of infinite length is shown in Fig. 1.1. The only consideration that needs to be specified in relation to this line is that it maintains a constant cross-section throughout its length. At any frequency, a potential difference applied to the line causes some current to flow into

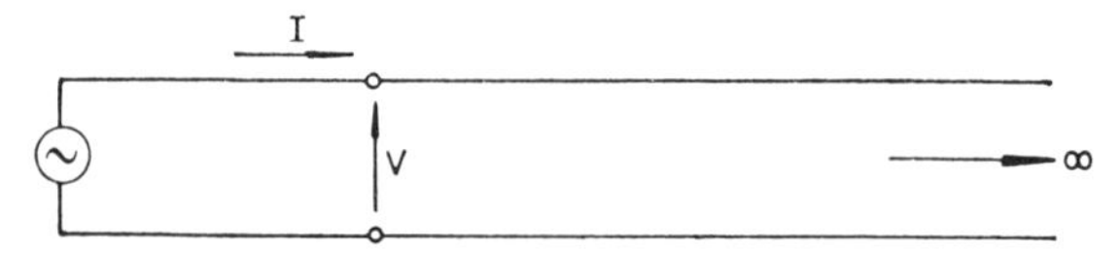

FIG. 1.1. Two-wire transmission line of infinite length.

the line, because even if there is no leakage conductance between the conductors, there is capacitance between them which provides a path for an alternating current. The current flow gives the line an equivalent impedance which is called the *characteristic impedance* of the line. It is given by the relationship

$$Z_0 = \frac{V}{I} \tag{1.1}$$

where Z_0, V and I are all phasor quantities. For this two-conductor line which has no losses, it is found that the current and potential difference between the conductors are in phase and the characteristic impedance is a resistive quantity. It will be shown in section 1.4 that this is true for all lossless transmission lines.

If a short transmission line is terminated by an infinite line, it is the same as an infinite line and its input impedance is the characteristic impedance. The infinite line may be replaced by its characteristic impedance without disturbing the electrical conditions on the short line so that a short transmission line terminated in its characteristic impedance behaves like an infinite line. This is shown diagrammatically in Fig. 1.2.

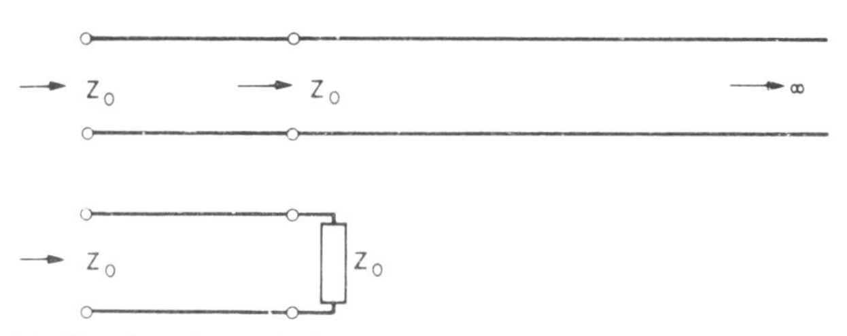

FIG. 1.2. Showing the equivalence between an infinite line and a short line terminated in its characteristic impedance.

Any two-port circuit can be replaced by an equivalent-T circuit and the equivalent-T of a finite transmission line terminated in its characteristic impedance is shown in Fig. 1.3. Its input impedance is also the characteristic impedance, so that the input impedance of the equivalent-T is given by

$$Z_0 = Z_1 + \frac{Z_2(Z_1 + Z_0)}{Z_1 + Z_2 + Z_0} \tag{1.2}$$

whence

$$Z_0^2 = Z_1^2 + 2Z_1Z_2 \tag{1.3}$$

If the characteristic impedance of the line is not known, the line cannot be correctly terminated, but its input impedance can be measured with the end of the line either open or short circuited. From the equivalent-T circuit of Fig. 1.3 it can be shown that the open- and short-circuit impedances respectively of the line are

$$Z_{oc} = Z_1 + Z_2 \tag{1.4}$$

$$Z_{sc} = Z_1 + \frac{Z_1Z_2}{Z_1 + Z_2} = \frac{Z_1^2 + 2Z_1Z_2}{Z_1 + Z_2} \tag{1.5}$$

and substitution from eqns. (1.3) and (1.4) into eqn. (1.5) gives

$$Z_0 = \sqrt{(Z_{sc}Z_{oc})} \tag{1.6}$$

The characteristic impedance of a line is the geometric mean of the open- and short-circuit impedances.

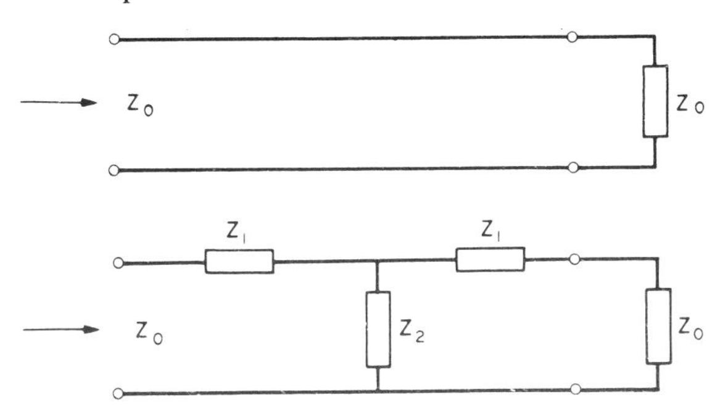

FIG. 1.3. The equivalent-T circuit of a short transmission line.

1.3. Transmission Line Equation

Consider the infinite two-wire transmission line. The current flowing in the line and the potential difference between the two wires of the line are functions of distance along the line, which is defined as the dimension z. The line has an effective series impedance Z ohm per unit length and an effective shunt admittance Y siemens per unit length. The effect of a short length of line δz is shown in Fig. 1.4. Hence we see that

$$\delta V = -IZ\delta z \tag{1.7}$$

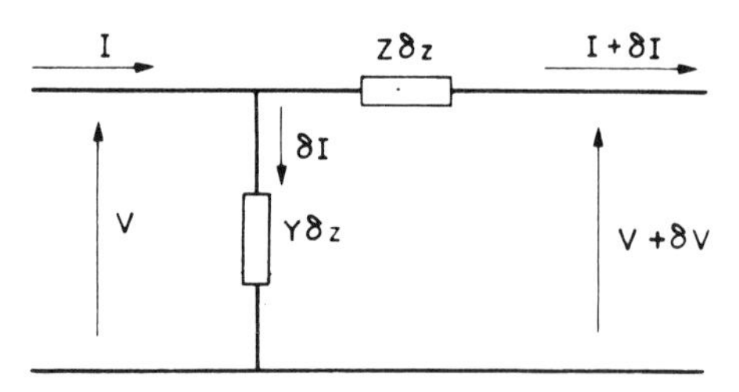

FIG. 1.4. Equivalent circuit of a short element, length δz, of an infinite transmission line.

and

$$\delta I = -VY\delta z \tag{1.8}$$

where the negative signs show that the line voltage and current decrease with increasing z. In the limit of small δz, eqns. (1.7) and (1.8) become

$$\frac{dV}{dz} = ZI \tag{1.9}$$

$$\frac{dI}{dz} = -YV \tag{1.10}$$

If eqns. (1.9) and (1.10) are both differentiated and substituted into the original equations, we obtain

$$\frac{d^2V}{dz^2} = \gamma^2 V \tag{1.11}$$

and

$$\frac{d^2I}{dz^2} = \gamma^2 I \tag{1.12}$$

where γ is defined by

$$\gamma^2 = ZY \tag{1.13}$$

These are linear second order equations having two independent solutions $\exp -\gamma z$ and $\exp \gamma z$. The general solution is obtained by forming a combination of these two solutions with arbitrary constants A and B so that a complete solution of eqn. (1.11) is

$$V = A \exp - \gamma z + B \exp \gamma z \tag{1.14}$$

Substitution into eqn. (1.9) gives the expression for the current

$$I = \frac{\gamma}{Z}(A \exp - \gamma z - B \exp \gamma z) \tag{1.15}$$

In general terms, the impedance and admittance of the line are complex, hence eqn. (1.13) shows that γ is complex. Any complex number can be split into its real and imaginary parts so that

$$\gamma = \alpha + j\beta \tag{1.16}$$

If the harmonic time dependence $\exp j\omega t$ is introduced into eqn. (1.14), the expression for the line potential becomes

$$V = A\exp[-\alpha z + j(\omega t - \beta z)] + B\exp[\alpha z + j(\omega t + \beta z)] \tag{1.17}$$

Inspection of eqn. (1.17) shows that it represents two waves travelling along the line. The first term represents a wave travelling in the positive z-direction and the second represents one travelling in the opposite direction. The factor $\exp j(\omega t - \beta z)$ is a function which remains unchanged for an observer travelling with the speed of the wave. Hence the speed is given by differentiating

$$\omega t - \beta z = \text{constant}$$

Therefore,

$$v = \frac{dz}{dt} = \frac{\omega}{\beta} \tag{1.18}$$

and this is called the *phase velocity* of the wave. The second term in eqn. (1.17) has a phase velocity that is the negative of that given in eqn. (1.18). That is, it represents a wave travelling with the same speed but in the opposite direction. The first part of each exponential expression in eqn. (1.17) represents an exponential decay in amplitude of the wave as the wave travels along the line. Hence α is called the *attenuation constant* of the wave. β is called the *phase constant* and γ is the *propagation constant*. The phase constant is related to the wavelength by the expression

$$\beta = \frac{2\pi}{\lambda} \tag{1.19}$$

[See section 2.10 for a detailed derivation of eqn. (1.19).]

1.4. Line Constants

For a low-frequency two-wire transmission line, the primary line constants are impedance properties that can be measured. They are:

Series resistance of the line	R ohm/metre (Ω/m)
Leakage conductance of the line	G siemens/metre (S/m)
Series inductance of the line	L henry/metre (H/m)
Capacitance of the line	C farad/metre (F/m)

Hence

$$Z = R + j\omega L \tag{1.20}$$

$$Y = G + j\omega C \tag{1.21}$$

and the propagation constant is given by

$$\begin{aligned}\gamma^2 &= (R + j\omega L)(G + j\omega C)\\ &= (RG - \omega^2 LC) + j\omega(GL + RC)\end{aligned} \tag{1.22}$$

In most practical transmission lines the series resistance and shunt leakage conductance are small so that the losses on the line are small and

$$\frac{R}{\omega L} \ll 1 \quad \text{and} \quad \frac{G}{\omega C} \ll 1$$

The expression for γ can be factorized to give

$$\gamma = j\omega\sqrt{(LC)}\left[1 - \frac{RG}{\omega^2 LC} - j\left(\frac{G}{\omega C} + \frac{R}{\omega L}\right)\right]^{1/2} \tag{1.23}$$

The centre of the three terms in eqn. (1.23) is very small and may be neglected. The last term is also small so that the equation may be expanded by Taylor's theorem to give

$$\gamma = j\omega\sqrt{(LC)}\left[1 - j\left(\frac{G}{2\omega C} + \frac{R}{2\omega L}\right)\right] \tag{1.24}$$

Therefore by comparison with eqn. (1.16)

$$\beta = \omega\sqrt{(LC)} \tag{1.25}$$

$$\alpha = \frac{G}{2}\sqrt{\left(\frac{L}{C}\right)} + \frac{R}{2}\sqrt{\left(\frac{C}{L}\right)} \tag{1.26}$$

Referring back to the equivalent-T circuit of the transmission line shown in Fig. 1.3, the elements for a line of length l are

$$Z_1 = \tfrac{1}{2}(R + j\omega L)l$$

$$Z_2 = \frac{1}{(G + j\omega C)l}$$

and substitution of these values into eqn. (1.3) gives

$$Z_0^2 = \frac{(R + j\omega L)^2 l^2}{4} + \frac{(R + j\omega L)l}{(G + j\omega C)l} \tag{1.27}$$

But the substitution of the primary line constants for the elements of the equivalent-T is only strictly valid for an infinitely short line. Hence $l \to 0$ and eqn. (1.27) becomes

$$Z_0 = \sqrt{\left(\frac{R + j\omega L}{G + j\omega C}\right)} = \sqrt{\left[\frac{L}{C}\left(\frac{1 - jR/\omega L}{1 - jG/\omega C}\right)\right]} \tag{1.28}$$

Therefore the low loss approximation is

$$Z_0 = \sqrt{\left(\frac{L}{C}\right)} \tag{1.29}$$

1.5. Lossless Transmission Line

For many practical purposes the losses of a transmission line may be neglected; for example, the losses of a low loss line are negligible when the length is of the order met in the laboratory. For the lossless line, eqns. (1.14) and (1.15) become

$$V = A \exp - j\beta z + B \exp j\beta z \tag{1.30}$$

$$I = \frac{1}{Z_0}(A \exp - j\beta z - B \exp j\beta z) \tag{1.31}$$

where the time dependence of $\exp j\omega t$ has been assumed for all the voltages and currents. Comparison between eqns. (1.15) and (1.31) and substitution from eqn. (1.13) shows that

$$Z_0 = \frac{Z}{\gamma} = \sqrt{\left(\frac{Z}{Y}\right)}$$

and this relationship will be found to be self-consistent with the lossless conditions $R = 0$ and $\alpha = 0$ and with eqns. (1.16), (1.20), (1.25) and (1.29).

The characteristic impedance is purely resistive for a lossless line. A mathematical proof will now be given of the effect of terminating a lossless transmission line with its characteristic impedance. Consider a line of length l terminated in a resistor of value Z_0. The voltage and current at the termination are given by substitution into eqns. (1.30) and (1.31),

$$V_l = A \exp - j\beta l + B \exp j\beta l \tag{1.32}$$

$$I_l = \frac{1}{Z_0}(A \exp - j\beta l - B \exp j\beta l) \tag{1.33}$$

and the ratio of these is the terminating resistance so that

$$\frac{V_l}{I_l} = Z_0 = Z_0 \frac{A \exp - j\beta l + B \exp j\beta l}{A \exp - j\beta l - B \exp j\beta l} \tag{1.34}$$

Solving eqn. (1.34) for the ratio B/A gives the result

$$\frac{B}{A} = -\frac{B}{A}$$

which can only be true if $B = 0$. This means that there is no wave travelling in the reverse direction. There is no reflected wave on a transmission line terminated in its characteristic impedance. A line with no reflected wave on it is called a *matched line*. Any line terminated in its characteristic impedance is matched. Similarly an infinitely long line produces no reflected wave so that it presents an impedance equal to its characteristic impedance everywhere along its length. This is a mathematical proof of the conditions described in section 1.1.

1.6. Voltage Standing Wave Ratio

If a line is terminated in some arbitrary impedance other than its characteristic impedance, it is necessary to have both forward and reverse wave components of the line voltage in order to satisfy the boundary conditions at the ends of the line. If eqn. (1.17) is simplified to give the conditions for a lossless line, and if it is written in the sinusoidal rather than in the exponential form, it gives

$$V = A\sin(\omega t - \beta z) + B\sin(\omega t + \beta z) \tag{1.35}$$

where A is the amplitude of the forward wave and B is the amplitude of the reflected wave. Expansion of eqn. (1.35) gives an expression for the amplitude of this voltage

$$\begin{aligned} V &= \sqrt{[(A + B)^2\cos^2\beta z + (A - B)^2\sin^2\beta z]}\sin(\omega t + \phi) \\ &= \sqrt{(A^2 + B^2 + 2AB\cos 2\beta z)}\sin(\omega t + \phi) \end{aligned}$$

which means that the amplitude of the voltage oscillates between the values

$$V_{\max} = A + B$$

$$V_{\min} = A - B$$

with a wavelength of $\frac{1}{2}\lambda$ where λ is the wavelength of both the forward and reverse waves.

The *voltage standing wave ratio* (*VSWR*) is a measure of the relative amplitude of the forward and reverse waves. It is defined as the ratio of the maximum to the minimum values of the voltage in the standing wave, which is always greater than one, or as the inverse of this ratio which is always less than one. There is never any doubt as to which definition is being used as the VSWR is always less than or greater than one. In this book, VSWR will always be greater than one. Hence the VSWR is given by

$$S = \frac{V_{max}}{V_{min}} = \frac{A+B}{A-B} = \frac{1+B/A}{1-B/A} = \frac{1+|\rho|}{1-|\rho|} \tag{1.36}$$

where $|\rho|$ is defined as

$$|\rho| = \frac{B}{A}$$

which is the ratio of the amplitude of the reflected wave to the forward wave. It is the modulus of the *reflection coefficient*. From eqn. (1.36) it is given by

$$|\rho| = \frac{S-1}{S+1} \tag{1.37}$$

1.7. Impedance Transformation

In microwave systems, the operating frequency is so high that the individual line potentials and currents cannot be measured easily. The standing wave in the line can be measured both in amplitude and in phase. Although the properties of the wave on a transmission line could be solved in terms of the voltage and current on the line, it would not be profitable and solutions will be sought in terms of the VSWR.

For convenience, in the rest of this chapter all impedances are normalized to the characteristic impedance of the transmission line in which measurements are being made. That is, the relative or normalized value of an impedance is the absolute value of that impedance divided by the characteristic impedance of the line.

Consider a line terminated in some impedance different from its characteristic impedance. Distances are measured in the negative z-direction with zero at the termination. Then at some length l in front of the termination, the line voltage and current are

$$V_l = A \exp j(\omega t + \beta l) + B \exp j(\omega t - \beta l) \tag{1.38}$$

$$I_l = \frac{1}{Z_0}[A \exp j(\omega t + \beta l) - B \exp j(\omega t - \beta l)] \tag{1.39}$$

where now A and B are both phasor quantities.

The normalized impedance of the terminated line at a distance l in front of the termination is given by

$$Z_l = \frac{V_l}{Z_0 I_l} = \frac{A \exp j(\omega t + \beta l) + B \exp j(\omega t - \beta l)}{A \exp j(\omega t + \beta l) - B \exp j(\omega t - \beta l)}$$

which simplifies to

$$Z_l = \frac{1 + \rho \exp - 2j\beta l}{1 - \rho \exp - 2j\beta l} \tag{1.40}$$

where ρ has phase as well as amplitude. If the terminating impedance is Z_t then

$$Z_t = Z_l \quad \text{when} \quad l = 0$$

and from eqn. (1.40)

$$Z_t = \frac{1 + \rho}{1 - \rho}$$

or

$$\rho = \frac{Z_t - 1}{Z_t + 1}$$

Substitution for ρ into eqn. (1.40) gives the relationship

$$Z_l = \frac{Z_t + j \tan \beta l}{1 + jZ_t \tan \beta l} \tag{1.41}$$

which relationship enables us to find the effective impedance anywhere on a uniform transmission line due to its terminating impedance. Alternatively eqn. (1.41) can be used to find the effective impedance on the transmission line in terms of the effective impedance at some distance l nearer the load. There is a similar relationship between the admittances on the line. It is left as an exercise for the reader to start with eqns. (1.38) and (1.39) to prove

$$Y_l = \frac{Y_t + j \tan \beta l}{1 + jY_t \tan \beta l} \tag{1.42}$$

1.8. Smith Chart

The performance of calculations involving eqns. (1.41) and (1.42) is simplified by the use of graphical methods. One of the most used of these is the *Smith chart* or *circle diagram* shown in Fig. 1.5. Equation (1.40) may be written in the form

$$Z = \frac{1 + W}{1 - W} \tag{1.43}$$

where W is given by

$$W = |\rho| \exp j(\phi - 2\beta l)$$

and ϕ is the phase angle of the reflection coefficient. The transformation

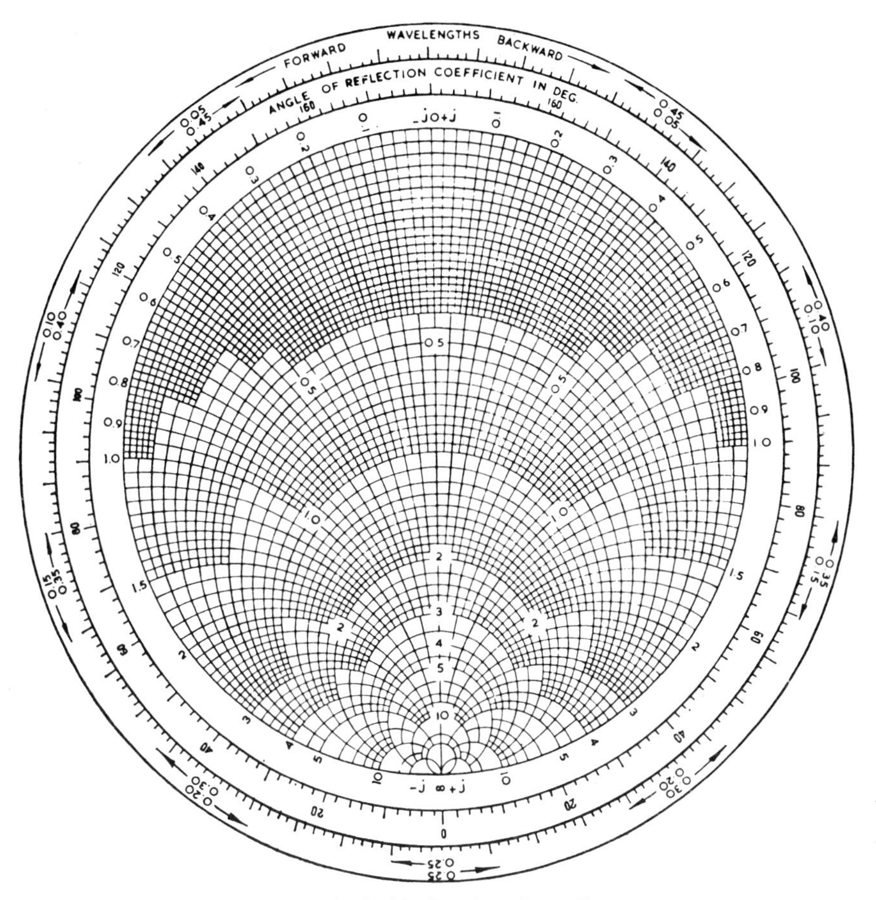

FIG. 1.5. The Smith chart impedance diagram.

between the z-plane and the w-plane given in eqn. (1.43) maps the rectangular coordinates $Z = R + jX$ into the coordinate lines of the circle diagram shown in Fig. 1.5. The resistive and reactive components of the normalized impedance are circles and segments of circles respectively. The circle diagram provides a polar plot of thc function W in the w-plane. It is the reflection coefficient in magnitude and phase transferred along a length l away from the termination. The centre of the circle diagram corresponds to the condition of zero reflection coefficient for which $|W| = |\rho| = 0$. This is the perfectly matched line, terminated in its characteristic impedance. For any mismatched condition, the modulus of the reflection coefficient remains constant but the effective impedance varies with distance along the line. A constant reflection coefficient is equivalent to a constant radius from the centre of the circle diagram. The locus of the effective impedance is a circle about the centre of the circle diagram. Since the amplitude of the reflection coefficient is equal to the distance from the centre of the diagram, eqn. (1.36) shows that any particular VSWR lies on a circle about the centre of the diagram.

1.9. Impedance Measurement

The impedance connected to the end of a transmission line can be determined by measuring the amplitude and phase of the forward and reflected waves on the transmission line, using a vector voltmeter and directional couplers. In the network analyser, such measurements are processed to provide readings of impedance. Alternatively, transmission line impedances may be measured by deduction from the VSWR in the line. Reference to eqn. (1.41) and the circle diagram shows that the resistive part of the effective impedance varies periodically in step with the standing wave pattern. Hence the minimum on the standing wave pattern is appropriate to an impedance located on the real axis between 0 and 1. This is the position of the zero of the phase angle measurement round the outside of the diagram, and reference to eqn. (1.43) shows that it is π in the phase angle of the reflection coefficient. Movement round the diagram at a constant radius from the centre is equivalent to movement along the transmission line. Movement along the line is best measured in wavelengths and it is seen that the outside of the circle diagram is calibrated in wavelengths. Comparison of eqns. (1.41) and (1.42) shows that, if the circle diagram can plot the relationships of eqn. (1.41), it can also plot the relationships of eqn. (1.42). Hence the circle diagram can also be used as an admittance diagram. Instead of plotting R and X, it plots G and B and the minimum of the standing wave pattern corresponds to a maximum value of admittance.

Consider a transmission line connected to an unknown impedance so that the only source of reflected power on the transmission line is from the unknown impedance. The impedance can be measured by finding the VSWR and the position of the first minimum of the standing wave pattern nearer to the generator.

Let the VSWR be S and the distance between the unknown impedance and the first minimum of the standing wave pattern be δ. Then the effective impedance at the position of the minimum of the standing wave pattern is known because the VSWR is at a constant radius on the diagram and the minimum is on the resistance axis. In Fig. 1.6, the effective impedance is shown as the point A, corresponding to $S = 2.0$. It is now necessary to move the effective impedance forward along the transmission line a distance $\delta = 0.18\,\lambda$ towards the load. This is shown to give the point B in Fig. 1.6. The coordinates of the point B give the value of the unknown impedance, so that $Z_t = 1.3 - j0.75$.

1.10. Stub Matching

A shorted stub is a short length of transmission line with a short circuit at the end. The standing wave circle of a shorted lossless line is the outer

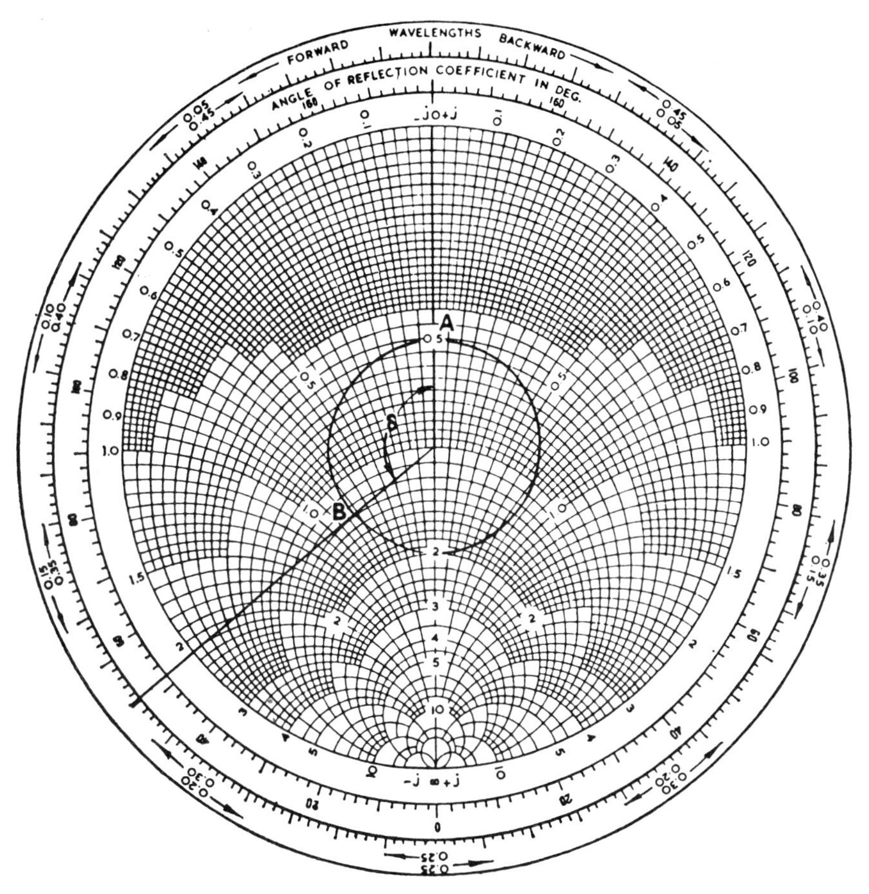

FIG. 1.6. Smith chart plot of an impedance giving a VSWR of 2 with the minimum a distance $\delta = 0.18\,\lambda$ from the plane of the impedance.

circumference of the circle diagram. Hence its effective impedance passes through the values of zero and infinite impedance and through the values $\pm j$. Since the impedance of a short circuit is zero, the effective impedance of a short-circuited stub of length l is seen from eqn. (1.41) to be

$$Z = j \tan \beta l \tag{1.44}$$

Such a shorted stub can be used to match any impedance on a transmission line. Consider the impedance $Z_t = 0.45 + j0.32$ given by the point A in Fig. 1.7. If a point on the transmission line in front of the impedance is chosen so that the resistive component of the effective impedance is unity, this corresponds to the point B on the diagram with the effective impedance $1 + j0.95$. A shorted stub may be added in series with the main transmission line at this point with an effective impedance of $-j0.95$ so that the total impedance is now unity and the line is matched. If the stub is to be added

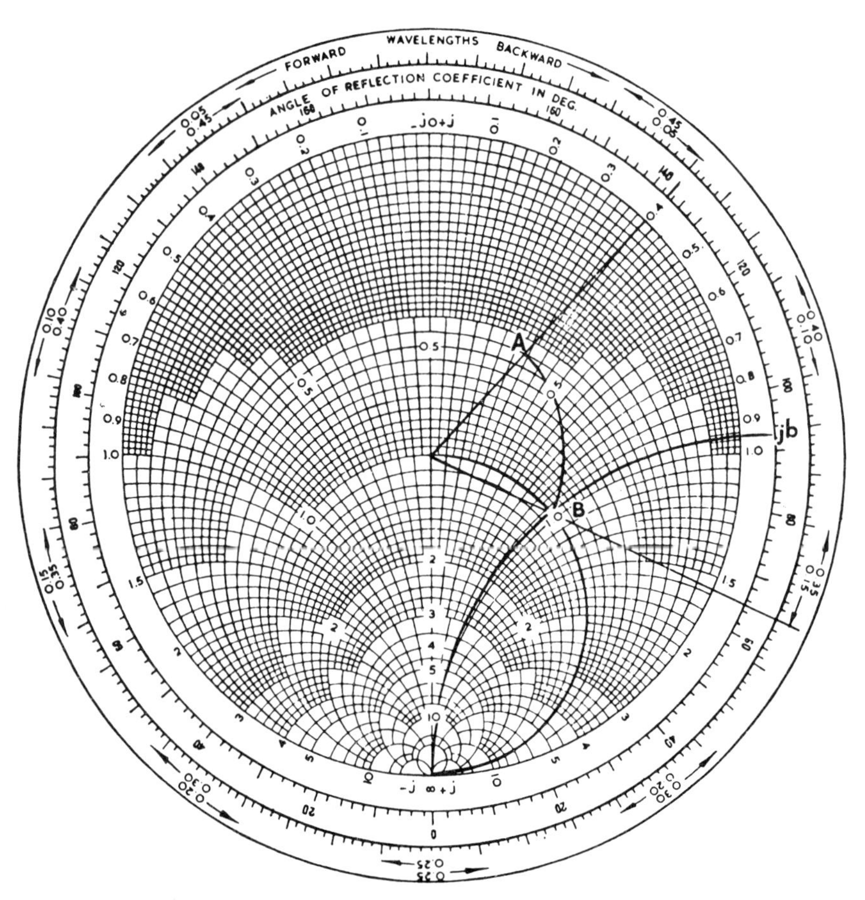

FIG. 1.7. Showing the construction to find the impedance and position of a shorted stub to match the impedance A.

in shunt, then it is easier to use admittances in the calculation. The Smith chart is used as an admittance diagram and the matching procedure is the same as that described for impedances.

It is often difficult to add shorted stubs exactly where they are needed, so three stubs a fixed distance apart are used. This has the advantage of allowing for experimental determination of the optimum matched condition by trial and error.

1.11. Impedance Transformer

A length of transmission line may be used as an impedance transformation device. In effect, this process has already been carried out in the discussion of stub matching, but a quarter wavelength section of transmission line may also be used to match two unlike impedances. It may be seen from eqn.

(1.41) that a quarter wavelength line transforms the terminating impedance into its inverse. Consider the transmission line system shown in Fig. 1.8. The

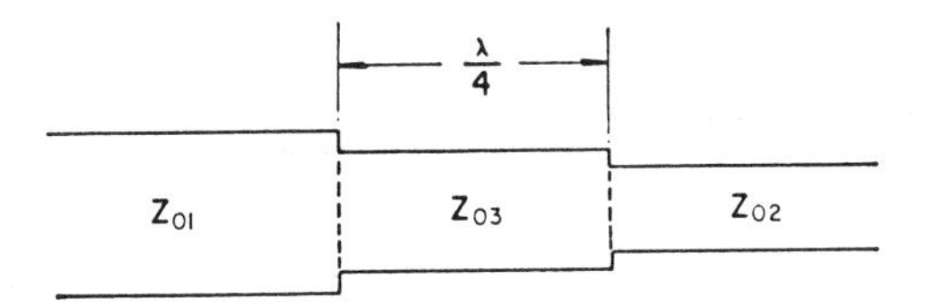

FIG. 1.8. A quarter wavelength matching section between two lines of different characteristic impedance.

characteristic impedances of the different lines are shown on the diagram. The centre section of line is a quarter wavelength long and is being used to match the differing impedances of the two lines so that there is no reflected wave from the change of line impedance. The impedances of the two end sections of line need to be normalized to that of the centre section of line. From the centre section of line, the impedances at the two ends are

$$\frac{Z_{01}}{Z_{03}} \quad \text{and} \quad \frac{Z_{02}}{Z_{03}}$$

The impedance at the left-hand end of the line when transformed along the length of the line to the right-hand end is

$$\frac{Z_{03}}{Z_{01}}$$

which for matching will now have to be the same as that at the right-hand end. Therefore the necessary characteristic impedance of the intermediate matching section between the two lines is given by

$$Z_{03} = \sqrt{(Z_{01}Z_{02})} \tag{1.45}$$

It must be equal to the geometric mean of the impedances of the two lines being matched.

1.12. Summary

1.2. Z_0, the ***characteristic impedance***, is the input impedance of an infinitely long uniform transmission line.

The input impedance of a short line terminated in Z_0 is Z_0.

The characteristic impedance of a short line can be measured because it is the geometric mean of the open- and short-circuit impedances.

1.3. The ***transmission line equations*** are:

$$\frac{d^2V}{dz^2} = \gamma^2 V \tag{1.11}$$

$$\frac{d^2 I}{dz^2} = \gamma^2 I \tag{1.12}$$

$$\gamma^2 = ZY \tag{1.13}$$

$$V = A\exp[-\alpha z + j(\omega t - \beta z)] + B\exp[\alpha z + j(\omega t + \beta z)] \tag{1.17}$$

$$I = \sqrt{\left(\frac{Y}{Z}\right)}\Big\{A\exp[-\alpha z + j(\omega t - \beta z)] - B\exp[\alpha z + j(\omega t + \beta z)]\Big\} \tag{1.46}$$

Propagation constant,

$$\gamma = \alpha + j\beta \tag{1.16}$$

Phase constant,

$$\beta = \frac{2\pi}{\lambda} \tag{1.19}$$

Attenuation constant, α

1.4. The ***primary line constants*** are the impedance properties that can be measured. The secondary line constants are given by the relationships

$$\beta = \omega\sqrt{(LC)} \tag{1.25}$$

$$\alpha = \frac{G}{2}\sqrt{\left(\frac{L}{C}\right)} + \frac{R}{2}\sqrt{\left(\frac{C}{L}\right)} \tag{1.26}$$

$$Z_0 = \sqrt{\left(\frac{R + j\omega L}{G + j\omega C}\right)} \approx \sqrt{\left(\frac{L}{C}\right)} \tag{1.28}$$

1.6. ***Voltage standing wave ratio,***

$$S = \frac{V_{max}}{V_{min}} = \frac{1 + |\rho|}{1 - |\rho|} \tag{1.36}$$

Reflection coefficient,

$$|\rho| = \frac{S - 1}{S + 1} \tag{1.37}$$

1.7. The ***impedance transformation*** is

$$Z_l = \frac{Z_t + j\tan\beta l}{1 + jZ_t\tan\beta l} \tag{1.41}$$

where Z_t and Z_l are both normalized with respect to Z_0. Similarly the admittance transformation is

$$Y_l = \frac{Y_t + j\tan\beta l}{1 + jY_t\tan\beta l} \qquad (1.42)$$

1.8. Equation (1.41) is solved graphically using the **Smith chart** (Fig. 1.5). The radius from the centre of the chart is the amplitude of the reflection coefficient. A circle about the centre is a locus of constant VSWR and constant amplitude reflection coefficient. Distance along the transmission line is equivalent to movement round the diagram. Rotation through 360° is equivalent to movement of half a wavelength along the line.

1.9. ***Impedance is measured*** by finding the VSWR and the position of the first minimum of the standing wave pattern. The VSWR gives the radial distance from the centre of the Smith chart. The distance of the measured minimum from the plane of the impedance gives the angular position on the Smith chart.

1.10. The impedance of a ***shorted stub*** is

$$Z = j\tan\beta l \qquad (1.44)$$

A shorted stub in the correct position may be used to cancel the reflected wave from an incorrect termination of a line. Alternatively three stubs at fixed positions may be used.

1.11. A quarter wavelength section of line may be used to match two unlike impedances. The characteristic impedance of the intermediate section must be equal to the geometric mean of the impedances of the two lines being matched.

Problems

1.1. If a short transmission line can be represented by the equivalent circuit of Fig. 1.9, prove eqns. (1.25) and (1.29). What assumptions are necessary and why are they valid?

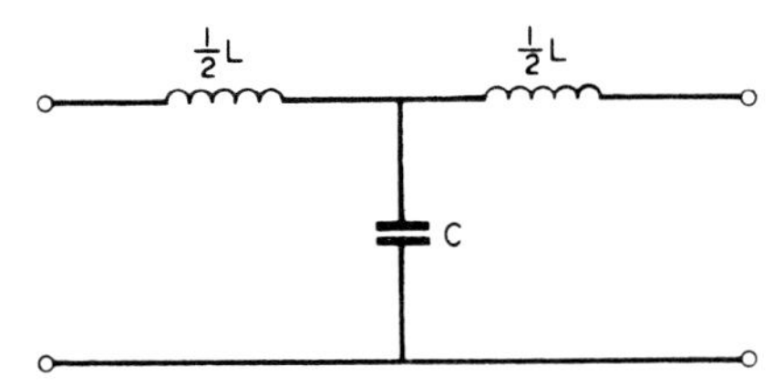

FIG. 1.9. Equivalent circuit of a short transmission line.

1.2. The primary line constants of a coaxial transmission line are given by

$$L = \frac{1}{2\pi}\mu_0\mu_r \log_e\frac{b}{a} \text{ H/m}$$

$$C = \frac{2\pi\varepsilon_0\varepsilon_r}{\log_e b/a} \text{ F/m}$$

where a and b are the radii of the inner and outer conductors respectively. Find the characteristic impedance of the line and show that the velocity of propagation of the wave is $1/\sqrt{(\mu_0\mu_r\varepsilon_0\varepsilon_r)}$, the same as the speed of light for an air-filled line where $\mu_r = \varepsilon_r = 1$. What is the ratio of the radii required to give a 50 Ω line, (a) air-filled, (b) using PTFE filling with $\varepsilon_r = 2$, $\mu_r = 1$?

[2.30:1; 3.25:1]

1.3. The VSWR on a transmission line is 2. Plot the shape of the standing wave pattern in the line. Repeat the exercise when the VSWR is 11.

1.4. Derive eqn. (1.42) from eqns. (1.38) and (1.39).

1.5. For a uniform air-filled transmission line, the phase velocity is the same as the speed of light. Calculate from eqn. (1.41) the effective impedance at a distance of (a) 2 cm and (b) 1 cm along the line from a terminating resistance of value twice the characteristic impedance of the line. The frequency is 3.75 GHz. Confirm the results using the Smith chart.

[$\frac{1}{2}Z_0$; $(0.8 - j0.6)Z_0$]

1.6. Draw the locus on the Smith chart circle diagram of VSWR values 1, 2, 3, 4, 5, 10 and ∞. Calculate the amplitude of the reflection coefficient appropriate to these loci.

If the termination of a transmission line is resistive and variable, plot on the impedance diagram the locus of the effective impedance at 0.125 of a wavelength in front of the termination as its resistance value is changed.

If the termination of a transmission line is reactive, capacitative and variable, plot on the impedance diagram the locus of the effective impedance at 0.25 of a wavelength in front of the termination as its capacitance value is changed.

Repeat these exercises using an admittance diagram.

1.7. Using the Smith chart, find the impedance of the terminations giving rise to the following VSWR readings at a frequency of 3 GHz (λ = 10 cm):

VSWR	Distance to position of 1st minimum	
1.5	2.50 cm	[1.50 Z_0]
2	1.52 cm	[$(1.0 - j0.7)Z_0$]
2	4.38 cm	[$(0.55 + j0.3)Z_0$]
3	0.68 cm	[$(0.4 - j0.4)Z_0$]
4	2.08 cm	[$(2.0 - j1.8)Z_0$]
5	2.82 cm	[$(2.6 + j2.4)Z_0$]
10	3.74 cm	[$(0.2 + j1.0)Z_0$]

1.8. A swept frequency measurement (see section 14.8) gives a plot of reflected power ratio (in dB) against frequency. The table gives some readings from such a measurement of the effect of a line terminated in an inductor whose series resistance is 50 Ω at all frequencies. Find the value of its inductance when the characteristic impedance of the line is 50 Ω.

f GHz	Reflected power dB
0.60	9.5
1.00	6.0
1.30	4.4
1.55	3.5
1.90	2.5
2.30	1.9

[9.1 nH]

1.9. (a) Calculate the length and position of a shorted stub to be added in series with a uniform transmission line to cancel the mismatch due to a terminating impedance of $(0.5 - j0.9)\,Z_0$ at 3 GHz (λ = 10 cm). Plot on the Smith chart the total effective impedance at the plane of the stub for a ± 5 per cent change in frequency. [1.56 cm, 4.52 cm]

(b) A transmission line system gives the following reading for VSWR at 3 GHz ($\lambda = 10$ cm): VSWR = 5 and position of minimum = 5.2 cm from some arbitrary datum. Find the length and position of a shorted stub to be added in shunt to cancel the mismatch of the system at 3 GHz. [4.53 cm, 0.81 cm]

1.10. Design a transformation to provide a matched junction between two similar transmission lines having characteristic impedances of 50 Ω and 75 Ω.

[Quarter wavelength section of 61.2 Ω]

2

Electromagnetic Fields

2.1. Electromagnetic Field Components

Any system of electric charges gives rise to corresponding potential differences and to electric and magnetic fields. As far as electromagnetic waves are concerned, it is the electric and magnetic fields that are important. In this chapter the basic mathematical relationships that enable electromagnetic theory to be such a precise science are discussed. Historically these relationships were derived by deductive reasoning from experimental observations. It is not felt to be part of the object of this book to detail these experiments and the reasoning leading to the basic electromagnetic relationships, but the relationships are stated and discussed in terms of elementary electric and magnetic field theory. These relationships are the basis of all the rest of the mathematical analysis given in this book. Any reader who is unfamiliar with these basic relationships is referred to one of the books listed in the Bibliography as relating to this chapter.

These precise mathematical relationships between the different electromagnetic field components and the electric charges and currents enable us to derive expressions for the electromagnetic fields for every precisely defined situation. As the fields, currents and charges exist in the body of a medium, they are all defined in terms of some space distribution. The electromagnetic field components together with their notation and units of measurement are:

Electric field	$\boldsymbol{E}$	volt/metre
Electric flux density	$\boldsymbol{D}$	coulomb/metre^2
Magnetic field	$\boldsymbol{H}$	ampere/metre
Magnetic flux density	$\boldsymbol{B}$	tesla = weber/metre^2
Charge density	ρ	coulomb/metre^3
Current density	$\boldsymbol{J}$	ampere/metre^2

Two of these units are gradients of some scalar quantity, $\boldsymbol{E}$ and $\boldsymbol{H}$, three of them are area density functions of a vector field, $\boldsymbol{D}$, $\boldsymbol{B}$ and $\boldsymbol{J}$, and the charge is a volume density function. Some of the terms may be unfamiliar to the reader now. They are all defined here for completeness. A full discussion of their implications is postponed until later in this chapter.

2.2. Material Properties

Some of the field components are related by the properties of the medium in which they exist:

$$\boldsymbol{B} = \mu \boldsymbol{H} \tag{2.1}$$

$$\boldsymbol{D} = \varepsilon \boldsymbol{E} \tag{2.2}$$

$$\boldsymbol{J} = \sigma \boldsymbol{E} \tag{2.3}$$

Equation (2.1) is the well-known relationship between the applied magnetic field and the resultant magnetic flux density. The constant μ is usually called the permeability of the medium. In the S.I. system of units it is a dimensional constant. With reference to eqn. (2.1), its dimensions are given by the relationship

$$\mu = \frac{\boldsymbol{B}}{\boldsymbol{H}} = \frac{\text{weber}}{\text{metre}^2} \times \frac{\text{metre}}{\text{ampere}}$$

(a changing flux of 1 weber/second generates 1 volt)

$$\mu = \frac{(\text{volt})(\text{second})}{(\text{ampere})(\text{metre})} = \text{henry/metre}$$

The permeability is a measure of both the relative effect of having a particular material in the path of the field and also a dimensional constant, so that it can be divided into two parts:

$$\mu = \mu_0 \mu_r$$

μ_0 is the *permeability constant* which is dimensional. It is defined to be of value $4\pi \times 10^{-7}$ henry/metre, and it is sometimes called the permeability of free space. μ_r is the *relative permeability* which is dimensionless and makes allowance for the effect of the material relative to vacuum or free space.

Equation (2.2) provides a similar relationship between the applied electric field and the electric flux density, $\boldsymbol{D}$. ε is the permittivity of the medium and is also a dimensional constant, its dimensions being given by

$$\varepsilon = \frac{\boldsymbol{D}}{\boldsymbol{E}} = \frac{\text{coulomb}}{\text{metre}^2} \times \frac{\text{metre}}{\text{volt}}$$

$$= \frac{(\text{ampere})(\text{second})}{(\text{volt})(\text{metre})} = \text{farad/metre}$$

The permittivity may also be divided into two parts:

$$\varepsilon = \varepsilon_0 \varepsilon_r$$

ε_0 is the *permittivity constant* or permittivity of free space. It is defined from the relationship given below for the speed of light and from the value of the permeability constant. It has the approximate value $1/(36\pi \times 10^9)$

farad/metre. ε_r is the dimensionless *relative permittivity* which takes account of the effect of the medium on the electric fields. It is the same as the dielectric constant of the medium.

In the solution of the equations for the electromagnetic fields, these two dimensional constants are most often met in combination to provide two more dimensional constants. The derivation and usefulness of these constants arises as we seek solutions for the expressions describing the fields of an electromagnetic wave. Their significance appears as they are obtained in the mathematics. They are

$$\text{The velocity of light } c = \frac{1}{\sqrt{(\mu_0 \varepsilon_0)}} \approx 3 \times 10^8 \text{ m/s}$$

$$\text{The impedance of free space } \eta = \sqrt{\left(\frac{\mu_0}{\varepsilon_0}\right)} \approx 120\pi = 377\,\Omega$$

The expression for the velocity of light is used in the definition for the permittivity constant given above.

Equation (2.3) is the conductivity relationship for the material. When the conductivity is a constant, it is an expression of Ohm's law. σ is the conductivity of the material. It is the reciprocal of the resistivity. From eqn. (2.3) its dimensions are seen to be

$$\sigma = \frac{J}{E} = \frac{\text{ampere}}{\text{metre}^2} \times \frac{\text{metre}}{\text{volt}} = \text{siemens/metre}$$

2.3. Vector Analysis

Electromagnetic fields are most easily described in terms of vectors and the definitions of section 2.1 already assume that the components of the fields are vector quantities. Vector analysis gives a convenient shorthand for the manipulation of differential equations incorporating vectors. It is assumed that the reader is conversant with the use of vectors and vector analysis. There are many mathematical textbooks which give the theory of vector analysis. One specializing in the subject is given in the Bibliography. This section defines the vector notation to be used and also gives a summary of the properties of the vector and the vector differential operator.

The three systems of three-dimensional space coordinates to be used in this book are shown in Fig. 2.1. The diagram shows an arbitrary vector resolved into three mutually perpendicular components labelled by reference to the coordinate system used. Throughout this book, subscripts are used to denote these components of the vectors. The subscripts are x, y and z in rectangular coordinates, r, θ and z in cylindrical polar coordinates and r, θ and ϕ in spherical polar coordinates.

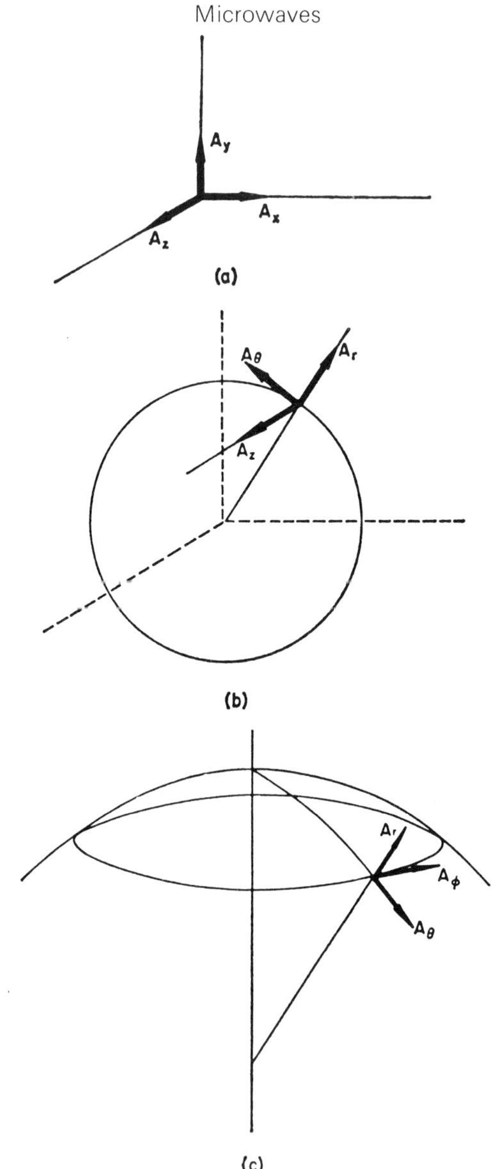

FIG. 2.1. The components of a vector in three coordinate systems. (a) Rectangular. (b) Cylindrical polar. (c) Spherical polar.

The properties of a vector are:

Addition: the addition of vectors requires the addition of each component of the vectors individually.

If

$$\boldsymbol{C} = \boldsymbol{A} + \boldsymbol{B}$$

then

$$C_x = A_x + B_x$$
$$C_y = A_y + B_y$$
$$C_z = A_z + B_z$$

Multiplication by a scalar just increases the amplitude of the vector. If

$$\boldsymbol{C} = k\boldsymbol{A}$$

then

$$C_x = kA_x$$
$$C_y = kA_y$$
$$C_z = kA_z$$

Scalar multiplication of two vectors gives a quantity which is the product of the amplitude of one vector with the projection of the other vector upon it.

$$k = \boldsymbol{A} \cdot \boldsymbol{B} = A_x B_x + A_y B_y + A_z B_z$$

Vector multiplication of two vectors gives another vector acting in a direction perpendicular to the two original vectors.

If

$$\boldsymbol{C} = \boldsymbol{A} \times \boldsymbol{B}$$

then

$$C_x = A_y B_z - A_z B_y$$
$$C_y = A_z B_x - A_x B_z$$
$$C_z = A_x B_y - A_y B_x$$

The differential operations on a vector are as follows.

Scalar differentiation. The ordinary differentiation of a vector $\boldsymbol{F}$ has the form as follows:

If

$$\frac{\partial \boldsymbol{F}}{\partial t} = \boldsymbol{Y}$$

then

$$Y_x = \frac{\partial F_x}{\partial t}; \qquad Y_y = \frac{\partial F_y}{\partial t}; \qquad Y_z = \frac{\partial F_z}{\partial t}.$$

Vector operator. The vector operator $\boldsymbol{\nabla}$ involves differentiation with regard to each of three perpendicular space dimensions multiplied by a unit vector in the direction of the dimension. If $\boldsymbol{U}_x$ is a unit vector in the x-direction and $\boldsymbol{U}_y$ and $\boldsymbol{U}_z$ are similarly defined, then

$$\boldsymbol{\nabla} \equiv \boldsymbol{U}_x \frac{\partial}{\partial x} + \boldsymbol{U}_y \frac{\partial}{\partial y} + \boldsymbol{U}_z \frac{\partial}{\partial z}$$

Its use is illustrated in the three vector differential operations.

Gradient of a scalar. The gradient of any scalar quantity is the vector differentiation of a scalar and is a vector.

If

$$\text{grad}\, F = \boldsymbol{\nabla} F = \boldsymbol{Y}$$

then

$$Y_x = \frac{\partial F}{\partial x}; \qquad Y_y = \frac{\partial F}{\partial y}; \qquad Y_z = \frac{\partial F}{\partial z}$$

Divergence of a vector. The divergence at a point is a measure of the total field flowing into or out of that point. Hence it may be considered as a measure of the sources of field at that point. It is the vector differentiation of a vector having a scalar result.

$$\text{div}\, \boldsymbol{F} = \boldsymbol{\nabla} . \boldsymbol{F} = \frac{\partial F_x}{\partial x} + \frac{\partial F_y}{\partial y} + \frac{\partial F_z}{\partial z}$$

Curl or rotation of a vector. Each component of the curl of a vector is a measure of the rotation or vorticity of the field round the component direction.

If

$$\text{curl}\, \boldsymbol{F} = \text{rot}\, \boldsymbol{F} = \boldsymbol{\nabla} \times \boldsymbol{F} = \boldsymbol{Y}$$

then

$$Y_x = \frac{\partial F_z}{\partial y} - \frac{\partial F_y}{\partial z}$$

$$Y_y = \frac{\partial F_x}{\partial z} - \frac{\partial F_z}{\partial x}$$

$$Y_z = \frac{\partial F_y}{\partial x} - \frac{\partial F_x}{\partial y}$$

Vector operators in the two polar coordinate systems (see Fig. 2.1 for details of the vector components):

Gradient, if $\nabla F = \boldsymbol{Y}$

$$Y_r = \frac{\partial F}{\partial r}; \qquad Y_\theta = \frac{1}{r}\frac{\partial F}{\partial \theta}; \qquad Y_z = \frac{\partial F}{\partial z}$$

$$Y_r = \frac{\partial F}{\partial r}; \qquad Y_\theta = \frac{1}{r}\frac{\partial F}{\partial \theta}; \qquad Y_\phi = \frac{1}{r\sin\theta}\frac{\partial F}{\partial \phi}$$

Divergence,

$$\nabla . \boldsymbol{F} = \frac{1}{r}\frac{\partial(rF_r)}{\partial r} + \frac{1}{r}\frac{\partial F_\theta}{\partial \theta} + \frac{\partial F_z}{\partial z}$$

$$\nabla . \boldsymbol{F} = \frac{1}{r^2}\frac{\partial(r^2F_r)}{\partial r} + \frac{1}{r\sin\theta}\frac{\partial(F_\theta \sin\theta)}{\partial \theta} + \frac{1}{r\sin\theta}\frac{\partial F_\phi}{\partial \phi}$$

Curl, if $\nabla \times \boldsymbol{F} = \boldsymbol{Y}$

$$Y_r = \frac{1}{r}\frac{\partial F_z}{\partial \theta} - \frac{\partial F_\theta}{\partial z}$$

$$Y_\theta = \frac{\partial F_r}{\partial z} - \frac{\partial F_z}{\partial r}$$

$$Y_z = \frac{1}{r}\left[\frac{\partial(rF_\theta)}{\partial r} - \frac{\partial F_r}{\partial \theta}\right]$$

$$Y_r = \frac{1}{r^2\sin\theta}\left[\frac{\partial(r\sin\theta F_\phi)}{\partial \theta} - \frac{\partial(rF_\theta)}{\partial \phi}\right]$$

$$Y_\theta = \frac{1}{r\sin\theta}\left[\frac{\partial F_r}{\partial \phi} - \frac{\partial(r\sin\theta F_\phi)}{\partial r}\right]$$

$$Y_\phi = \frac{1}{r}\left[\frac{\partial(rF_\theta)}{\partial r} - \frac{\partial F_r}{\partial \theta}\right]$$

The scalar Laplacian in all three coordinate systems:

$$\nabla^2 F = \frac{\partial^2 F}{\partial x^2} + \frac{\partial^2 F}{\partial y^2} + \frac{\partial^2 F}{\partial z^2}$$

$$\nabla^2 F = \frac{\partial^2 F}{\partial r^2} + \frac{1}{r}\frac{\partial F}{\partial r} + \frac{1}{r^2}\frac{\partial^2 F}{\partial \theta^2} + \frac{\partial^2 F}{\partial z^2}$$

$$\nabla^2 F = \frac{\partial^2 F}{\partial r^2} + \frac{2}{r}\frac{\partial F}{\partial r} + \frac{1}{r^2}\frac{\partial^2 F}{\partial \theta^2} + \frac{1}{r^2\tan\theta}\frac{\partial F}{\partial \theta} + \frac{1}{r^2\sin^2\theta}\frac{\partial^2 F}{\partial \phi^2}$$

Some vector identities:

$$\nabla \times \nabla F = 0$$

$$\nabla . \nabla \times \boldsymbol{F} = 0$$

$$\nabla \times (\nabla \times F) = \nabla(\nabla . F) - \nabla^2 \boldsymbol{F}$$

$$\nabla . (A \times B) = \boldsymbol{B} . \nabla \times A - A . \nabla \times \boldsymbol{B}$$

2.4. Maxwell's Equations

The electromagnetic field equations were derived by deductive reasoning from the results of experiments. They will not be derived here, but their derivation from the well-known laws of elementary electricity and magnetism is indicated. The equations are

$$\nabla . \boldsymbol{D} = \rho \tag{2.4}$$

$$\nabla . \boldsymbol{B} = 0 \tag{2.5}$$

$$\nabla \times \boldsymbol{E} = -\frac{\partial \boldsymbol{B}}{\partial t} \tag{2.6}$$

$$\nabla \times \boldsymbol{H} = \boldsymbol{J} + \frac{\partial \boldsymbol{D}}{\partial t} \tag{2.7}$$

Equation (2.4) arises from the fact that any stored electric charge gives rise to an electric field or conversely any discontinuous electric field gives rise to electric charge. It is a mathematical statement of Gauss's law which states that the integration of the perpendicular component of the flux density over any closed surface is equal to the total charge enclosed by that surface. Equation (2.5) is the magnetic form of Gauss's law where there are no isolated magnetic charges. It states that magnetic field must exist in closed loops. There are no magnetic charges or single magnetic poles in nature. Most of the fundamental particles appear to be magnetic dipoles, but the total magnetic flux integrated over a surface enclosing a dipole is zero; hence eqn. (2.5) is valid. Equation (2.6) is a statement of Faraday's law of electromagnetic induction, that the e.m.f. induced in a closed circuit (the curl of the electric field) is proportional to the rate of change of the magnetic flux threading the circuit.

Equation (2.7) is partly a statement of the Biot–Savart law, often commonly called Ampere's law. A current gives rise to a closed loop of magnetic field. However the conduction current is not the only source of magnetic field, for it is found that a magnetic field exists around the air gap between the plates of a parallel plate capacitor which has a time-varying current flowing in its leads. This difficulty was resolved by Maxwell who postulated that the circuit containing the capacitor must have a continuum of current flowing round it. Since no conduction current is flowing between the plates of the capacitor, the two halves of the circuit are linked by the electric field between the plates of the capacitor. Then the current is equal to the rate of change of charge on the plates of the capacitor which in its turn is equal to the rate of change of flux density, the $\boldsymbol{D}$ field, between the plates of the capacitor. Consequently he postulated a displacement current density which consisted of the time derivative of the electric flux density in the air gap between the plates of the capacitor. Hence the term $\partial \boldsymbol{D}/\partial t$ is added to the conduction current on the right-hand side of eqn. (2.7). Maxwell first formulated the electromagnetic field relationships into the form given in eqns. (2.4) to (2.7) and these equations are called by his name.

2.5. The Solution of Maxwell's Equations

The relationships between the components of any electromagnetic field are given by Maxwell's equations, eqns. (2.4) to (2.7), and by the equations representing the properties of the medium in which the electromagnetic field exists, eqns. (2.1) to (2.3). In its most general form, all the components of the field are functions of three space dimensions and time. Initially, consideration is given to a situation where the medium is a perfect insulator with no stored charges, i.e. $\rho = 0$, $\boldsymbol{J} = 0$, $\sigma = 0$, and the effect of conduction is considered in Chapter 7. μ and ε are considered to be scalar constants, which is true for most materials; there are a few situations of interest in microwave engineering where it is not so and these are considered in Chapters 8 and 9. The time dependence of any signal of interest is sinusoidal. If the angular frequency is ω, all fields have a time dependence of $\exp j\omega t$ and the partial differentiation $\partial/\partial t$ is equivalent to multiplication by $j\omega$. Performing this substitution and also substituting for $\boldsymbol{B}$ and $\boldsymbol{D}$ from eqns. (2.1) and (2.2), eqns. (2.4) to (2.7) become

$$\nabla . \boldsymbol{E} = 0 \tag{2.8}$$

$$\nabla . \boldsymbol{H} = 0 \tag{2.9}$$

$$\nabla \times \boldsymbol{E} = -j\omega\mu \boldsymbol{H} \tag{2.10}$$

$$\nabla \times \boldsymbol{H} = j\omega\varepsilon \boldsymbol{E} \tag{2.11}$$

In order to find a characteristic solution to these equations, it is necessary to eliminate one of the two remaining field vectors. A similar solution is obtained whichever field is eliminated. The mathematics is performed to eliminate $\boldsymbol{H}$ and to find an expression in $\boldsymbol{E}$ and the similar expression in $\boldsymbol{H}$ is quoted. It is a good exercise for the reader to satisfy himself that this expression for $\boldsymbol{H}$ in eqn. (2.13) is correct.

First take the curl of both sides of eqn. (2.10), that is, operate on both sides with $\nabla \times$.

$$\nabla \times \nabla \times \boldsymbol{E} = -j\omega\mu(\nabla \times \boldsymbol{H})$$

and substituting from eqn. (2.11)

$$\nabla \times \nabla \times \boldsymbol{E} = \omega^2 \mu\varepsilon \boldsymbol{E}$$

Using a vector identity for the left-hand side of this equation gives

$$\nabla(\nabla . \boldsymbol{E}) - \nabla^2 \boldsymbol{E} = \omega^2 \mu\varepsilon \boldsymbol{E}$$

and eqn. (2.8) makes the first term of this expression zero so that

$$\nabla^2 \boldsymbol{E} + \omega^2 \mu\varepsilon \boldsymbol{E} = 0 \tag{2.12}$$

and similarly

$$\nabla^2 \boldsymbol{H} + \omega^2 \mu\varepsilon \boldsymbol{H} = 0 \tag{2.13}$$

Equations (2.12) and (2.13) are general solutions of Maxwell's equations in terms of the material constants and the angular frequency of the electromagnetic signals. It is now necessary to use these equations to obtain solutions for the field quantities in terms of particular systems of space coordinates and particular physical constraints.

Using the expansion of ∇^2 in the rectangular coordinates x, y, and z, eqn. (2.12) becomes

$$\frac{\partial^2 \boldsymbol{E}}{\partial x^2} + \frac{\partial^2 \boldsymbol{E}}{\partial y^2} + \frac{\partial^2 \boldsymbol{E}}{\partial z^2} = -\omega^2 \mu\varepsilon \boldsymbol{E} \tag{2.14}$$

In an orthogonal rectangular coordinate system, eqn. (2.14) is separable into its individual component parts so that there is an equivalent differential equation for each of the components of the vector.

$$\frac{\partial^2 E_x}{\partial x^2} + \frac{\partial^2 E_x}{\partial y^2} + \frac{\partial^2 E_x}{\partial z^2} = -\omega^2 \mu\varepsilon E_x$$

$$\frac{\partial^2 E_y}{\partial x^2} + \frac{\partial^2 E_y}{\partial y^2} + \frac{\partial^2 E_y}{\partial z^2} = -\omega^2 \mu\varepsilon E_y$$

$$\frac{\partial^2 E_z}{\partial x^2} + \frac{\partial^2 E_z}{\partial y^2} + \frac{\partial^2 E_z}{\partial z^2} = -\omega^2 \mu\varepsilon E_z$$

2.6. Plane Wave Solution

Consider a situation where there is a variation of field quantities in only one direction. Let this be in the direction of the dimension z. Then

$$\frac{\partial}{\partial x} = \frac{\partial}{\partial y} = 0$$

and eqn. (2.14) becomes

$$\frac{\partial^2 \boldsymbol{E}}{\partial z^2} = -\omega^2 \mu\varepsilon \boldsymbol{E} \tag{2.15}$$

Equation (2.15) may now be separated into three similar equations involving the three component parts of the electric field. They are

$$\frac{\partial^2 E_x}{\partial z^2} = -\omega^2 \mu\varepsilon E_x \tag{2.16}$$

$$\frac{\partial^2 E_y}{\partial z^2} = -\omega^2 \mu\varepsilon E_y \tag{2.17}$$

$$\frac{\partial^2 E_z}{\partial z^2} = -\omega^2 \mu\varepsilon E_z \tag{2.18}$$

Equation (2.16) is a differential equation which has a solution of the form

$$E_x = A \exp(\pm\gamma z)$$

where

$$\gamma^2 = -\omega^2 \mu\varepsilon \tag{2.19}$$

There are apparently two solutions to eqn. (2.16) and the full solution is the sum of both possible solutions.

$$E_x = A \exp - \gamma z + B \exp \gamma z \tag{2.20}$$

It is seen that this result is the same as eqn. (1.14), the expression for the voltage on a transmission line. Hence the electromagnetic field equations have a solution for a wave that only varies in one direction, that is the same as a transmission-line wave where the direction of variation is the same as the direction of propagation. As a transmitting wave, this electromagnetic wave has all the properties of a wave on a transmission line. At the moment we are only interested in the fields of the wave and in its propagation properties and we are not interested in demonstrating the properties of transmission-line waves. That has already been adequately explained in Chapter 1. Further analysis is confined to the forward wave alone.

2.7. Propagation Properties

Introducing the time dependence, the full expression for the forward wave of the field is

$$E_x = E_0 \exp(j\omega t - \gamma z) \tag{2.21}$$

As already explained in section 1.3, γ is called the *propagation constant* of the wave. Provided μ and ε are both real, the use of β, the *phase constant*, serves to eliminate the negative in eqn. (2.19) so that

$$\gamma = j\beta$$

and

$$\beta = \omega\sqrt{(\mu\varepsilon)} \tag{2.22}$$

However, if there are any losses incurred by the wave as it propagates through the medium, the wave is attenuated. γ has to have a real part α, the *attenuation constant*, so that we have eqn. (1.16) which is rewritten here

$$\gamma = \alpha + j\beta \tag{1.16}$$

Equation (2.21) is an expression for a quantity which appears to be travelling in the direction z. If γ is imaginary the field is travelling without changing at the *phase velocity*,

$$v = \frac{\omega}{\beta} = \frac{1}{\sqrt{(\mu\varepsilon)}} \tag{2.23}$$

If γ has a real component, the field suffers an exponential decay with distance. Normally all waves suffer slight attenuation as they propagate even if the attenuation is so small that it may be ignored. The field quantities of all electromagnetic waves are modified by three terms:

$\exp j\omega t$ denotes a sinusoidal oscillation with regard to time,
$\exp -j\beta z$ denotes a sinusoidal oscillation with distance, and
$\exp -\alpha z$ denotes an exponential decay with distance.

The electromagnetic field here described is a plane wave. All the components of its fields have similar variations with respect to time and one direction. That direction is also the direction in which the wave appears to be travelling. The speed of travel is $1/\sqrt{(\mu\varepsilon)}$. If the electromagnetic wave is travelling through vacuum (or air which for electromagnetic waves may be considered to be a vacuum) its speed is $1/\sqrt{(\mu_0\varepsilon_0)}$, the velocity of light, c, which is 3×10^8 m/s.

2.8. Field Components of a Plane Wave

To find the individual components of an electromagnetic wave it is necessary to return to the two curl equations, eqns. (2.10) and (2.11), in order to obtain relationships between the field component that is known ($E_x = E_0 \exp(j\omega t - \gamma z)$) and the other components of the field. First eqns. (2.10) and (2.11) are given with each vector quantity separated into three orthogonal components.

Equation (2.10) becomes

$$\left.\begin{aligned} \frac{\partial E_z}{\partial y} - \frac{\partial E_y}{\partial z} &= -j\omega\mu H_x \\ \frac{\partial E_x}{\partial z} - \frac{\partial E_z}{\partial x} &= -j\omega\mu H_y \\ \frac{\partial E_y}{\partial x} - \frac{\partial E_x}{\partial y} &= -j\omega\mu H_z \end{aligned}\right\} \tag{2.24}$$

Equation (2.11) becomes

$$\left.\begin{aligned} \frac{\partial H_z}{\partial y} - \frac{\partial H_y}{\partial z} &= j\omega\varepsilon E_x \\ \frac{\partial H_x}{\partial z} - \frac{\partial H_z}{\partial x} &= j\omega\varepsilon E_y \\ \frac{\partial H_y}{\partial x} - \frac{\partial H_x}{\partial y} &= j\omega E_z \end{aligned}\right\} \tag{2.25}$$

Substitution of the conditions for a plane wave into eqns. (2.24) and (2.25) gives the relationships between the components of an electromagnetic plane wave. The conditions are that there is no variation of field in two dimensions and exp $-\gamma z$ variation in the third dimension giving

$$\frac{\partial}{\partial x} = \frac{\partial}{\partial y} = 0; \qquad \frac{\partial}{\partial z} = -\gamma$$

These conditions may be substituted into eqns. (2.24) and (2.25) to give

$$\left.\begin{aligned} \gamma E_y &= -j\omega\mu H_x \\ -\gamma E_x &= -j\omega\mu H_y \\ 0 &= -j\omega\mu H_z \end{aligned}\right\} \tag{2.26}$$

$$\left.\begin{aligned} \gamma H_y &= j\omega\varepsilon E_x \\ -\gamma H_x &= j\omega\varepsilon E_y \\ 0 &= j\omega\varepsilon E_z \end{aligned}\right\} \tag{2.27}$$

If it is a lossless wave, there is no attenuation and

$$\gamma = j\omega\sqrt{(\mu\varepsilon)}$$

Substituting for γ and using the shorthand

$$\eta = \sqrt{\left(\frac{\mu}{\varepsilon}\right)}$$

eqns. (2.26) and (2.27) simplify to

$$\begin{aligned} E_y &= -\eta H_x \\ E_x &= \eta H_y \\ E_z &= H_z = 0 \end{aligned} \tag{2.28}$$

Equation (2.28) gives an interesting insight into some of the properties of a plane wave. We have already specified that the z-direction is the direction of propagation and now we can see that there is no field acting in this direction. The ratio of the electric field strength to the magnetic field strength is η. η has the dimensions of impedance (ohms) and is called the *intrinsic impedance* of the medium through which the electromagnetic wave is propagating. Equations (2.28) also show that there are two completely separate sets of field components, E_y and H_x or E_x and H_y, with no relationship between the two sets. This means that the field associated with either set may be zero without affecting the strength of the fields associated with the other set. Hence we may consider that each set constitutes a separate wave. It is found in practice that a wave propagating in space a great distance from its source is a plane wave and has the properties specified by eqn. (2.28). If a situation occurs where more complicated fields are propagated, it is found that the wave may be analysed in terms of the sum of a number of separate plane waves in the same way as Fourier analysis enables any periodic waveform to be analysed as the sum of a number of sinusoidal waves of different frequency.

2.9. Plane Wave

Summary of the properties of a plane wave. The foregoing analysis shows that a plane wave has:

- no fields acting in the direction of propagation;
- no variation of field in the plane perpendicular to the direction of propagation;
- an electric field normal to the magnetic field;
- both fields act in a direction along the plane of the wave, that is in a direction perpendicular to the direction of propagation;
- the electric and magnetic fields are in phase with one another and directly related by the intrinsic impedance.

A plane wave is the electromagnetic wave that propagates in unbounded free space. It continues to infinity in all directions, i.e. it starts from infinity, it goes to infinity and it extends to infinity all round. No boundaries have been postulated and a plane wave cannot exist when any boundary conditions have to be considered. A plane wave is obtained in practice when the source is a long distance away (so that any curvature of the wave front may be neglected) and when any boundaries to the space are a long distance away. In this context, the expressions long and large are both relative to the wavelength of the electromagnetic wave. Light may usually be considered to be a plane wave in most practical situations, but at microwave frequencies, although many situations give rise to an electromagnetic wave approximating to a plane wave, a true plane wave would only exist for signals originating from a satellite or another planet.

2.10. Wavelength of a Propagating Wave

The wavelength of any periodic waveform is the distance in which the waveform repeats itself. As the field strength of a propagating wave is a function of both distance and time, in the above definition of wavelength it is necessary to specify that the distance is measured at any one instant of time. This definition is similar to that of the period of a wave which is the time in which the waveform repeats itself at any one position in space.

For a wave propagating with a sinusoidal waveform, $\gamma = j\beta$ and eqn. (2.21) shows that the waveform repeats itself, for $t = 0$, when $\beta z = 2\pi$ and the distance z is the wavelength. But wavelength is denoted by λ, so that

$$\lambda = \frac{2\pi}{\beta} \tag{2.29}$$

Substituting the value of phase constant for a plane wave into eqn. (2.29) shows that the wavelength of a plane wave is a function only of frequency and the material constants. If a plane wave is propagating in free space, it has a wavelength which is characteristic of electromagnetic radiation of that frequency. That wavelength is called the *characteristic wavelength* or *free space wavelength* and denoted by the symbol λ_0. Then

$$\lambda_0 = \frac{2\pi}{\omega\sqrt{(\mu_0\varepsilon_0)}} = \frac{2\pi c}{\omega} \tag{2.30}$$

but

$$\omega = 2\pi f$$

where f is the frequency of the electromagnetic wave. Therefore

$$\lambda_0 f = c \tag{2.31}$$

Equation (2.31) is the well-known relationship whereby the frequencies of many broadcast radio stations used to be known by their characteristic wavelength.

2.11. Power Flow

Energy is carried by an electromagnetic wave. The derivation of a simple mathematical relationship for the rate of power flow in the wave is given. The instantaneous electric and magnetic energy densities in a lossless medium are given by

$$W_e = \int \boldsymbol{E} . \frac{\partial \boldsymbol{D}}{\partial t} dt \tag{2.32}$$

$$W_m = \int \boldsymbol{H} . \frac{\partial \boldsymbol{B}}{\partial t} dt \tag{2.33}$$

Hence the rate of change of stored energy is given by

$$\frac{\partial}{\partial t}(W_e + W_m) = \boldsymbol{E}.\frac{\partial \boldsymbol{D}}{\partial t} + \boldsymbol{H}.\frac{\partial \boldsymbol{B}}{\partial t} \tag{2.34}$$

In a conducting medium, the resistive power loss is given by $\boldsymbol{E}.\boldsymbol{J}$. If the expression for the resistive power loss is added to the expression for the rate of change of stored energy given in eqn. (2.34), an expression is obtained for the total rate of change of electromagnetic energy in any small volume. Hence

$$\frac{\partial W}{\partial t} = \boldsymbol{H}.\frac{\partial \boldsymbol{B}}{\partial t} + \boldsymbol{E}.\boldsymbol{J} + \boldsymbol{E}.\frac{\partial \boldsymbol{D}}{\partial t} \tag{2.35}$$

Substitution for appropriate expressions from eqns. (2.6) and (2.7) into eqn. (2.35) gives

$$\frac{\partial W}{\partial t} = -\boldsymbol{H}.\nabla \times \boldsymbol{E} + \boldsymbol{E}.\nabla \times \boldsymbol{H} \tag{2.36}$$

By the application of a vector identity given in section 2.3, the expression in eqn. (2.36) becomes

$$\frac{\partial W}{\partial t} = -\nabla.\boldsymbol{E} \times \boldsymbol{H} \tag{2.37}$$

The divergence of the vector product of the two fields of the electromagnetic wave is equal to the rate of change of energy in the wave. This vector product has been given a name. It is called the *Poynting vector* and is defined by

$$\boldsymbol{S} = \boldsymbol{E} \times \boldsymbol{H} \tag{2.38}$$

Hence the time average of the divergence of the Poynting vector over any closed volume is equal to the rate of change of energy in that volume. For an electromagnetic wave flowing through a non-conducting medium, there is no dissipation of power in the medium, there is no resistive power loss in the medium and the time average of the divergence of the Poynting vector over any closed volume is equal to the power flow through the surface enclosing the volume.

Gauss's theorem gives a relationship between the divergence of any vector over any closed volume and the integral of that vector over the surface enclosing the volume. The theorem gives

$$\iiint_{\text{volume}} \nabla.\boldsymbol{S}\,dv = \iint_{\text{surface}} \boldsymbol{S}.d\boldsymbol{a} \tag{2.39}$$

where dv represents a small element of volume and $d\boldsymbol{a}$ is a vector representing a small element of the surface area and directed normal to that surface. Hence, the power flow through the surface is equal to the time average of the integral of the normal component of the Poynting vector over that surface.

For a plane wave in infinite space, the total power flow through any surface is also given by

$$\iint_{\text{surface}} \boldsymbol{S}.d\boldsymbol{a}$$

As $\boldsymbol{E}$ and $\boldsymbol{H}$ lie in the plane of the wave, eqn. (2.38) shows that $\boldsymbol{S}$ is a vector in the direction of propagation of the wave. If $\boldsymbol{a}$ represents a plane area in the plane of the wave, $\boldsymbol{S}$ and $\boldsymbol{a}$ are parallel vectors so that the power flow is given by

$$p = E_x H_y a$$

and the power flow density by

$$\boldsymbol{S} = E_x H_y \tag{2.40}$$

For sinusoidally time-varying fields, the time average of the product between two quantities is the same as half the real part of the product of one quantity and the complex conjugate of the other. Hence the complex Poynting vector is useful when considering time-varying fields,

$$\boldsymbol{S} = \tfrac{1}{2}\boldsymbol{E} \times \boldsymbol{H}^* \tag{2.41}$$

where the asterisk indicates the complex conjugate. Power is equal to the real part of the complex Poynting vector.

2.12. Boundary Conditions

Consider any arbitrary boundary between two media labelled 1 and 2 as in Fig. 2.2. The components of the fields at the boundary are categorized

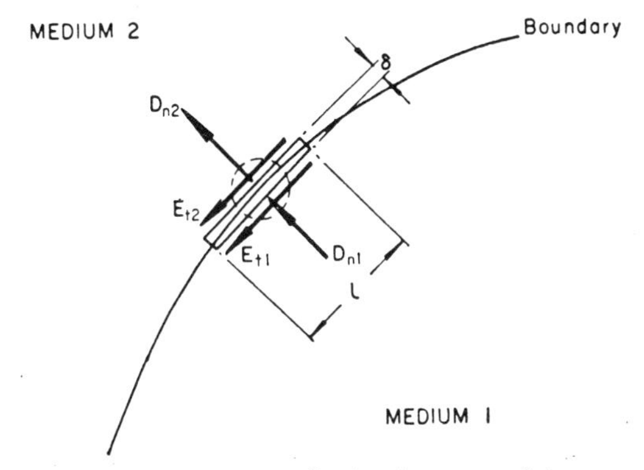

FIG. 2.2. A boundary between two media showing some of the components of the field at a point on the boundary.

into those parallel with the boundary, called the tangential component and having the subscript t, and those perpendicular to the boundary, called the normal component and having the subscript n. Some of the field components acting at any point on the boundary are shown in Fig. 2.2. Consider the divergence of the fields from an arbitrarily small volume enclosing a point on the boundary. Then Maxwell's equations give us from eqns. (2.4) and (2.5)

$$\nabla . \boldsymbol{D} = \rho \tag{2.4}$$

$$\nabla . \boldsymbol{B} = 0 \tag{2.5}$$

If the elemental boundary enclosing the point on the boundary is so small that there is no change in the tangential components of the field across the volume, eqns. (2.4) and (2.5) give us

$$-B_{n1} + B_{n2} = 0$$

and

$$-D_{n1} + D_{n2} = \rho$$

where the first subscript denotes the direction of the component of the field and the second denotes the medium in which the field is acting. Hence for the components of the field normal to the surface,

$$B_{n1} = B_{n2} \tag{2.42}$$

$$D_{n2} = D_{n1} + \rho \tag{2.43}$$

where ρ is the surface charge density at the boundary. In most practical situations there is no surface charge when both the media are non-conducting, hence for a charge-free surface

$$D_{n2} = D_{n1} \tag{2.44}$$

To find a relationship between the tangential components of the fields, it is necessary to consider the Maxwell curl equations, (2.6) and (2.7):

$$\nabla \times \boldsymbol{E} = -\frac{\partial \boldsymbol{B}}{\partial t} \tag{2.6}$$

$$\nabla \times \boldsymbol{H} = \boldsymbol{J} + \frac{\partial \boldsymbol{D}}{\partial t} \tag{2.7}$$

Consider a small rectangular area enclosing the surface as shown in Fig. 2.2 of length l and width δ. If we take the curl of the electric field around the circumference of this area and make δ so small that the contribution from the field at the ends of the rectangle may be ignored, then from eqn. (2.6)

$$(E_{t2} - E_{t1})l = j\omega B_p l\delta$$

where B_p is the average field perpendicular to the area $l \times \delta$. If δ is made so small that the right-hand side of the equation vanishes, then

$$E_{t2} - E_{t1} = 0$$

or

$$E_{t2} = E_{t1} \tag{2.45}$$

Similarly from eqn. (2.7)

$$(H_{t2} - H_{t1})l = -\left(J_p + \frac{\partial D_p}{\partial t}\right)l\delta$$

where the subscript p is again used to denote the average component of the fields perpendicular to the area $l \times \delta$. If δ is made sufficiently small, the electric displacement term in the equation will vanish, but the current density term will not necessarily be zero. For a good conductor, the current only exists in a thin skin close to the surface so that the product $J_p\delta$ is finite. It is shown in section 7.6 that a large current density exists in a thin skin close to the surface and the product of the current density and distance into the surface is equivalent to the total surface current.

Hence for a perfect conductor

$$J_p\delta = I_s$$

and since no electromagnetic fields can exist inside a perfect conductor, $H_{t2} = 0$ and

$$H_{t1} = I_s \tag{2.46}$$

However, the general relationship between the tangential magnetic field components on each side of the boundary is

$$H_{t2} = H_{t1} - J_p\delta \tag{2.47}$$

and for a non-conducting medium

$$H_{t2} = H_{t1} \tag{2.48}$$

Summary. A general statement relating the fields across a boundary in non-conducting and charge-free media:

The normal components of $\boldsymbol{B}$ and $\boldsymbol{D}$ and the tangential components of $\boldsymbol{H}$ and $\boldsymbol{E}$ are the same on each side of the boundary.

For a conducting medium the normal components of $\boldsymbol{D}$ and the tangential components of $\boldsymbol{H}$ are not necessarily the same, but the other two relationships still apply.

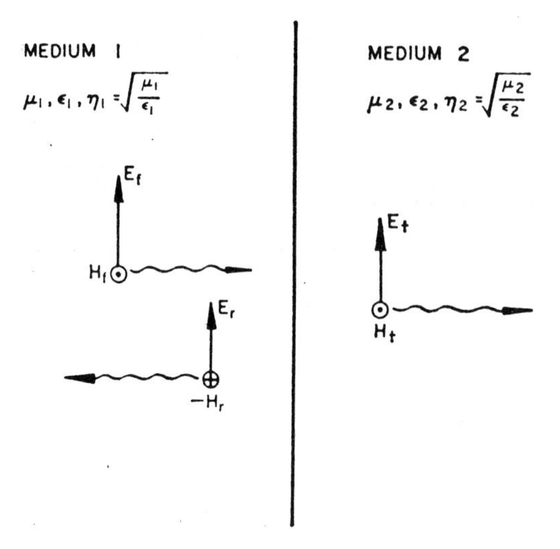

FIG. 2.3. Plane boundary between two semi-infinite media showing the material properties of the two media and the field components of the forward, transmitted and reflected plane waves.

2.13. Reflection from a Plane Boundary

As an example of the application of boundary conditions, consider the effect of a plane boundary between two semi-infinite blocks of non-conducting material as shown in Fig. 2.3. Let there be an incident plane wave in medium 1 normal to the boundary. Then there is a plane wave reflected from the boundary in medium 1 and there is a plane wave transmitted through the boundary into medium 2. The components of the three waves are denoted by the subscripts f, r and t for the incident or forward wave, the reflected wave and the transmitted wave respectively. The electric field components of the three waves are all parallel as shown in Fig. 2.3. It is assumed that the magnetic fields of the three waves are also parallel and all acting in the same direction. Hence the normally incident wave will have fields

$$E_f \quad \text{and} \quad H_f \qquad \text{where} \qquad E_f = \eta_1 H_f$$

which will give rise to a transmitted wave having fields

$$E_t \quad \text{and} \quad H_t \qquad \text{where} \qquad E_t = \eta_2 H_t$$

and a reflected wave having fields

$$E_r \quad \text{and} \quad H_r \qquad \text{where} \qquad E_r = -\eta_1 H_r$$

The negative sign occurs in the previous equation because all the magnetic fields are defined as acting in the same direction, but the reflected wave is propagating in the reverse direction.

At the boundary the tangential fields are the same, therefore

$$H_f + H_r = H_t \tag{2.49}$$

$$E_f + E_r = E_t \tag{2.50}$$

Substituting the relationships between E and H into eqn. (2.50) gives

$$\eta_1 H_f - \eta_1 H_r = \eta_2 H_t \tag{2.51}$$

and solving eqns. (2.49) and (2.51) gives

$$2H_f = \left(1 + \frac{\eta_2}{\eta_1}\right) H_t$$

$$2H_r = \left(1 - \frac{\eta_2}{\eta_1}\right) H_t$$

whence

$$\frac{H_f}{H_r} = \frac{\eta_1 + \eta_2}{\eta_1 - \eta_2} \tag{2.52}$$

and

$$\frac{E_f}{E_r} = \frac{\eta_1 + \eta_2}{\eta_1 - \eta_2} \tag{2.53}$$

The reflected wave causes a standing wave pattern in the medium 1 where

$$E_{\text{max}} = E_f + E_r$$

$$E_{\text{min}} = E_f - E_r$$

and the *voltage standing wave ratio* (VSWR) is given by

$$S = \frac{E_{\text{max}}}{E_{\text{min}}} \tag{2.54}$$

Substitution of values for E_{max} and E_{min} from eqn. (2.53) gives

$$S = \frac{\eta_2}{\eta_1} \tag{2.55}$$

2.14. Summary

2.2. Material relationships:

$$\boldsymbol{B} = \mu \boldsymbol{H} \tag{2.1}$$

$$\boldsymbol{D} = \varepsilon \boldsymbol{E} \tag{2.2}$$

$$\boldsymbol{J} = \sigma \boldsymbol{E} \tag{2.3}$$

Permeability $\mu = \mu_0 \mu_r$

Permittivity $\varepsilon = \varepsilon_0 \varepsilon_r$

$$\mu_0 = 4\pi \times 10^{-7}\ \mathrm{H/m}$$

$$\varepsilon_0 = \frac{1}{c^2 \mu_0} \approx \frac{1}{(36\pi \times 10^9)}\ \mathrm{F/m}$$

Velocity of light

$$c = \frac{1}{\sqrt{(\mu_0 \varepsilon_0)}} \approx 3 \times 10^8\ \mathrm{m/s}$$

Impedance of free space

$$\eta = \sqrt{\left(\frac{\mu_0}{\varepsilon_0}\right)} \approx 120\pi = 377\ \Omega$$

2.3. Section 2.3 gives a summary of the useful vector relationships.

2.4. ***Maxwell's equations*** are

$$\nabla . \boldsymbol{D} = \rho \tag{2.4}$$

$$\nabla . \boldsymbol{B} = 0 \tag{2.5}$$

$$\nabla \times \boldsymbol{E} = -\frac{\partial \boldsymbol{B}}{\partial t} \tag{2.6}$$

$$\nabla \times \boldsymbol{H} = \boldsymbol{J} + \frac{\partial \boldsymbol{D}}{\partial t} \tag{2.7}$$

2.5. The general solutions for a non-conducting medium are

$$\nabla^2 \boldsymbol{E} + \omega^2 \mu \varepsilon \boldsymbol{E} = 0 \tag{2.12}$$

$$\nabla^2 \boldsymbol{H} + \omega^2 \mu \varepsilon \boldsymbol{H} = 0 \tag{2.13}$$

2.6. For a ***plane wave*** propagating in the z-direction, the differential equation is

$$\frac{\partial^2 E_x}{\partial z^2} = -\omega^2 \mu \varepsilon E_x \tag{2.16}$$

2.7.

$$E_x = E_0 \exp(j\omega t - \gamma z) \tag{2.21}$$

$$\gamma = j\beta = j\omega\sqrt{(\mu\varepsilon)} \tag{2.22}$$

$$\text{Speed} = \frac{1}{\sqrt{(\mu\varepsilon)}}$$

2.8. The field relationships of a plane wave propagating in the z-direction are

$$E_y = -\eta H_x \tag{2.28}$$

and all other components of the field are zero, or

$$E_x = \eta H_y \tag{2.28}$$

and all other components of the fields are zero.

2.9. The properties of a plane wave are summarized at the beginning of section 2.9.

2.10. The wavelength

$$\lambda = \frac{2\pi}{\beta} \tag{2.29}$$

The ***characteristic wavelength***

$$\lambda_o = \frac{2\pi}{\omega\sqrt{(\mu_0\varepsilon_0)}} = \frac{2\pi c}{\omega} \tag{2.30}$$

$$\lambda_0 f = c \tag{2.31}$$

2.11. ***Poynting vector***

$$\boldsymbol{S} = \boldsymbol{E} \times \boldsymbol{H} \tag{2.38}$$

Power flow through any surface is the integral over the surface of the time average of the Poynting vector component normal to the surface.

2.12. At any boundary between non-conducting media, the normal components of $\boldsymbol{B}$ and $\boldsymbol{D}$ and the tangential components of $\boldsymbol{H}$ and $\boldsymbol{E}$ are the same on each side of the boundary.

Problems

2.1. If a free magnetic pole were possible, eqn. (2.5) would be of the form

$$\nabla . \boldsymbol{B} = \rho_m$$

Find the dimensions of the magnetic charge density ρ_m. Similarly, a magnetic current density term would appear in eqn. (2.6). What would be its dimensions?

2.2. Derive the expressions for the vector operators in the two polar coordinate systems given in section 2.3.

2.3. By writing eqn. (2.7) in its component parts in a cylindrical polar coordinate system, derive an expression for the magnetic field due to a direct current flowing in a wire.

2.4. By writing eqns. (2.8) to (2.11) in their component parts in a rectangular coordinate system, derive eqn. (2.14) without using any vector analysis.

2.5. Calculate the speed of an electromagnetic plane wave through the following materials and its wavelength compared with air in each case.

	ε_r	μ_r
Air	1	1
PTFE	2.0	1
Alumina	10	1
Titanium dioxide	90	1

2.6. A circularly polarized plane wave is described by the equation $E_x = -jE_y$. Write expressions for all the components of the fields of this wave and describe it as the sum of a number of normal plane waves. [See section 6.7 for a discussion of this problem.]

2.7. Calculate the frequencies of the following different types of electromagnetic radiation, where the characteristic wavelength is given:

X-rays	30 pm
Visible light	0.6 μm
Infrared	100 μm
Microwave	3 cm
Radio	300 m

2.8. Making the assumption that a spherical wavefront is an approximation to a plane wave, find an expression for the microwave power density (W/m^2) at any distance from a 1 kW isotropic source of microwave radiation. Hence calculate the field strength at distances of 1 m and 1 km from the source. [173 V/m, 0.173 V/m]

2.9. The boundary condition at a conducting surface is that the tangential electric field components of an electromagnetic wave are zero at the plane of the surface. A plane wave is normally incident onto a plane conducting sheet. Find an expression for the field components of the reflected wave and the position of the first minimum of the standing wave pattern.

2.10. Calculate the VSWR in the air space for a plane wave in air normally incident onto the plane surface of a dielectric medium of relative permittivity 2.

The boundary in Fig. 2.3 is matched by means of a quarter wavelength thickness of material separating the two media. Derive expressions for the material properties of the new medium.

3

Waveguide Transmission

3.1. Waveguide

So far mathematical results have been obtained which describe the properties of an electromagnetic wave propagating in a uniform medium of infinite extent. The plane wave, which is the fundamental form of wave propagating in such a medium, cannot exist when there are any boundaries to space. In practice, at microwave frequencies these conditions are not even approximately true and the effect of boundaries has to be considered. It is often necessary to control the electromagnetic radiation and to channel it from one point to another without allowing it to escape as radiation into the surrounding space.

Any system of multiple conductor transmission line guides and controls the electromagnetic radiation with varying amounts of containment. Some of those most used at microwave frequencies are discussed in sections 3.6 and 3.8. For microcircuits, the dimensions are small, any losses associated with the insulating material of the transmission line are also small and the complete circuit may be enclosed in a sealed metal box. However, for high power, microcircuits are unsuitable and a larger all metal line is required. Only a totally enclosed conductor system provides no radiation loss. Hollow metal pipe is one of the most convenient totally enclosed systems and provides the lowest attenuation.

It is only under certain conditions that electromagnetic radiation propagates freely along the inside of a hollow metal pipe. The pipe used for this purpose is called a *waveguide*. Chapters 5 and 6 are devoted to the solution of Maxwell's equations inside a hollow metal pipe and the determination of the necessary conditions of shape and size. Before considering the special properties of rectangular and circular waveguide, in Chapters 5 and 6, this chapter develops those relationships which are common to a number of different types of waveguide. First in this chapter, propagation in parallel plate waveguide is described by means of a pictorial description of the superposition of plane waves, that has helped many to a better visualization of waveguide propagation.

3.2. Parallel Plate Waveguide

Consider a waveguide system consisting of two parallel plates as shown in Fig. 3.1. The figure shows only a section of the system since both the plates ought to be considered to extend to infinity in both directions. The plates are the only boundaries to the electromagnetic wave propagating between them. The boundary condition is that each metal plate acts as a perfectly conducting sheet. No metal is a perfect conductor, but to a first approximation the metal forming the waveguide wall may be considered to be a perfect conductor. No electric field can exist parallel to the wall at the surface of the wall. Hence the electric field of the electromagnetic wave must be perpendicular to the plane of the walls as shown in Fig. 3.1. A

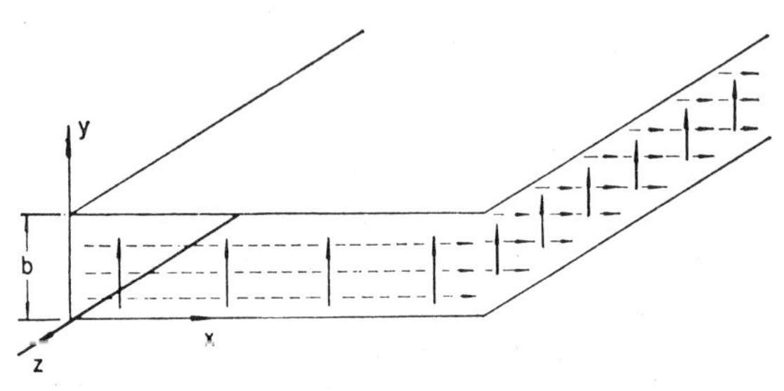

FIG. 3.1. A portion of an infinite parallel plate waveguide. ——→ electric field. — — — → magnetic field.

perfectly conducting sheet is also a perfectly reflecting mirror so that the electromagnetic field patterns shown in Fig. 3.1 would be reflected to make the field patterns of an infinite plane wave in an infinite medium propagating in the z-direction. Hence the parallel plate waveguide can be considered to have taken a slice out of a plane wave which is propagating in a direction in the plane of the plates and whose electric field is perpendicular to the plane of the plates. This mode of propagation exhibits all the properties of a plane wave and is called a transmission line mode. Two other systems of conductors which support transmission line modes are discussed in section 3.6. However, there are other possible configurations for the fields inside parallel plate waveguide and these exist for other modes of propagation called waveguide modes.

3.3. Waveguide Modes

There is a pictorial description of the parallel plate waveguide modes which is very helpful in understanding some of the properties of waveguide propagation. Some results are developed intuitively here, and their strict mathematical derivation is postponed until Chapter 5 or 6. In Fig. 3.2, a plane wave is impinging at an angle onto a plane conducting sheet which is acting as a perfect reflector. The lines on the diagram denote positions of equal phase, both on the incident wave and on the reflected wave. The direction of propagation of the wave is perpendicular to the lines of equal

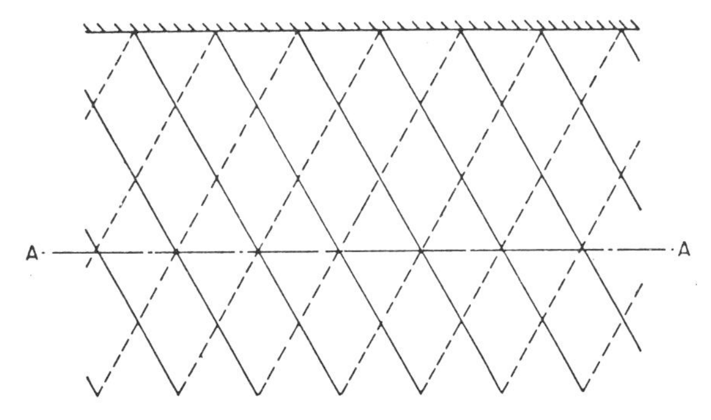

FIG. 3.2. Plane wave reflection from a plane conducting sheet, showing lines of equal phase for both the incident and reflected wave.

phase which are a wavelength apart. It is seen that a second conducting sheet could be positioned at A–A, parallel to the first sheet, without affecting the wave pattern between the sheets. Such a plane wave propagating by reflection from two parallel reflectors is shown in Fig. 3.3 together with a single ray along the direction of propagation of the plane wave. It is seen that the ray is reflected without loss at an angle θ from the faces of the reflectors.

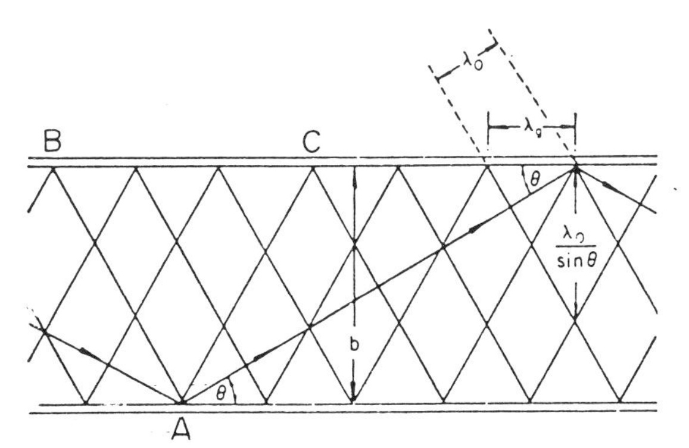

FIG. 3.3. Ray and phase front representation of an electromagnetic wave propagating between parallel plate waveguide. One ray is shown and it is perpendicular to the lines of equal phase. The separation between the plates is b.

A construction such as that shown in Fig. 3.3 is dependent upon there being a formal relationship between the spacing between the plates and the angle and wavelength of the plane wave. The condition is that there should be a whole number of the projections of half a wavelength in the spacing between the plates. If the wavelength is λ_0 the projection of a wavelength is $\lambda_0/\sin\theta$ and the condition is given by

$$b = \frac{n\lambda_0}{2\sin\theta} \tag{3.1}$$

where n is an integer. In Fig. 3.3 the direction of propagation is in the plane of the paper parallel to the plane of the plates and it is seen that the wavelength in the direction of propagation is not the same as the wavelength of the plane wave. The wavelength in the direction of propagation along the waveguide is called the *waveguide wavelength*, denoted by λ_g. It is the projection of the plane wave wavelength in the direction of propagation between the plates, hence

$$\lambda_g = \frac{\lambda_0}{\cos\theta} \tag{3.2}$$

3.4. Cut-off Conditions

If the spacing between the plates is such that

$$b = \tfrac{1}{2}n\lambda_0 \tag{3.3}$$

then $\theta = 90^\circ$ and the waveguide wavelength is infinite, which means that the wave does not propagate along the waveguide. This situation is termed *cut off* and the condition is given by eqn. (3.3). At a constant frequency, which means that the plane wave wavelength or characteristic wavelength λ_0 remains constant, let the spacing between the plates of the parallel plate waveguide be varied. As the spacing is reduced, the angle θ increases and the waveguide wavelength increases. Eventually the spacing is reduced so much that the cut-off conditions are reached and propagation in that particular waveguide mode ceases. For smaller spacing between the plates, there cannot be any propagation in that particular waveguide mode. A fuller mathematical description of cut-off conditions is given in section 5.3.

For a fixed size of waveguide which is a more normal condition, there is a frequency given by substituting eqn. (2.31) into the cut-off condition, eqn. (3.3), to give the *cut-off frequency*

$$f_c = \frac{cn}{2b}$$

The cut-off frequency has a characteristic wavelength associated with it called the *cut-off wavelength*. The symbols associated with both these quantities have the subscript c. Hence

$$\lambda_c = \frac{2b}{n}$$

Combining eqns. (3.1) and (3.2) to eliminate θ gives

$$\lambda_g = \frac{\lambda_0}{\sqrt{\left[1 - \left(\frac{n\lambda_0}{2b}\right)^2\right]}}$$

and substituting for λ_c gives

$$\lambda_g = \frac{\lambda_0}{\sqrt{\left[1 - \left(\frac{\lambda_0}{\lambda_c}\right)^2\right]}} \tag{3.4}$$

It is seen from eqn. (3.4) that the waveguide wavelength is determined by the characteristic wavelength of the electromagnetic radiation and the cut-off wavelength of the waveguide. The cut-off wavelength is a function of the size and, as will be seen in Chapters 5 and 6, shape of the waveguide and of the integer n. The integer is determined by the pattern of the fields of the electromagnetic wave in the waveguide and is constant for any particular mode of waveguide propagation. A detailed discussion of waveguide modes is given in Chapter 5 and in particular in section 5.8 on mode nomenclature.

Consider the modes of propagation in any particular parallel plate waveguide. At low frequencies, only the transmission line mode shown in Fig. 3.1 is able to propagate. The frequency is below the cut-off frequency of all the possible waveguide modes. At frequencies greater than that given by

$$\lambda_0 = 2b$$

the first of the waveguide modes is able to propagate as well as the transmission line mode. At frequencies greater than that given by

$$\lambda_0 = b$$

two waveguide modes are able to propagate. The waveguide is able to support more and more modes as the frequency is increased. The parallel plate waveguide consists of two separate conductors and so it is able to support the transmission line mode. Any transmission line, consisting of two or more parallel conductors, can support the transmission line mode. The hollow metal waveguide, however, consists of only one conductor and cannot support the transmission line mode; this idea is developed mathematically in section 6.8. Consider Fig. 3.1; if this parallel plate waveguide is changed into a closed waveguide which is not infinite in the x-direction, then somewhere there must be a conducting wall parallel to the electric field which then could not exist. The plane wave type of transmission line mode cannot exist inside a closed waveguide. However, the waveguide modes exhibiting the properties of cut off and having a waveguide wavelength different from the plane wave wavelength can exist inside any hollow conducting pipe.

3.5. Wave Velocities

The phase velocity has already been mentioned in sections 1.3 and 2.7 as the speed of a wave, but it will be derived formally for completeness. All the field components of the electromagnetic waves propagating through any system have a time and distance dependence of

$$\exp j(\omega t - \beta z)$$

where the z-direction is the direction of propagation. The velocity of propagation of the wave is the velocity taken by an observer who could remain at a point of constant phase on the wave. This velocity is called the *phase velocity*, denoted by v_p, and is such that the expression $(\omega t - \beta z)$ is a constant. The velocity is given by

$$v_p = \frac{dz}{dt} = \frac{\omega}{\beta} \tag{3.5}$$

Reference to Fig. 3.3 shows that the waveguide wavelength is longer than the plane wave wavelength and hence the phase velocity of the waveguide wave is faster than that of the plane wave. Since the plane wave and the waveguide wave both propagate a distance of one wavelength in the same time, if the plane wave propagates with the velocity c then the waveguide mode will propagate with the velocity

$$v_p = \frac{c}{\cos\theta}$$

or

$$v_p = \frac{c\lambda_g}{\lambda_0} \tag{3.6}$$

The phase constant of a propagating wave is related to the waveguide wavelength by the relationship (similar to eqn. (2.29))

$$\lambda_g = \frac{2\pi}{\beta} \tag{3.7}$$

Substituting for c from eqn. (2.31) and λ_g from eqn. (3.7) into eqn. (3.6) gives

$$v_p = \frac{\omega}{\beta} \tag{3.5}$$

hence showing, as would be expected, that the phase velocity derived from the pictorial description of parallel plate waveguide is the same as that derived theoretically. The phase velocity here specified is faster than the speed of light and it may appear to contradict the theory of relativity. Phase velocity, however, is only an apparent velocity. It is the speed of movement of a point of constant phase in the centre of a continuous wave and it is not the speed at which information travels along the waveguide. Consider water waves on the surface of an otherwise still sea. The waves appear to be moving across the surface of the water. However, the water molecules are only moving up and down in a vertical direction perpendicular to the direction of propagation of the wave. Therefore, the phase velocity of the wave can be considerably faster than the actual velocity of movement of the water. In our electromagnetic wave, the initial point of any disturbance travels along the waveguide with the speed of light, but the main content of the disturbance arrives later at a slower speed called the *group velocity*, denoted by v_g.

To return to the pictorial description of the waveguide, any information could be considered to be travelling along a path shown by the ray in Fig. 3.3. Consider the point A in Fig. 3.3 where the ray is reflected from the wall of the waveguide. The incident ray at A is associated with the plane wave phase front AC, but the reflected ray is associated with the phase front AB. Therefore, on reflection, BC, which is three waveguide wavelengths in this diagram, is effectively lost. This gives another explanation why the phase velocity can be considerably faster than the effective velocity of the ray along the waveguide. Then its speed along the waveguide is

$$v_g = c\cos\theta$$

which by substitution from eqn. (3.2) gives

$$v_g = c\frac{\lambda_0}{\lambda_g}$$

Substituting for c from eqn. (2.31) and λ_g from eqn. (3.7) gives

$$v_g = c^2\frac{\beta}{\omega} \tag{3.8}$$

Information must be propagated by means of pulses or by modulation of a continuous wave. Let us consider the simplest form of modulation of a continuous wave, that consisting of two equal waves with a small frequency difference between them. The angular frequencies will be ω and $\omega + \delta\omega$ and the corresponding phase constants will be β and $\beta + \delta\beta$. The expressions for the electric field strength of the two waves are

$$E_1 = E_0\sin(\omega t - \beta z)$$
$$E_2 = E_0\sin(\omega t + \delta\omega t - \beta z - \delta\beta z)$$

and the field strength of the combined wave is

$$E = 2E_0\sin(\omega t + \tfrac{1}{2}\delta\omega t - \beta z - \tfrac{1}{2}\delta\beta z)\cos(\tfrac{1}{2}\delta\omega t - \tfrac{1}{2}\delta\beta z)$$

This is an amplitude modulated wave and the velocity of propagation of the modulation envelope is given by

$$v_g = \frac{dz}{dt} = \frac{\delta\omega}{\delta\beta} \tag{3.9}$$

In the limit of small changes, eqn. (3.9) becomes

$$v_g = \frac{\partial \omega}{\partial \beta}$$

The relationship for v_g for a normal waveguide will be derived from eqn. (3.9) and will be shown to be the same as that given by eqn. (3.8). Substitution for the different wavelengths in eqn. (3.4) gives

$$\frac{1}{\beta} = \frac{c}{\omega\sqrt{[1 - (\omega_c/\omega)^2]}}$$

whence

$$\omega^2 = c^2\beta^2 + \omega_c^2 \tag{3.10}$$

Differentiation of eqn. (3.10) gives the expression for v_g which is the same as that given by eqn. (3.8):

$$v_g = \frac{\partial \omega}{\partial \beta} = \frac{c^2\beta}{\omega}$$

Here it is seen that for waveguides in general

$$v_p > c > v_g$$

and that from eqns. (3.5) and (3.8)

$$v_p v_g = c^2$$

For the plane wave propagating in free space or for the two-conductor transmission line

$$v_p = v_g = c$$

3.6. Transmission line modes

Any two-conductor or multiple-conductor transmission line supports propagation of the transmission line mode where the components of the fields exist only in a plane perpendicular to the direction of propagation similar to the plane wave. Consider the coaxial line having a central conductor of circular cross-section and outer conducting cylindrical sheath, as shown in Fig. 3.4. The field distribution in the coaxial line may be developed by inference from the parallel plate waveguide field shown in Fig. 3.1. If the parallel plate waveguide is bent to simulate the fields in a coaxial line as shown in Fig. 3.5, the fields remain entirely in the plane perpendicular to the direction of propagation. The electric field remains perpendicular to the plane of the conductors and the magnetic field remains parallel to them and perpendicular to the electric field. When the bending is completed, the fields in a coaxial line as shown in Fig. 3.6 are obtained. For the transmission line mode, with any shape of conductor, the fields are always in a plane

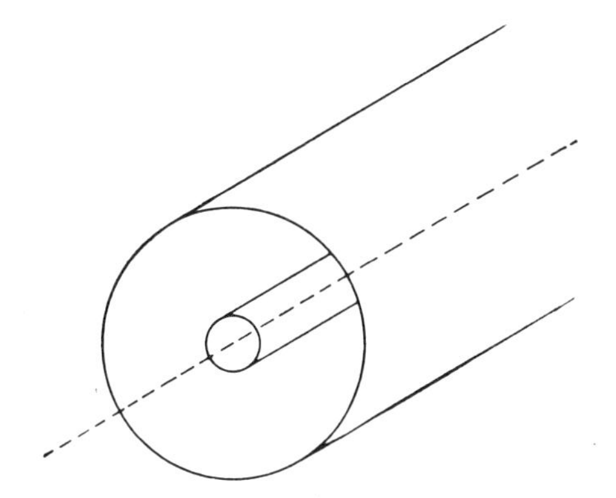

FIG. 3.4. Coaxial transmission line.

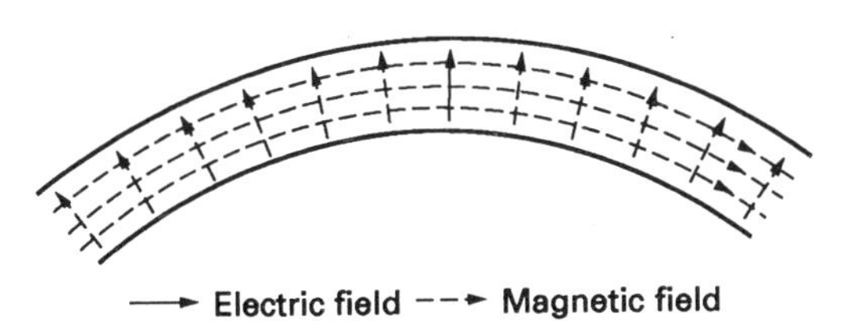

FIG. 3.5. Part of the field in a parallel plate waveguide bent to simulate a coaxial line.

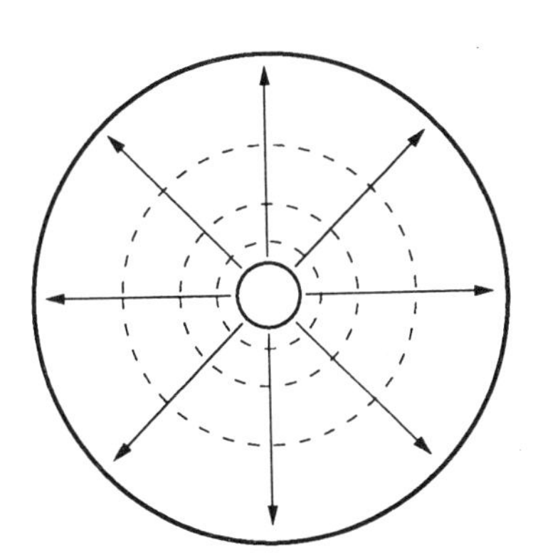

FIG. 3.6. The electric and magnetic fields in a coaxial transmission line.

perpendicular to the plane of propagation. The electric field is always perpendicular to the magnetic field and at any point their ratio is always the intrinsic impedance. The fields in the coaxial line are derived mathematically in section 6.8 and show that, for the coaxial line, the previous statement is true.

The field distribution in any two-conductor or multi-conductor transmission line which is propagating a transmission line mode is the same as the d.c. field distribution due to a potential difference between the conductors and a current in the conductors. The only difference is that the electric and magnetic fields are related by Maxwell's equations and at any point their ratio is the intrinsic impedance. In the same way, the d.c. potential difference and current on the line are controlled by external factors, whereas the a.c. potential difference and current are related by the characteristic impedance of the line. From the properties of the d.c. fields, the electric and magnetic fields at any radius r in a coaxial line are given by

$$E = \frac{V}{r \log_e(a/b)} \tag{3.11}$$

$$H = \frac{I}{2\pi r} \tag{3.12}$$

where a and b are the radial dimensions of the coaxial line. At d.c. these two equations are independent but at a.c. E and H are related by Maxwell's equations to give

$$\frac{E}{H} = \eta \tag{3.13}$$

Substitution from eqns. (3.11) to (3.13) show that the characteristic impedance of a coaxial line is given by

$$Z_0 = \frac{V}{I} = \frac{\eta}{2\pi} \log_e \frac{a}{b}$$

Another simple shape of transmission line is the two-wire line, whose fields are shown in Fig. 3.7. All the field lines are circles or segments of circles and again they are directly related to the d.c. electric and magnetic fields.

3.7. Impedance

For a uniform two-wire transmission line, the *characteristic impedance* of the line, eqn. (1.1), is given by

$$Z_0 = \frac{V}{I}$$

which is the ratio of the potential and current in a line of infinite length. Provided the line is uniform, the ratio is constant and the impedance is a characteristic of the line.

For a plane wave in a uniform homogeneous medium, the *intrinsic*

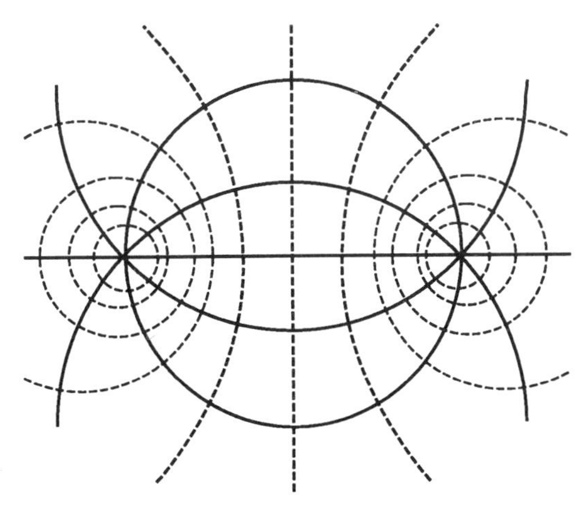

FIG. 3.7. The electric and magnetic fields in a two-wire transmission line.

impedance of the medium is given by eqn. (2.28) or eqn. (3.13)

$$\eta = \frac{E}{H} \tag{3.13}$$

which is the ratio of the electric and magnetic fields in the wave. In the plane wave, the fields both lie in a plane perpendicular to the direction of propagation. Hence for any travelling wave, a *wave impedance* is defined as the ratio of the transverse components of the electric and magnetic fields. The wave impedance of a propagating model will use the same notation as the characteristic impedance of a two-wire transmission line so that

$$Z_0 = \frac{E_t}{H_t} \tag{3.14}$$

where the subscript t is used to denote the components of the field perpendicular to the direction of propagation. For the transmission line mode, discussed in section 3.6, the wave impedance is the same as the intrinsic impedance. For the other possible modes, which are waveguide modes exhibiting cut off, the wave impedance is different from the intrinsic impedance. For hollow metal waveguides, the wave impedance is derived in section 5.12.

3.8. Stripline and Microstrip

One transmission line which is very useful in the design of microwave miniature circuits and integrated circuits is *microstrip* which is one form of the planar conductor system called *stripline.* It consists of a thin planar conductor supported on a dielectric sheet with an earthed metallic plane on the opposite side of the dielectric. The symmetrical stripline has the conductor sandwiched between two dielectric sheets and two earthed metal conductors.

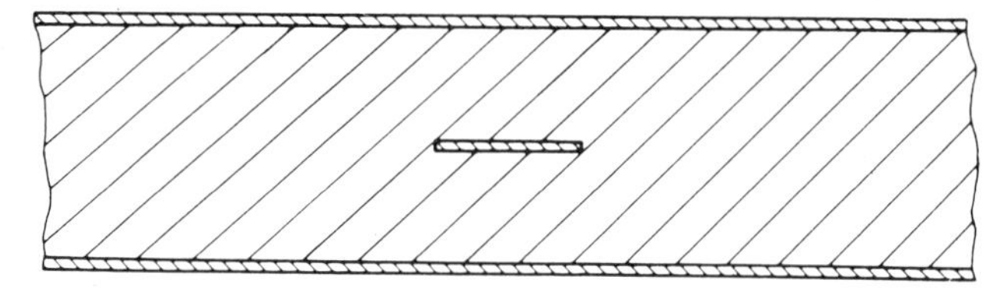

FIG. 3.8. Triplate stripline.

It is shown in Fig. 3.8 and is sometimes called triplate line. It has the advantage that the microwave field is enclosed as in a waveguide and there is little radiation of the field but the disadvantage that, being enclosed, it is difficult to adjust and to add active devices. The unsymmetrical stripline is shown in Fig. 3.9 and is usually called *microstrip*. It has the advantage that

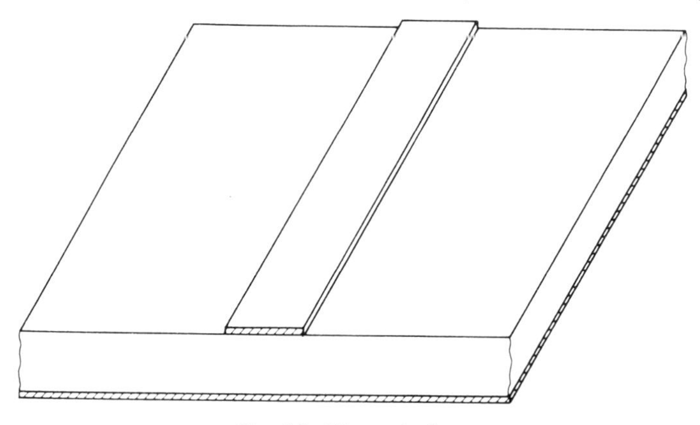

FIG. 3.9. Microstrip line.

it can easily be constructed using techniques similar to those used to make printed circuits, and lumped devices can easily be added to the circuit. In a microminiature form, it is used with unencapsulated semiconductor devices to make oscillator or amplifier circuits. It has the disadvantage that part of the field is in the air space above the dielectric so that the line must be enclosed in a box. Obviously some waveguide theory must be applied to the design of the box. Another disadvantage of microstrip is that the wave is not a pure transmission line mode, because the field is partly in a dielectric and partly in air.

In order to help in the design of stripline circuits, it is useful to know the waveguide wavelength and characteristic impedance of the line. The triplate line shown in Fig. 3.8 supports a pure transmission line mode of propagation. The waveguide wavelength is the same as the free space wavelength of a plane wave in an infinite dielectric medium having a relative permittivity the same as that of the dielectric filling the stripline. The capacitance and inductance of the line can be calculated from electrostatics and magnetostatics and the characteristic impedance found from eqn. (1.29). However, the transmission medium of the microstrip line shown in Fig. 3.9 is not uniform. While most of the microwave power is contained in the dielectric between the strip and the groundplane some microwave power exists in the air space above the conducting strip. Also calculations based on electrostatics and magnetostatics do not give entirely correct answers for the fields in the stripline. An approximate impression of the fields on microstrip is shown in Fig. 3.10. From electrostatic calculations, the form of the electric field is

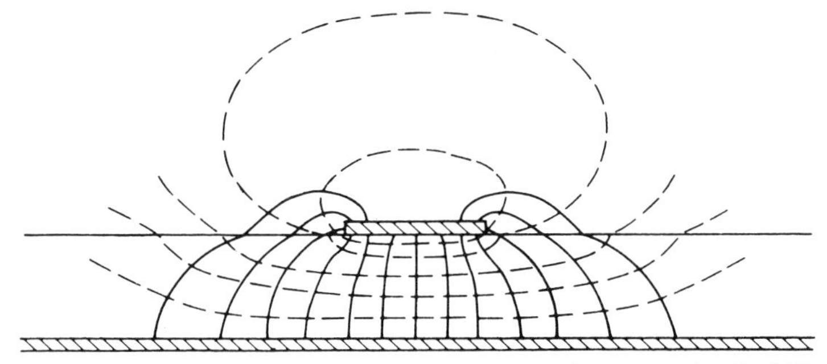

FIG. 3.10. Approximate form of the fields on microstrip. ——— electric field, - - - - magnetic field.

approximately that shown in Fig. 3.10 and there is a kink in the electric field lines at the air dielectric boundary. However, for magnetostatic calculations the air dielectric boundary does not affect the magnetic field and the magnetic field lines are continuous smooth curves surrounding the stripline conductor. Maxwell's equations for time varying fields relate the electric and magnetic fields at each point in space, so that for the microwave field the magnetic field is affected by the electric field and the magnetic field lines also kink the air dielectric boundary. Pure transmission line mode propagation does not occur and calculations based on purely static field considerations are not exact. At lower frequencies, however, static field calculations give satisfactory results. There is a quasi transmission line mode of propagation which is a good static approximation to the true dynamic fields.

For the quasi transmission line mode, the characteristic impedance of microstrip against various values of the ratio, width of the strip to thickness of the dielectric, w/h, is given in Fig. 3.11. A number of curves are given for different values of the relative permittivity of the dielectric of the line. The curves are calculated from an empirical formula* devised to be the best approximation to theoretical results. The theoretical calculations assume that the strip is of negligible thickness; formulae for the small correction to

* H. A. Wheeler, Transmission-line Properties of a Strip on a Dielectric Sheet on a Plane. *IEEE Trans. on Microwave Theory and Techniques*, Vol. **MTT-25**, pp. 631–647 (August 1977). (See particularly eqn. (10) and Figs. 2 and 3.)

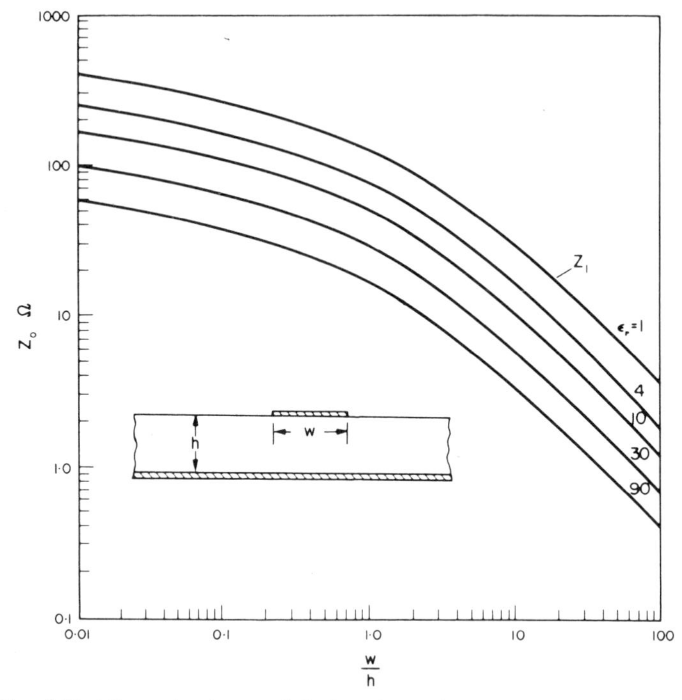

FIG. 3.11. Microstrip characteristic impedance for various values of relative permittivity of the dielectric, calculated from a formula given by Wheeler.*

allow for strip thickness are also available.* The microstrip line for which $\varepsilon_r = 1$ is the theoretical air-filled line which does support a transmission line mode of propagation and whose waveguide wavelength is the same as the characteristic wavelength. For any other dielectric material in the line, the waveguide wavelength is less than the characteristic wavelength in the same ratio as the characteristic impedance Z_0 is less than Z_1, the characteristic impedance of the same shape of air-filled line. Then the waveguide wavelength in the microstrip line is given by

$$\lambda = \lambda_0 \frac{Z_0}{Z_1} \tag{3.15}$$

At higher frequencies, there is a small increase in characteristic impedance and a small decrease in phase velocity with increase in frequency. This deviation from transmission line mode propagating conditions is dependent on both the frequency and the size of the microstrip circuit. Much work has been done on the propagating conditions in microstrip line and design formulae making allowance for high-frequency effects have been published. However, the correction for high-frequency effects is small, and the results for the quasi transmission line mode are satisfactory for most purposes.

* See footnote, p. 63.

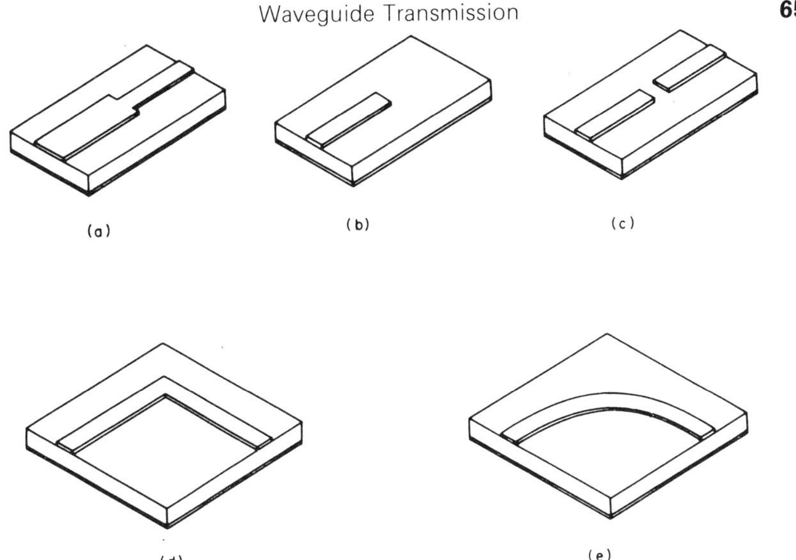

FIG. 3.12. Some stripline discontinuities, (a) step changes in width, (b) open end, (c) gap, (d) abrupt corner and (e) radius bend.

3.9. Discontinuities and stubs

Any discontinuity in a stripline triplate or microstrip conductor introduces some reactive impedance at the point of the discontinuity. This section gives the effects of the discontinuities shown in Fig. 3.12.

Step change in width of the conductor: Fig. 3.12 (a), characteristic impedance changes in a line can be effected by changing the width of the conductor. As already discussed in Chapter 1, such a change in impedance causes a mismatch on the line or may be used as described in section 1.11 to match an existing mismatch. However, the step change in width of the conductor also introduces a small inductive reactance in series with the line. The effect of this inductance is small and may usually be neglected.

Open end: Fig. 3.12 (b), as with all microwave circuits, the open ended line is not a perfect open circuit because some of the microwave power is radiated from the end of the line. The effect may usually be neglected unless the open circuit is part of a high Q resonant length of line; the power radiated from the end of the line is of the same order as the power dissipated in the line. Also electrically the open circuit is slightly beyond the physical end of the line. The equivalent increase in length is less than $0.5\,h$ for microstrip.

Gap: Fig. 3.12 (c). A gap in the conductor of stripline provides a capacitance in series with the conductor.

Corner: the abrupt right angle corner shown in Fig. 3.12 (d) introduces a capacitive shunt reactance at the point of the corner. However, for a radius bend, Fig. 3.12 (e), having a radius greater than $3\,w$ this reactance becomes so small that it may be neglected.

Stubs: such as those used for matching in section 1.10, are easily realized in stripline. The impedance of a shorted stub of length l is given by eqn. (1.44). Therefore the reactive impedance of such a stub is given by

$$X_L = Z_0 \tan \beta l = Z_0 \tan (\omega l/v) \tag{3.16}$$

For a short line, $l \ll v/\omega$, and the impedance becomes

$$X_L = \omega L \approx Z_0 \omega l/v \tag{3.17}$$

Therefore the inductance is given by

$$L = Z_0 l/v \tag{3.18}$$

Equation (3.18) is also valid for calculating the inductance of a short length of stripline conductor and then the inductance per unit length is Z_0/v.

Similarly, from eqn. (1.42), the reactive admittance of an open circuited line is given by

$$B_C = Y_0 \tan \beta l = Y_0 \tan \omega l/v \tag{3.19}$$

Similarly, for a short line, where $l \ll v/\omega$,

$$B_C = \omega C \approx Y_0 \omega l/v \tag{3.20}$$

Therefore the capacitance is given by

$$C \approx Y_0 l/v \tag{3.21}$$

However, eqn. (3.21) is less accurate than eqn. (3.18) because the equivalent open circuit is not exactly at the end of the line due to the capacitance effect of the end of the line.

3.10. Waveguide Attenuation

The theory to be developed in Chapters 5 and 6 assumes that there are no losses associated with propagation along hollow metal waveguide. In practice, no waveguide is entirely lossless although for most purposes the effect of small losses may be neglected. Large losses modify the propagating conditions in the waveguide compared with those calculated for lossless waveguide, but the consideration of large losses is beyond the scope of this book. Losses attenuate the electromagnetic wave as it propagates along the waveguide. It is assumed if the losses are small that their effect is only secondary; the propagating conditions in the waveguide may be calculated on the assumption of no loss and then the effect of losses is calculated on the assumption that the losses do not modify either the phase constant or the field pattern in the waveguide. Attenuation due to losses in the material filling the waveguide is considered here, but attenuation due to the finite conductivity of the waveguide walls is deferred until chapter 5 (section 5.11).

The attenuation constant is the real part of the propagation constant, given by eqn. (1.16),

$$\gamma = \alpha + j\beta$$

If the material through which the wave is propagating is absorbing power from the wave, it is causing attenuation of the wave. The attenuating properties of the material may be allowed for by specifying complex values for the permeability and permittivity:

$$\mu = \mu' - j\mu''$$

$$\varepsilon = \varepsilon' - j\varepsilon''$$

If this concept seems strange, the reader is recommended to calculate the impedance of a capacitor whose dielectric is lossy and is represented by a complex permittivity. It is found that the imaginary part of the permittivity gives rise to a resistive component of the impedance of the capacitor.

The losses can be specified by *loss tangents*

$$\tan \delta_m = \frac{\mu''}{\mu'}$$

$$\tan \delta_e = \frac{\varepsilon''}{\varepsilon'}$$

For a plane wave eqn. (2.19) gives

$$\gamma^2 = -\omega^2 \mu \varepsilon$$

whence

$$\gamma^2 = -\omega^2(\mu' - j\mu'')(\varepsilon' - j\varepsilon'') \tag{3.22}$$

If the losses are small, $\mu'' \ll \mu'$ and $\varepsilon'' \ll \varepsilon'$, and eqn. (3.22) simplifies to

$$\gamma^2 = -\omega^2 \mu' \varepsilon'[1 - j(\tan \delta_m + \tan \delta_e)] \tag{3.23}$$

For propagation inside waveguide, the propagation constant will be different from that given in eqn. (3.23) which is for a plane wave in an unbounded medium. If β is the phase constant for propagation in lossless waveguide of the same shape and size, the propagation constant in the lossy waveguide is given by

$$\gamma^2 = -\beta^2[1 - j(\tan \delta_m + \tan \delta_e)] \tag{3.24}$$

Actual values of the attenuation constant may be calculated from expressions similar to those given in eqn. (7.5). Continuing the simplification for small losses, eqn. (3.24) becomes,

$$\gamma \approx j\beta + \tfrac{1}{2}\beta(\tan \delta_m + \tan \delta_e) \tag{3.25}$$

If the intrinsic impedance of the medium is defined from the real parts of the permeability and permittivity,

$$\eta = \sqrt{\left(\frac{\mu'}{\varepsilon'}\right)}$$

The phase constant and the attenuation constant may be specified in terms of the material properties in a relationship very similar to eqns. (1.25) and (1.26)

$$\beta = \omega\sqrt{(\mu'\varepsilon')} \tag{3.26}$$

$$\alpha = \tfrac{1}{2}\beta(\tan\delta_m + \tan\delta_e) = \frac{\omega}{2}\left(\frac{\mu''}{\eta} + \varepsilon''\eta\right) \tag{3.27}$$

Values of the loss tangents are given in tables of material data.

3.11. Microwave Resonators

The reader will be familiar with resonant circuits and the concept of resonant frequency and Q-factor. A typical resonant curve is shown in Fig. 3.13 on which are marked the resonant frequency f_0 and the bandwidth at the half power points δf. At low frequencies, a resonance curve similar to Fig. 3.13 is obtained by measuring either the current or voltage in the circuit when the other is maintained constant. At microwave frequencies, when current or voltage in a circuit does not have much practical significance, the resonance curve is measured by the variation of impedance or reflection coefficient of a circuit. The Q-factor of a microwave resonant circuit is usually measured from the width of the resonance curve. The well-known formulae are

$$\left.\begin{aligned} Q &= 2\pi f_0 \frac{\text{energy stored}}{\text{power dissipated}} \\ Q &= \frac{f_0}{\delta f} \end{aligned}\right\} \tag{3.28}$$

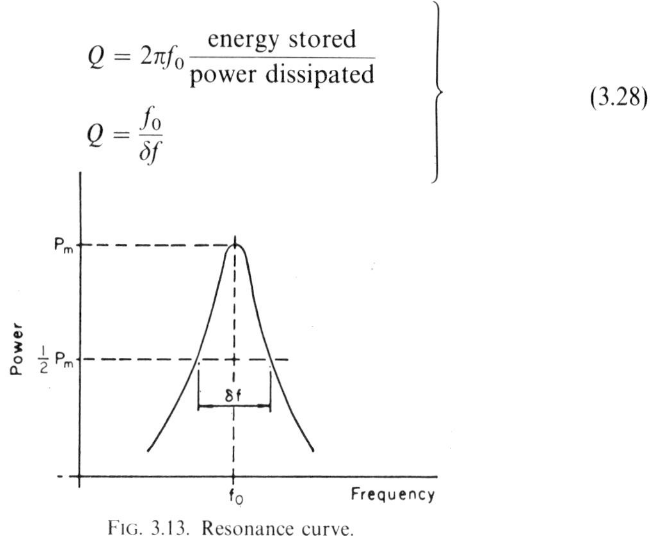

FIG. 3.13. Resonance curve.

In practice a microwave resonator consists of a cavity. This is an enclosure with perfectly conducting metal walls having a small aperture in one of the walls. If this cavity is connected to a microwave system by means of the aperture, it is found that at most frequencies there is very little microwave penetration into the cavity and the cavity has very little effect on the microwave system. But at certain distinct frequencies, the cavity abstracts appreciable power from the microwave system and the electromagnetic fields inside the cavity are of the same order as the fields in the rest of the system. These are the resonant frequencies of the cavity. For the purposes of analysis, the cavity may be considered to be a totally enclosed waveguide system. It is a length of waveguide or transmission line with a short circuit at each end. A short circuit creates a standing wave pattern which has a minimum of electric field at a half wavelength from the short circuit and at every subsequent additional half wavelength away from the short circuit. Neglecting losses, the electric field strength will be as shown in Fig. 3.14 and it is seen that a second short circuit could be placed at any of the minima of the electric field without affecting the fields in the waveguide. Hence a microwave resonator will be a length of transmission line or waveguide enclosed by two short circuits which is an integral number of half wavelengths long.

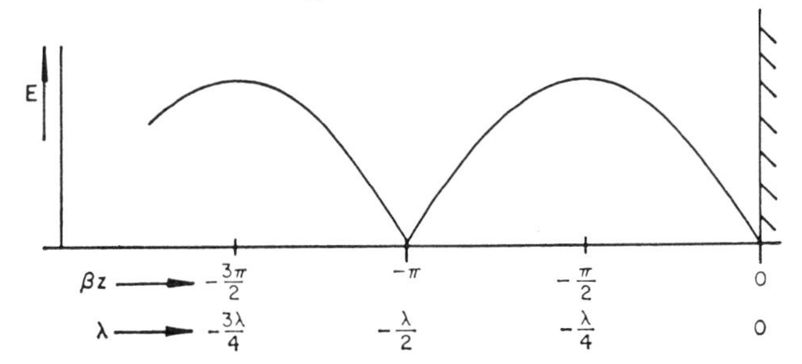

FIG. 3.14. Electric field intensity of a standing wave due to a short circuit.

3.12. Summary

3.2. ***Parallel plate waveguide*** propagates a transmission line mode similar to a plane wave.

3.3. It can also support waveguide modes.

3.4. In parallel plate waveguide, the ***cut-off wavelength*** is

$$\lambda_c = \frac{2b}{n}$$

and the ***waveguide wavelength***

$$\lambda_g = \frac{\lambda_0}{\sqrt{\left[1 - \left(\frac{\lambda_0}{\lambda_c}\right)^2\right]}} \tag{3.4}$$

A waveguide mode cannot exist as a propagating mode below the cut-off frequency.

3.5. ***Phase velocity*** $$v_p = \frac{\omega}{\beta} = c\frac{\lambda_g}{\lambda_0} \tag{3.5}$$

Group velocity $$v_g = \frac{\partial\omega}{\partial\beta} = c^2\frac{\beta}{\omega} = c\frac{\lambda_0}{\lambda_g} \tag{3.8}$$

For the transmission line mode, $v_p = v_g = c$.

3.6. A transmission line mode will propagate along any two conductor transmission line.

3.7. ***Characteristic impedance*** $$Z_0 = \frac{V}{I} \tag{1.1}$$

Intrinsic impedance $$\eta = \frac{E}{H} \tag{3.13}$$

Wave impedance $$Z_0 = \frac{E_t}{H_t} \tag{3.14}$$

3.8. ***Stripline*** consists of a thin conductor supported on a dielectric sheet with an earthed metallic backing. Symmetrical stripline has two dielectric sheets and earthed conductors and is sometimes called triplate. Unsymmetrical stripline is usually called ***microstrip***. The characteristic impedance of microstrip is shown in Fig. 3.11.

3.9. Discontinuities usually introduce an effective shunt capacitance across the line.

Step change in width of the conductor introduces a small inductive reactance in series with the line but it may usually be neglected.

Open end of the line: the electrical position of the open circuit is slightly beyond the end of the line.

A ***gap*** in the conductor introduces a capacitance in series with the conductor.

An abrupt ***corner*** introduces a capacitive shunt reactance at the point of the corner. A ***radius bend*** makes this capacitance negligible.

The ***shorted stub*** acts as an inductor

$$L = Z_0 l/v \tag{3.18}$$

which is also the inductance of a length of conductor.

The ***open circuited stub*** acts as a capacitor.

$$C = Y_0 l/v \tag{3.21}$$

3.10. Attenuating properties of a material are specified by complex values for the permeability and permittivity:

$$\mu = \mu' - j\mu'' \quad \text{and} \quad \varepsilon = \varepsilon' - j\varepsilon''$$

and loss tangents

$$\tan\delta_m = \frac{\mu''}{\mu'} \quad \text{and} \quad \tan\delta_e = \frac{\varepsilon''}{\varepsilon'}$$

Waveguide propagation constant is given by

$$\gamma^2 = -\beta^2[1 - j(\tan\delta_m + \tan\delta_e)] \tag{3.24}$$

for low-loss materials.

3.11. ***Q-factor*** of a ***resonator*** with δf as the half-power bandwidth is given by

$$Q = 2\pi f_0 \frac{\text{energy stored}}{\text{power dissipated}} = \frac{f_0}{\delta f} \tag{3.28}$$

The resonator consists of a length of waveguide or transmission line, closed at each end by short circuits and an integral number of half wavelengths long.

Problems

3.1. Strip transmission line could be considered as two parallel plate conductors on each side of a PFTE slab 2 mm thick. If the relative permittivity of PTFE is 2, calculate the maximum frequency at which this stripline might be expected to operate only in the transmission line mode. [53 GHz]

3.2. Calculate a few points and plot a graph of characteristic wavelength against waveguide wavelength in terms of cut-off wavelength.

3.3. Plot the variation of the phase and group velocities with frequency for the first waveguide mode in air-filled parallel plate waveguide, having 2 cm separation between the plates.

3.4. From a consideration of the phase front representation of an electromagnetic plane wave as shown in Fig. 3.2, prove the normal laws of reflection and refraction applying to a plane wave incident at an angle onto the plane surface between two media shown in Fig. 2.3.

3.5. Two coaxial cables need to be connected together in a microwave system. They have PTFE insulation ($\varepsilon_r = 2$) and the following inner and outer conductor diameters, 1.0 mm I.D., 3.25 mm O.D. and 0.75 mm I.D., 4.40 mm O.D. Suggest dimensions for an intermediate matching section of coaxial cable also using PTFE insulation.
[Quarter wavelength of cable, 0.875 mm I.D., 3.70 mm O.D.]

3.6. A coaxial cable of 50 Ω characteristic impedance has a conductor loss of 3.0 mΩ/m and negligible dielectric loss. Find its attenuation constant. [3.0×10^{-5} nepers/m]

3.7. A microcircuit is constructed in microstrip on an alumina substrate ($\varepsilon_r = 10$), 1.0 mm thick. Find the width of conductor needed to give a 50 Ω transmission line.

A microwave integrated circuit is constructed on a GaAs substrate ($\varepsilon_r = 14$), 100 μm thick. Find the width of a microstrip conductor to give a 100 Ω transmission line, and the length of that microstrip to provide an inductance of 0.1 nH. [1.0 mm; 10 μm, 30 μm]

3.8. Assuming that the inductance of a coil is directly proportional to the permeability of the core threading the coil, find an expression for the impedance of a coil when the core material permeability is $\mu' - j\mu''$. When the same coil is measured with no magnetic core, its impedance is $R + j\omega L$. Prove that the imaginary term in the permeability gives rise to a resistive term in the impedance.

3.9. A low-loss transmission line (where $\alpha = 0$) has short circuits applied to it at distances along it of 25 cm and 40 cm. By using the expression for the voltage on a transmission line, eqn. (1.17), find the lowest frequency at which an electromagnetic wave can exist on the transmission line between the short circuits. Are there other frequencies also at which these fields can exist? If so, derive an expression for their frequency. [1 GHz]

3.10. By considering rectangular waveguide as a parallel plate waveguide cavity with a short circuit at each end, derive an expression for the cut-off frequencies of rectangular waveguide of dimensions $a \times b$. [See eqn. (5.22)]

4

Microwave Circuits

4.1. Introduction

Microwave transistor amplifiers and oscillators are now commonplace. They are possible because the useful upper frequency of transistors has risen as their dimensions get smaller. The active components are incorporated into miniature microcircuits or integrated circuits which may be designed using lumped circuit techniques even at microwave frequencies. However, it is still necessary to appreciate the fields that are associated with microwave circuits and the necessity to consider all interconnections as transmission lines. Even when lumped circuit components are part of the microstrip conductor pattern in either miniature microcircuits or integrated circuits, they do not behave strictly as lumped components because of their parasitic capacitance and the inductance of any interconnecting wires. The capacitance and inductance is small but at microwave frequencies even a small capacitance or inductance produces an impedance that cannot be neglected.

For low power, microwave circuits are constructed in microstrip and interconnected with coaxial lines. The attenuation in microstrip line and in coaxial line is greater than that of hollow metal waveguide, but because microstrip circuits are small, the overall attenuation is usually no greater. Discrete components may be inserted into microstrip circuits. When the scale of the circuit is similar to that of a conventional printed circuit board, the devices are fully encapsulated and usually have beam lead output conductors which are similar in dimension to the stripline conductor. A sketch of a beam lead device mounted in a microstrip circuit is shown in Fig. 4.1.

In order that some of the advantages of miniaturisation may be used at microwave frequencies, unencapsulated active devices are mounted in miniature microstrip circuits. The circuit conductors are deposited as thin or thick film circuits on ceramic substrates about 5 or 10 cm square. In these miniature circuits, the active or semiconductor devices are used in the form of unencapsulated chips which are bonded directly into the microstrip circuit. Design consists of a combination of lumped circuit techniques and transmission line techniques. Further miniaturization can be achieved by forming the microwave circuit on a semiconductor chip to make a microwave

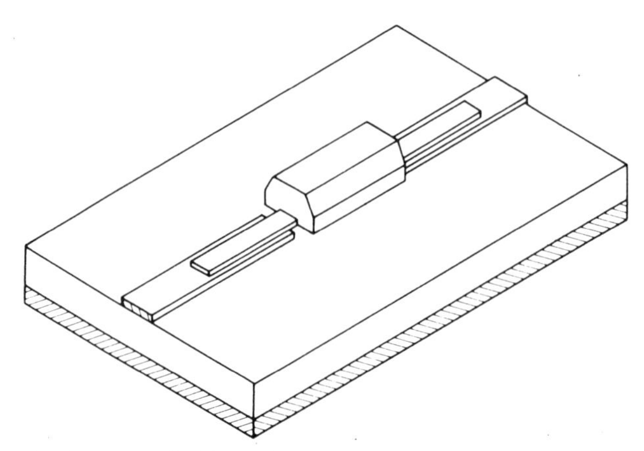

FIG. 4.1. A two-terminal beam lead device mounted in a microstrip circuit.

integrated circuit. Many microwave processing functions are being integrated onto the semiconductor chip. Microwave amplifiers, oscillators and mixers can be produced with all the circuit components on the same semiconductor chip. Further details are given in section 10.3.

Hollow metal waveguide has to be used to contain high power radiation and it is also used where minimum attenuation is required, but in many other applications miniature microstrip circuits are perfectly satisfactory. Before proceeding to develop waveguide theory in Chapters 5 and 6, the particular techniques necessary for the design of microwave circuits are given in this chapter. Most microwave circuits are constructed in microstrip, whose properties have been discussed already in sections 3.8 and 3.9.

4.2. Network Parameters

At low frequencies, the properties of any network may be specified in terms of the voltages and currents at its input and output terminals as shown in Fig. 4.2. Then the network is represented completely by its impedance parameters given by

$$\left.\begin{aligned} V_1 &= z_{11}I_1 + z_{12}I_2 \\ V_2 &= z_{21}I_1 + z_{22}I_2 \end{aligned}\right\} \tag{4.1}$$

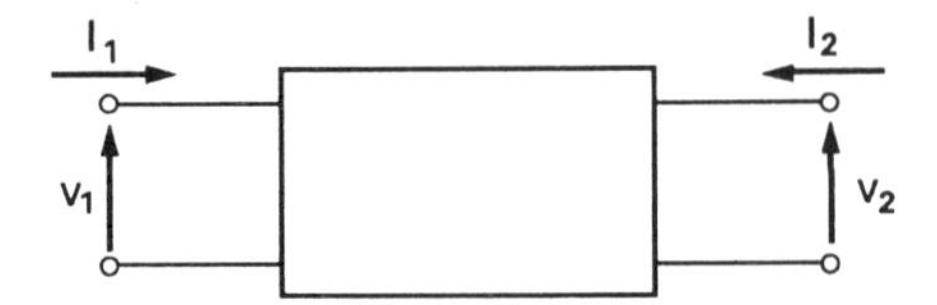

FIG. 4.2. A two-port network.

The parameters are often presented as a matrix,

$$\mathbf{Z} = \begin{vmatrix} z_{11} & z_{12} \\ z_{21} & z_{22} \end{vmatrix} \tag{4.2}$$

Similarly, the network may be represented by its admittance parameters

$$\left.\begin{aligned} I_1 &= y_{11}V_1 + y_{12}V_2 \\ I_2 &= y_{21}V_1 + y_{22}V_2 \end{aligned}\right\} \tag{4.3}$$

where

$$\mathbf{Y} = \begin{vmatrix} y_{11} & y_{12} \\ y_{21} & y_{22} \end{vmatrix} \tag{4.4}$$

At high frequencies, active devices tend to be capacitive, so that admittance parameters are often used to represent the properties of active devices at high frequencies.

At microwave frequencies, however, voltage and current have little meaning and wave amplitudes are a better way to describe microwave fields. The network of Fig. 4.2 is now represented by its incident and emergent wave amplitudes in Fig. 4.3. The wave amplitudes a represent the incident waves and the wave amplitudes b represent the resultant waves. They are related by the *scattering parameters*.

$$\left.\begin{aligned} b_1 &= s_{11}a_1 + s_{12}a_2 \\ b_2 &= s_{21}a_1 + s_{22}a_2 \end{aligned}\right\} \tag{4.5}$$

If **a** and **b** are column vectors representing the input and output wave respectively, eqn. (4.5) may be written in matrix form

$$\mathbf{b} = \mathbf{Sa} \tag{4.6}$$

where **S** is the *scattering matrix*,

$$\mathbf{S} = \begin{vmatrix} s_{11} & s_{12} \\ s_{21} & s_{22} \end{vmatrix} \tag{4.7}$$

Equation (4.6) is valid whatever the size of its matrices so that the same equation may be used to represent the conditions of any microwave network

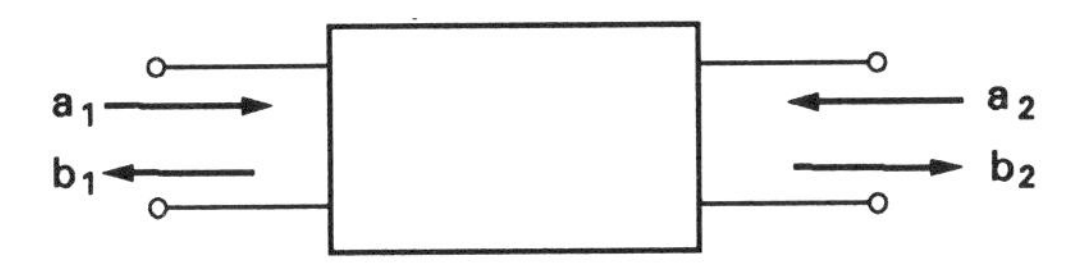

FIG. 4.3. A fields representation of a two-port network.

however many ports it may have. Similarly, there is always a scattering matrix to represent any microwave network whatever its number of ports.

For the network shown in Fig. 4.3, take port 1 as the input port and port 2 as the output port. If the two ports are terminated in their characteristic impedance, the scattering parameters for the network take on the following meanings,

$$s_{11} = \frac{b_1}{a_1}$$

is the input reflection coefficient.

$$s_{12} = \frac{b_1}{a_2}$$

is the reverse transfer coefficient.

$$s_{21} = \frac{b_2}{a_1}$$

is the forward transfer coefficient or gain.

$$s_{22} = \frac{b_2}{a_2}$$

is the output reflection coefficient. The properties of microwave circuits and microwave circuit components are measured and quoted in s-parameters. For a reciprocal passive device, the performance is reciprocal and $s_{12} = s_{21}$.

4.3. Lumped Components

Lumped components for use at microwave frequencies can be incorporated into miniature stripline circuits. Resistors can be realized by the controlled deposition of nichrome or a similar resistive material between conductors on the stripline or microstrip dielectric material. Relatively large values of capacitance have to be realized in the form of overlay capacitors or discrete capacitors which are inserted into the circuit. Overlay capacitors consist of a deposited layer of dielectric material on the surface of the conductor onto which a further conductor is also deposited. Smaller values of capacitance can be realized by having a small gap in the stripline conductor. The range of possible capacitance is increased if the gap is interdigital in shape as shown in Fig. 4.4. (a). The length of the fingers and the spacing may be varied to give a range of values of capacitance of the order of 0.01 to 1.0 pF.

Any length of conductor has an inductance which provides an appreciable impedance at microwave frequencies. Usually the inductance is incorporated into the transmission line properties of the stripline, but in lumped circuits it must be considered. The value of inductance is given approximately by eqn. (3.18). In a deliberately designed inductor the length and width of the conductor may be varied to control the inductance. Usually the conductor is formed into a loop as shown in Fig. 4.4 (b). Possible inductance values are of the order of 0.1 to 1.0 nH. An example of a lumped inductor and an interdigital capacitor combined in a parallel tuned circuit is shown in Fig. 4.4 (c).

Larger values of inductance up to about 10 nH may be provided by a planar spiral as shown in Fig. 4.5. However, some special overlay is needed to make connection to the centre of the spiral. This connection is easily supplied in microwave integrated circuits where a number of layers of metallization are used in the overall processing of the semiconductor chip.

As already mentioned, lumped circuit components do not behave as ideal components at microwave frequencies. Their true performance may be allowed for by incorporating their measured *s*-parameters into any network analysis or by representing them by an equivalent circuit. A simple equivalent circuit is shown in Fig. 4.6 and typical values of the circuit elements for a number of integrated circuit components on a gallium arsenide (GaAs) substrate are given in Table 4.1.

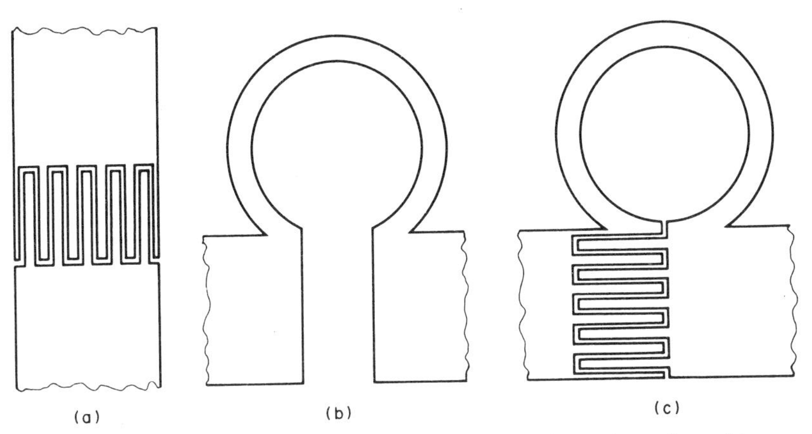

FIG. 4.4. A lumped capacitor and inductor. (a) A lumped capacitance formed by an interdigital gap in the stripline conductor. (b) A lumped stripline inductor. (c) A parallel tuned circuit formed from lumped components in stripline.

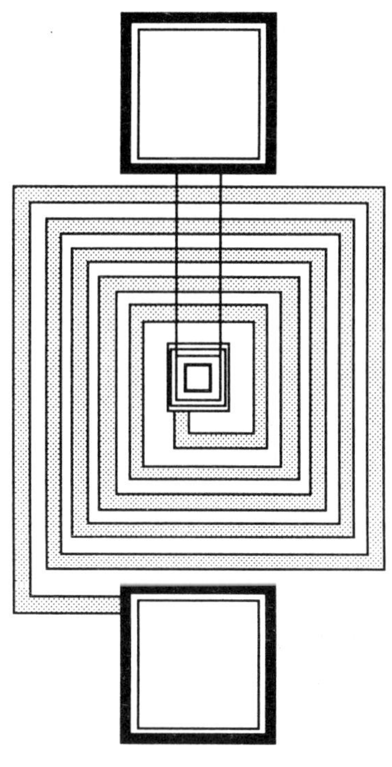

FIG. 4.5. A planar spiral inductor.

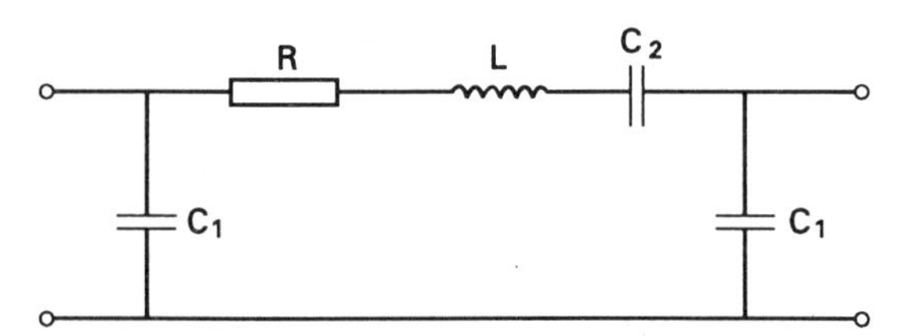

FIG. 4.6. Equivalent circuit of microstrip lumped components. Some typical values are given in Table 4.1.

TABLE 4.1. *Typical equivalent circuit elements for a number of integrated circuit components on GaAs.*

	$R(\Omega)$	L(nH)	C_1(pF)	C_2(pF)
Nichrome planar resistor	200	0.06	0.06	short circuit
Spiral inductor	9	9.0	0.27	short circuit
Overlay capacitor	0.8	0.26	0.10	5.6
Interdigital capacitor	4	0.26	0.10	0.65

4.4. Transistor Amplifier

The special techniques applicable to amplifier design at microwave frequencies are associated with the microwave properties of the circuit components. Silicon bipolar transistors may be used up to about 4 GHz and GaAs planar field effect transistors (FET) may be used above that frequency. Normal d.c. bias circuits are required, except that these connections are supplied through inductive low pass filters consisting of a series inductance and a shunt capacitance to earth. The microwave properties of the transistor are specified by its measured s-parameters or by an equivalent circuit. The s-parameters of an integrated circuit FET are shown in Fig. 4.7 and its

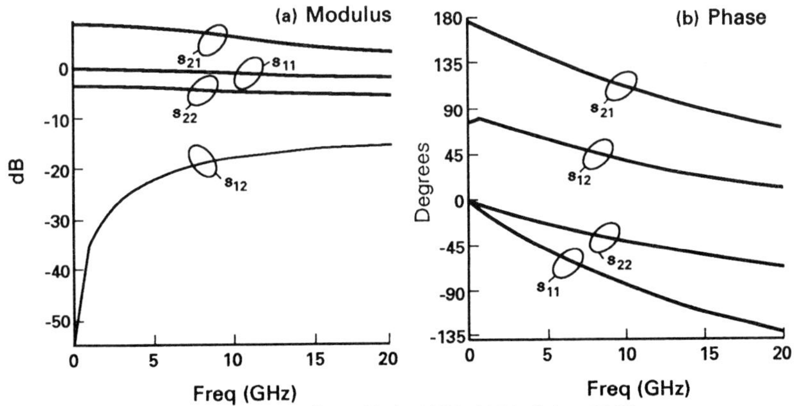

FIG. 4.7. S-parameters for a GaAs FET: (a) Modulus; (b) Phase.

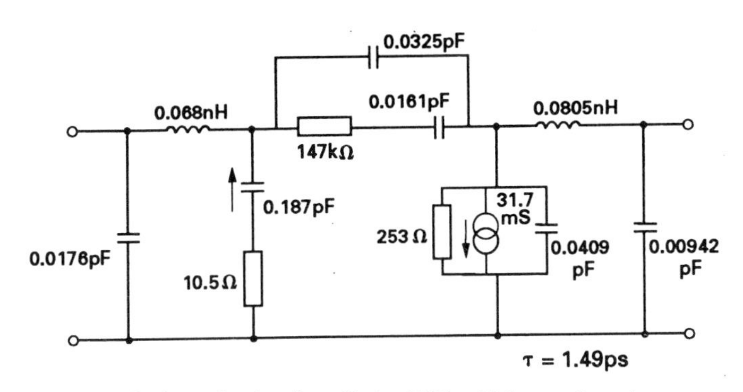

FIG. 4.8. Equivalent circuit of a GaAs FET which matches the measured performance shown in Fig. 4.7. τ is the phase delay in the current generator.

equivalent circuit in Fig. 4.8. These s-parameters are measured using small size microwave signals which assume that the transistor properties are linear and are called the *small signal s-parameters*.

Any transistor needs to be matched to the external circuit at its input and output by matching circuits as shown in Fig. 4.9. For such an amplifier, there are different measures of power gain. The transducer power gain is given by

$$G_T = \frac{P_L}{P_A} \tag{4.8}$$

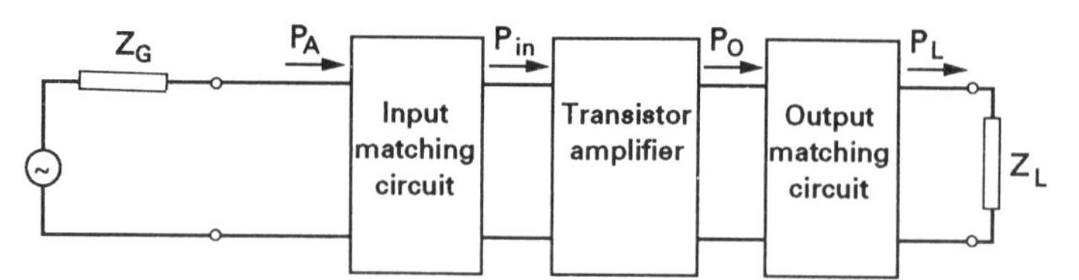

FIG. 4.9. A transistor as an amplifier with input and output matching circuits.

which is the overall gain of the amplifier in its operating circuit. The available power gain is

$$G_A = \frac{P_o}{P_A} \tag{4.9}$$

which is independent of the load match. The power gain is defined by

$$G = \frac{P_L}{P_{in}} \tag{4.10}$$

which is independent of the source match. When the source and load are both matched, then the transducer power gain is the maximum available gain.

A single-stage amplifier consists of a circuit similar to that shown in Fig. 4.9. The transistor usually consists of a common emitter or a common source stage for highest gain per stage. Simple input and output matching circuits

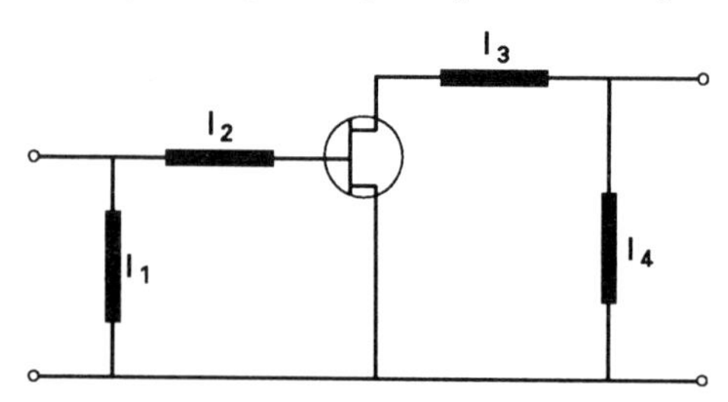

FIG. 4.10. Single-stage microwave transistor amplifier showing the microwave circuit elements only.

are shown in Fig. 4.10 to be single stub matching as described in section 1.10. The shorted stubs are also used to provide d.c. bias to the transistor as shown in Fig. 4.11. Most microwave transistor amplifier design is performed using a circuit analysis and optimisation program on a computer. One of the problems with microwave transistor amplifier design is that the transistor input and output impedances are a long way from the 50 Ω characteristic impedance usually used in microwave circuits. The microwave bipolar transistor is low impedance and the input gate of a microwave FET is capacitive.

A two-stage amplifier could consist of two single-stage amplifiers connected

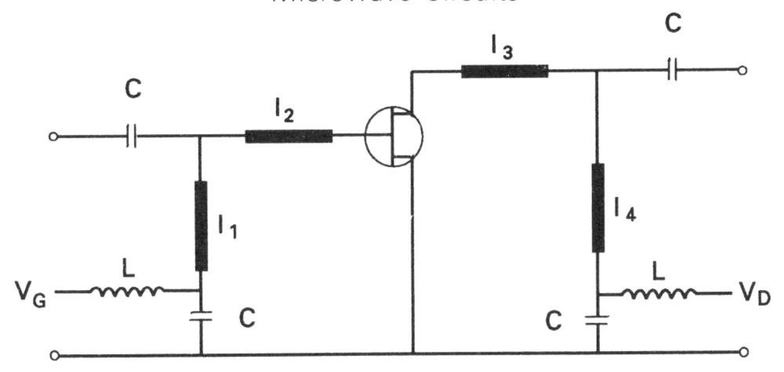

FIG. 4.11. The single-stage transistor amplifier of Fig. 4.10 with d.c. bias connections.

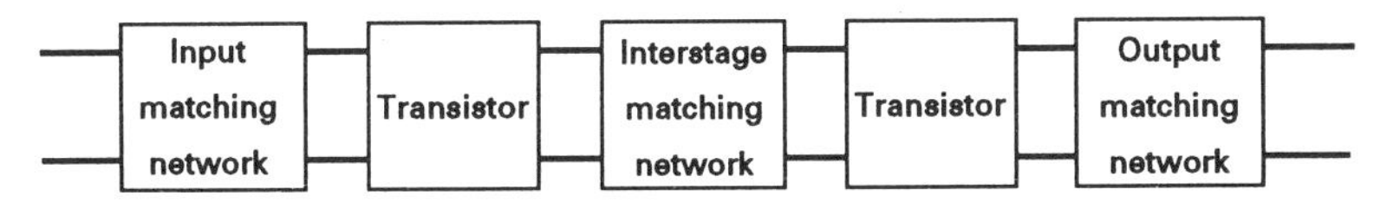

FIG. 4.12. A two-stage amplifier.

together. However, the two-stage amplifier as shown in Fig. 4.12 gives more freedom in the design of the interstage coupling network. The simplest two-stage design using transmission lines in the matching networks would still look like two stages from Fig. 4.11 connected together because of the need to provide d.c. bias to the transistors. With integrated circuits, there is a severe limitation on the length of transmission lines that is acceptable but alternatively lumped circuit components can be used. A computer generated design for a two-stage amplifier using lumped components to give 10 dB gain over the band 8–12 GHz is shown in Fig. 4.13.

4.5. Amplifier Stability

An amplifier is stable if it does not break into self oscillation. An unstable oscillator can often be prevented from oscillating by the correct matching

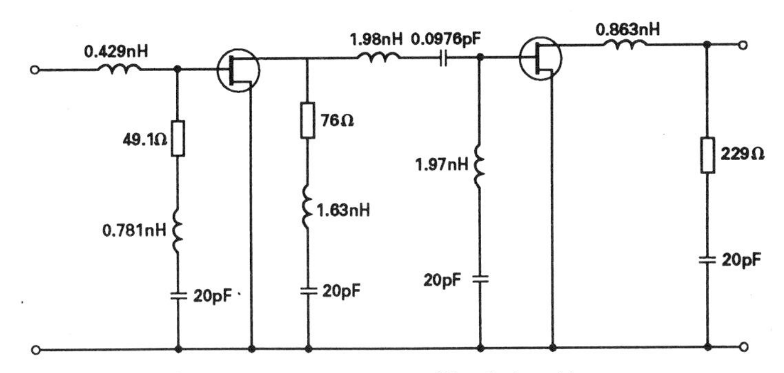

FIG. 4.13. Lumped component two-stage amplifier designed by computer program to have a gain of 10 dB over the band 8–12 GHz.

networks and terminating impedances. However, in amplifier design it is advisable to aim for an unconditionally stable amplifier, which will not oscillate whatever the amplitude and phase of the impedances connected to the input and output ports. When oscillation occurs, either the input or output reflection coefficient is greater than one. Therefore the condition for stability is $\rho_{in} < 1$ and $\rho_{out} < 1$, and this must occur for any condition of source and load impedances. Referring to Fig. 4.3,

$$\rho_{in} = \frac{b_1}{a_1} \tag{4.11}$$

and for the load impedance,

$$\rho_L = \frac{a_2}{b_2} \tag{4.12}$$

Substituting from eqns. (4.11) and (4.12) into eqn. (4.5) and after some algebraic manipulation we get

$$\rho_{in} = s_{11} + \frac{s_{12}s_{21}\rho_L}{1 - s_{22}\rho_L} \tag{4.13}$$

Similarly,

$$\rho_{out} = s_{22} + \frac{s_{21}s_{12}\rho_s}{1 - s_{11}\rho_s} \tag{4.14}$$

where ρ_s is the reflection coefficient from the source impedance. For stability $\rho_{in} < 1$ and $\rho_{out} < 1$ for any value of $\rho_L < 1$ and $\rho_s < 1$. Putting these conditions into eqns. (4.13) and (4.14), after a straightforward but lengthy algebraic manipulation, gives the conditions

$$\left.\begin{aligned} k &> 1 \\ |\Delta| &< 1 \end{aligned}\right\} \tag{4.15}$$

where

$$k = \frac{1 - |s_{11}|^2 - |s_{22}|^2 + |\Delta|^2}{2s_{12}s_{21}} \tag{4.16}$$

and

$$\Delta = s_{11}s_{22} - s_{12}s_{21}$$

k is often called the stability factor of the amplifier. Provided the conditions given by eqn. (4.15) are satisfied, any amplifier is unconditionally stable.

4.6. T-junction

When a line is added in shunt to another line, there is a T-junction between transmission lines. During the discussion of stub matching in section 1.10, it was assumed that the stub appeared as a pure inductance or capacitance across the main line. However, when the T-junction is realized in microstrip, there is a small inductance and capacitance associated with the junction as

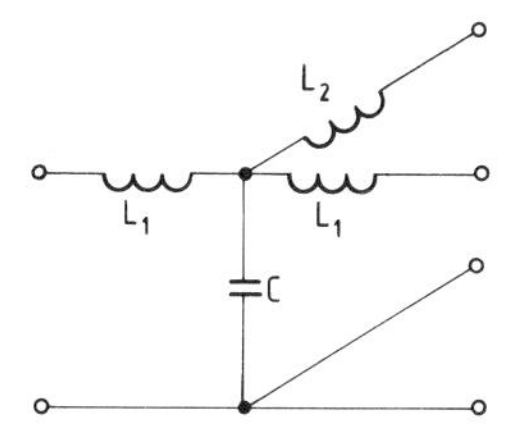

FIG. 4.14. Equivalent circuit of a microstrip T-junction.

shown in Fig. 4.14. The effect is small, but allowance is made for the effect in many microwave CAD programs. The T-junction operates satisfactorily for the addition of shorted stubs but is not suitable for power combining or splitting.

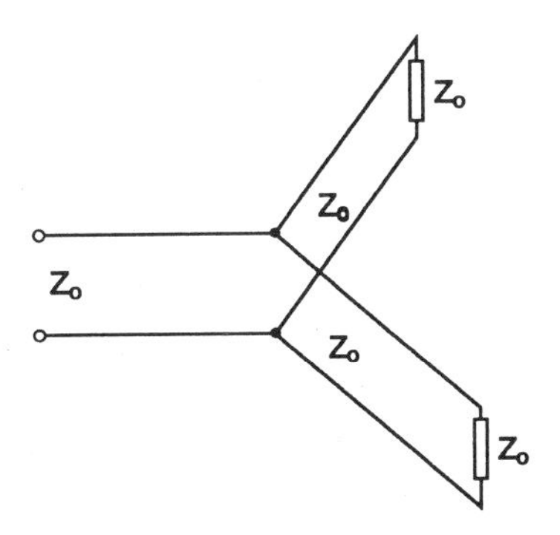

FIG. 4.15. Junction between three lines of equal characteristic impedance.

Consider the junction between three matched lines of equal characteristic impedance as shown in Fig. 4.15. The effective impedance presented by the junction is the characteristic impedance of two lines in parallel to each other, which is $\frac{1}{2}Z_0$. Therefore, it is impossible to provide a matched symmetrical T-junction. In terms of s-parameters, the parameters of a three port are

$$\mathbf{S} = \begin{vmatrix} s_{11} & s_{12} & s_{13} \\ s_{21} & s_{22} & s_{23} \\ s_{31} & s_{32} & s_{33} \end{vmatrix} \tag{4.17}$$

By symmetry, eqn. (4.17) becomes

$$\mathbf{S} = \begin{vmatrix} s_1 & s_2 & s_2 \\ s_2 & s_1 & s_2 \\ s_2 & s_2 & s_1 \end{vmatrix} \tag{4.18}$$

For a lossless network, it can be shown that

$$|s_1|^2 + 2|s_2|^2 = 1 \tag{4.19}$$

and

$$s_1 s_2^* + s_2 s_1^* + s_2 s_2^* = 0 \tag{4.20}$$

where the asterisk denotes the complex conjugate. From eqn. (4.19) it is seen that there is equal power output from the two exit ports. Also, eqn. (4.20) may be reduced to

$$2s_1 + s_2 = 0 \quad \text{or} \quad s_1 = -\tfrac{1}{2}s_2 \tag{4.21}$$

which confirms that it is impossible to provide a symmetrical matched T-junction.

4.7. Directional Coupler

A satisfactory system for diverting some power into a branch line is provided by the four-port network called a *directional coupler*. Consider two striplines parallel to each other with two high impedance lines coupling between them as shown in Fig. 4.16. The impedance of the coupling lines is chosen so that a portion of the field k in one line is coupled into the other line. It is assumed that the coupling lines are of equal length and create negligible disturbance to the field in either of the main lines. Then k will be small. The field strength in each of the lines takes the values indicated in the figure, where ϕ is the phase change due to the length l of main line between the two coupling lines. If the length l is a quarter of a wavelength, the phase change ϕ is $\pi/2$, and the output from port 4 cancels out and will be zero provided that $(1 - k) \approx 1$, whereas the output from port 3 is a maximum.

The directional coupler operates on the principle shown in Fig. 4.16, that power into port 1 gives an output in ports 2 and 3 but no output in port 4.

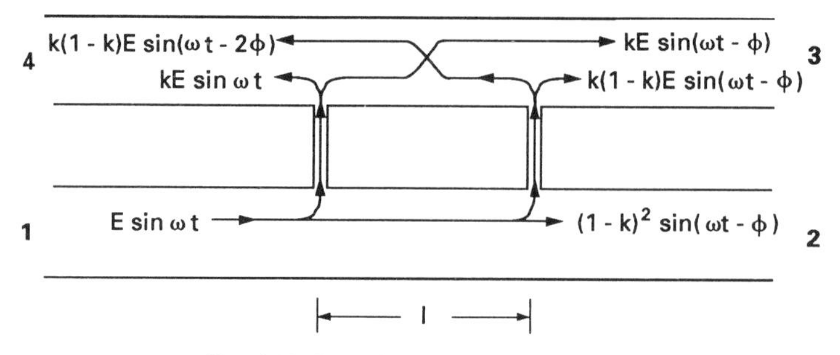

FIG. 4.16. Branch line directional coupler.

The directional coupler can be described by its scattering matrix,

$$\mathbf{S} = \begin{vmatrix} s_{11} & s_{12} & s_{13} & s_{14} \\ s_{21} & s_{22} & s_{23} & s_{24} \\ s_{31} & s_{32} & s_{33} & s_{34} \\ s_{41} & s_{42} & s_{43} & s_{44} \end{vmatrix} \tag{4.22}$$

Inspection of Fig. 4.16 and eqn. (4.22) shows that by symmetry,

$$s_{11} = s_{22} = s_{33} = s_{44} = s_1$$
$$s_{12} = s_{21} = s_{43} = s_{34} = s_2$$
$$s_{13} = s_{31} = s_{24} = s_{42} = s_3$$
$$s_{14} = s_{41} = s_{23} = s_{32} = s_4$$

so that the scattering matrix becomes

$$\mathbf{S} = \begin{vmatrix} s_1 & s_2 & s_3 & s_4 \\ s_2 & s_1 & s_4 & s_3 \\ s_3 & s_4 & s_1 & s_2 \\ s_4 & s_3 & s_2 & s_1 \end{vmatrix} \tag{4.23}$$

For an ideal device, there is no reflected power. Therefore $s_1 = 0$. For an ideal coupler there is no output in the reverse port so that $s_4 = 0$. Then s_3 gives a measure of the signal coupled into the side arm of the coupler and s_2 gives a measure of the signal left in the main line. The circuit symbol for a directional coupler is shown in Fig. 4.17. The performance of a directional coupler is usually quoted in terms of power ratio as,

$$\text{coupling factor} = s_3^2 = \frac{\text{power in port 3}}{\text{power in port 1}} \tag{4.24}$$

and the unwanted power in the reverse arm as,

$$\text{directivity} = \frac{\text{power in port 4}}{\text{power in port 3}} \tag{4.25}$$

Both of these power ratios are given in dB. The arrangement described only

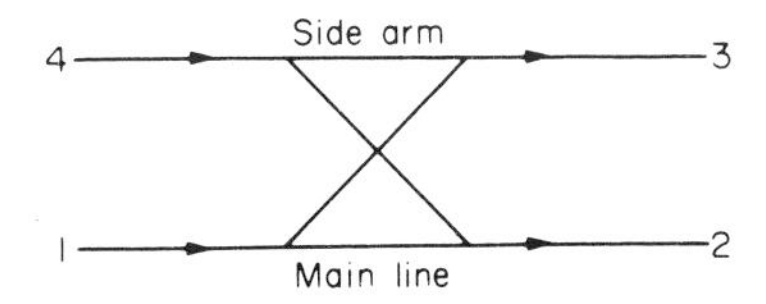

FIG. 4.17. Directional coupler.

works at one particular frequency when the coupling lines are a quarter of a wavelength apart. To increase the frequency band over which the coupler works, either a greater number of coupling lines are used or a method of continuous coupling between two lines is adopted.

4.8. Parallel Line Coupler

If two stripline conductors are brought sufficiently close together, power is coupled from one to the other and the device is a directional coupler. The

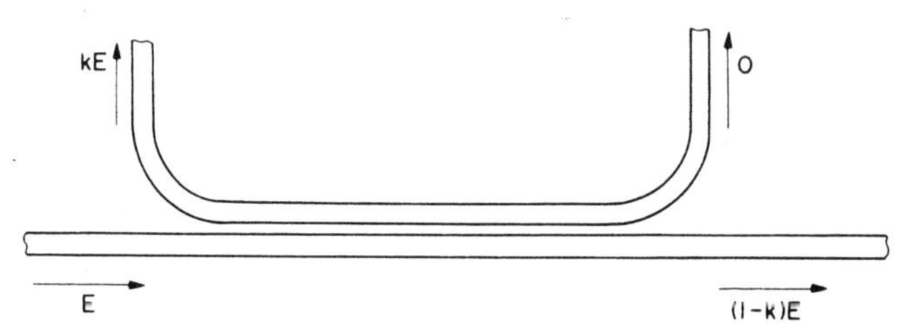

FIG. 4.18. Stripline directional coupler.

layout of the conductors is shown in Fig. 4.18. In this coupler energy is coupled backward instead of forward so that the output in the coupled arm occurs at the port adjacent to the input as shown in Fig. 4.18. Analysis of the stripline coupler is usually performed in terms of the odd and even mode impedances of the coupled section. The electric field pattern for the odd and

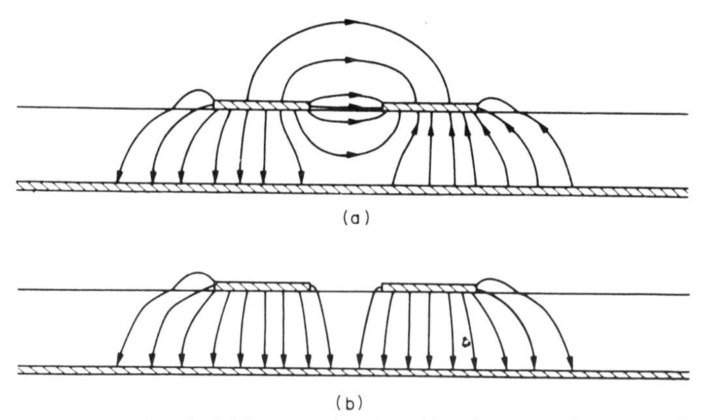

FIG. 4.19. The electric field pattern for the odd and even modes on a pair of coupled lines, (a) odd mode, (b) even mode.

even modes in a pair of coupled lines is shown in Fig. 4.19. The equivalent circuit for the odd and even excitation of the line is shown in Fig. 4.20. The conventional excitation of a directional coupler is shown in Fig. 4.20 (a) and it is seen that this can be composed of the sum of the odd excitation in Fig. 4.20 (b) and the even excitation in Fig. 4.20 (c). The characteristic impedances

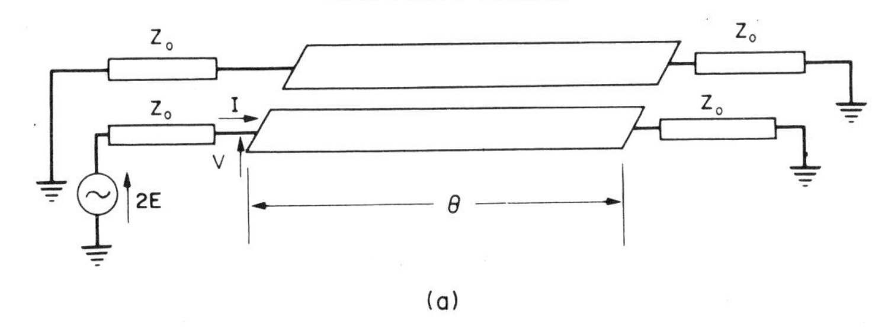

(a)

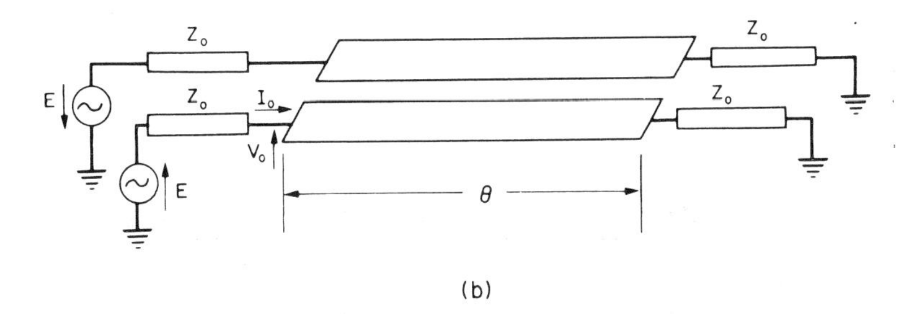

(b)

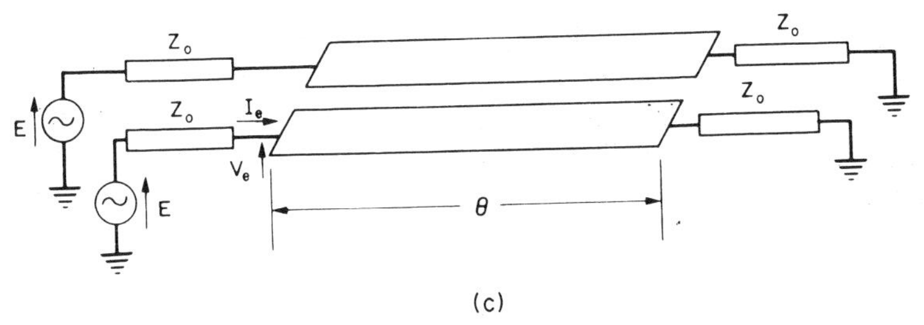

(c)

FIG. 4.20. (a) Equivalent circuit of a stripline directional coupler which can be considered as the sum of (b) odd excitation and (c) even excitation.

of the odd and even modes are different. The odd mode impedance is denoted Z_{0o} and the even mode impedance by Z_{0e}.

Analysis of these circuits shows that

$$\left.\begin{aligned} I_o &= \frac{E}{Z_0 + Z_{io}} \quad \text{and} \quad V_o = Z_{io} I_o \\ I_e &= \frac{E}{Z_0 + Z_{ie}} \quad \text{and} \quad V_e = Z_{ie} I_e \end{aligned}\right\} \qquad (4.26)$$

and the input impedance is given by the superposition of the effect of the two modes of excitation

$$Z_{in} = \frac{V_o + V_e}{I_o + I_e}$$
$$= \left(\frac{Z_{io}}{Z_0 + Z_{io}} + \frac{Z_{ie}}{Z_0 + Z_{ie}}\right) \Big/ \left(\frac{1}{Z_0 + Z_{io}} + \frac{1}{Z_0 + Z_{ie}}\right)$$
$$= \frac{Z_0(Z_{io} + Z_{ie}) + 2Z_{io}Z_{ie}}{Z_{io} + Z_{ie} + 2Z_0} \tag{4.27}$$

Using transmission line theory from section 1.7, the input impedance of the coupled line section under odd and even excitation as shown in Fig. 4.20 is given by eqn. (1.41). βl is the phase length of the coupled section, θ, and Z_0 is the terminating impedance. Then the input impedances become

$$Z_{io} = Z_{0o} \frac{Z_0 + jZ_{0o} \tan\theta}{Z_{0o} + jZ_0 \tan\theta} \tag{4.28}$$

$$Z_{ie} = Z_{0e} \frac{Z_0 + jZ_{0e} \tan\theta}{Z_{0e} + jZ_0 \tan\theta} \tag{4.29}$$

Substituting these values of impedance into eqn. (4.27) gives an expression for the input impedance of the coupled lines in terms of the odd and even mode impedances, and the overall characteristic impedance Z_0. The design aim is that when the coupler is terminated in its characteristic impedance on three of its ports, then the input impedance at the fourth port is also the same characteristic impedance. If the characteristic impedance is made to be the geometric mean of the odd and even mode impedances,

$$Z_0 = \sqrt{(Z_{0o}Z_{0e})} \tag{4.30}$$

then the input impedance given by eqn. (4.27) is the same as the characteristic impedance, $Z_{in} = Z_0$.

The maximum coupling occurs when the coupled section of line is a quarter wavelength long, i.e. $\theta = 90°$. It is found that the coupled output comes out of arm 4 and the output from arm 3 is zero, so that this form of coupler has sometimes been called a *backward coupler*.

The voltage coupling factor k is given by

$$k = \frac{V_{out}}{V_{in}} = \frac{Z_{0e} - Z_{0o}}{Z_{0e} + Z_{0o}} \tag{4.31}$$

For design purposes, the coupling factor is known, then rearranging eqn. (4.31) gives

$$\frac{Z_{0e}}{Z_{0o}} = \frac{1 + k}{1 - k} \tag{4.32}$$

Curves of the odd and even mode impedances of coupled striplines similar to Fig. 3.11 are given in the literature.* Theoretically, this simple quarter wavelength coupler has perfect match and isolation and it also has a useful bandwidth of operation of at least an octave.

Usually stripline conductors need to be quite close together for coupling to occur. Unless the lengths of line are in resonance, there is very little interaction between conductors that are more than the dielectric thickness h apart.

4.9. Hybrid Ring

Half power coupling can be realized in any transmission line by means of the hybrid ring, which is shown schematically in Fig. 4.21. The ring is $1\frac{1}{2}$

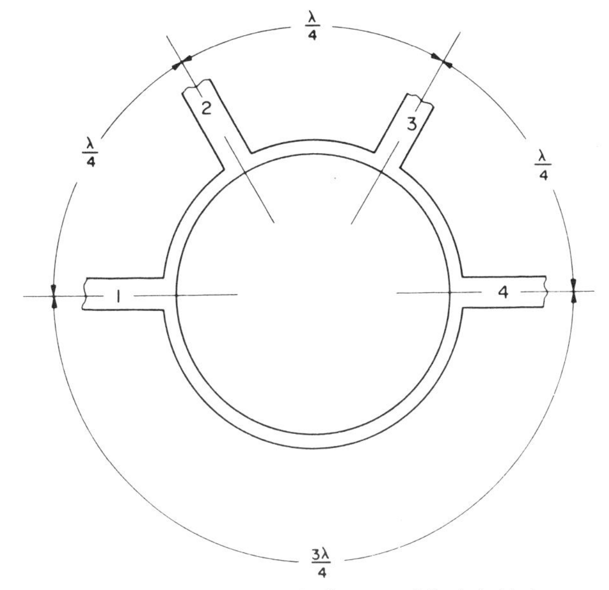

FIG. 4.21. A schematic diagram of the hybrid ring.

wavelengths long. A signal entering port 1 is divided into halves at the T-junction as it enters the ring. At port 2 and at port 4, the two signals travelling round the ring in opposite directions are in phase and half the power comes out of port 2 and half the power comes out of port 4 provided these two ports are terminated in their characteristic impedance. Conversely, the signals at port 3 are out of phase and nothing comes out of that port. Similarly, if two signals enter at ports 1 and 3, their sum comes out of port 2 and their difference comes out of port 4. The hybrid ring can be realised in any transmission line. In waveguide, the hybrid T-junction described in

* See, for example, T. S. Saad, *Microwave Engineers Handbook*. Airtech. 1971. Vol. 1. p. 132.

section 11.6 gives the same performance and is much more compact than the hybrid ring, although waveguide hybrid rings have been used. The hybrid T-junction cannot be realized in simple stripline so that the hybrid ring is used as a half power, or 3 dB, coupler in stripline.

4.10. Summary

4.2. *Impedance parameters*

$$\left.\begin{aligned} V_1 &= z_{11}I_1 + z_{12}I_2 \\ V_2 &= z_{21}I_1 + z_{22}I_2 \end{aligned}\right\} \quad (4.1)$$

Admittance parameters

$$\left.\begin{aligned} I_1 &= y_{11}V_1 + y_{12}V_2 \\ I_2 &= y_{21}V_1 + y_{22}V_2 \end{aligned}\right\} \quad (4.3)$$

Scattering parameters

$$\left.\begin{aligned} b_1 &= s_{11}a_1 + s_{12}a_2 \\ b_2 &= s_{21}a_1 + s_{22}a_2 \end{aligned}\right\} \quad (4.5)$$

$$\mathbf{b} = \mathbf{Sa} \quad (4.6)$$

s_{11} is the input reflection coefficient
s_{12} is the reverse transfer coefficient
s_{21} is the forward transfer coefficient or gain
s_{22} is the output reflection coefficient

4.3. ***Lumped components*** in microstrip can be used at microwave frequencies. The interdigital capacitor and the loop inductor can be produced by the conductor pattern.

4.4. For the ***transistor amplifier***

transducer gain $$G_T = \frac{P_L}{P_A} \quad (4.8)$$

available gain $$G_A = \frac{P_o}{P_A} \quad (4.9)$$

power gain $$G = \frac{P_L}{P_{\text{in}}} \quad (4.10)$$

A transistor amplifier needs input and output matching networks and intermediate matching networks between stages which are designed using transmission line matching techniques. Most microwave amplifiers are designed using CAD techniques.

4.5. ***Amplifier stability*** is unconditional provided

$$\left.\begin{aligned} k &> 1 \\ |\Delta| &< 1 \end{aligned}\right\} \quad (4.15)$$

where

$$k = \frac{1 - |s_{11}|^2 - |s_{22}|^2 + |\Delta|^2}{2s_{12}s_{21}} \tag{4.16}$$

and

$$\Delta = s_{11}s_{22} - s_{12}s_{21}$$

4.6. A ***T-junction*** cannot be matched for all three ports simultaneously, and is not suitable for use as a power splitter.

4.7. The ***directional coupler*** is used to couple a portion of the power from one transmission line into a side arm. It is a four-port network. Its performance is given by

$$\text{coupling factor} = \frac{\text{power in port 3}}{\text{power in port 1}} \tag{4.24}$$

$$\text{directivity} = \frac{\text{power in port 4}}{\text{power in port 3}} \tag{4.25}$$

4.8. In the ***backward coupler***, the characteristic impedance is the geometrical mean of the odd and even mode impedances.

$$\text{Voltage coupling factor, } k = \frac{Z_{0e} - Z_{0o}}{Z_{0e} + Z_{0o}} \tag{4.31}$$

4.9. The ***hybrid ring*** can be used as a half power split directional coupler.

Problems

4.1. Using the substitution, $V = a + b$ and $Z_0 I = a - b$ at each port, prove the following relationships between network parameters.

$$s_{11} = \{(1 - Z_0 y_{11})(1 + Z_0 y_{22}) + Z_0^2 y_{12} y_{21}\}/D$$
$$s_{12} = -2Z_0 y_{12}/D$$
$$s_{21} = -2Z_0 y_{21}/D$$
$$s_{22} = \{(1 + Z_0 y_{11})(1 - Z_0 y_{22}) + Z_0^2 y_{12} y_{21}\}/D$$
$$D = (1 + Z_0 y_{11})(1 + Z_0 y_{22}) - Z_0^2 y_{12} y_{21}$$

4.2. Starting with eqns. (1.30) and (1.31), find the network parameters of a lossless transmission line of length l.

[$z_{11} = z_{22} = Z_0 \cot \beta l$, $z_{12} = z_{21} = Z_0 \operatorname{cosec} \beta l$;
$y_{11} = y_{22} = Y_0 \cot \beta l$, $y_{12} = y_{21} = -Y_0 \operatorname{cosec} \beta l$;
$s_{11} = s_{22} = 0$, $s_{12} = s_{21} = \exp - j\beta l$]

4.3. Find the dimensions of a single loop of microstrip transmission line on alumina substrate 1.0 mm thick to provide 2.0 nH at 1 GHz. [$l = 3.0$ mm, $w = 0.3$ mm]

4.4. The s-parameters of a microwave transistor at 4.0 GHz are: $s_{11} = 0.96\angle -40°$, $s_{12} = 0.063\angle 65°$, $s_{21} = 2.45\angle 150°$ and $s_{22} = 0.65\angle -20°$, measured relative to 50 Ω. It is suitably biased and connected to 50 Ω transmission lines at input and output. Find its input impedance, the VSWR on the 50 Ω input transmission line and its gain at 4.0 GHz.

[$(10 - j140)$Ω; 49; 7.8 dB]

4.5(a). The transistor described in problem 4.4 is connected into the circuit shown in Fig. 4.10. Using the Smith chart, find the dimensions l_1 and l_2 at 4.0 GHz of the input matching circuit. The transmission lines all have 50 Ω characteristic impedance.

(b) An input matching circuit for the transistor at 4.0 GHz could consist of an inductor in series with the input and another inductor in shunt to earth at the opposite end of the series inductor. Using the Smith chart, find the values of the two inductors. It is all required to be matched to 50 Ω lines. [(a) 0.170λ, 0.025λ; (b) 4.8 nH, 1.0 nH]

4.6. Starting with eqns. (4.5), (4.11) and (4.12), derive eqn. (4.13).

4.7. Calculate the stability factor k at 4.0 GHz for the transistor of problem 4.4. [0.15—it is unstable]

4.8. Calculate the proportion k as defined in Fig. 4.16 to give a coupling factor of 40 dB. In these conditions, what is the directivity? [$k = 0.005$, 40 dB]

4.9. A backward coupler has odd and even mode impedances of 25 Ω and 100 Ω respectively. What is its power coupling factor? [4.4 dB]

4.10. For a characteristic impedance of 50 Ω find the odd and even mode impedances of a backward coupler to give a coupling factor of 10 dB. [69.4 Ω, 36.0 Ω]

5

Rectangular Waveguides

5.1. Introduction

The use of circuit techniques at microwave frequencies has been discussed in the last chapter. However, these circuits are only suitable for low-power applications. For high power, the electromagnetic radiation needs to be enclosed and guided by hollow metal waveguides. To eliminate the possibility of radiation and interference, microwave circuits are usually enclosed in a hollow metal box. Such a box is a section of rectangular waveguide. Therefore it is not satisfactory to study microwave circuit techniques alone, but it is also necessary to acquire a knowledge of the properties of hollow metal waveguide. This chapter and the next contain a detailed mathematical analysis of the properties of two simple shapes of hollow metal waveguide and provide a basis for a description of the properties of some common waveguide components and devices in Chapters 11 and 12.

In Chapter 2 consideration was given to the solution of Maxwell's equations in order to describe an electromagnetic wave in a uniform medium of infinite extent. In Chapter 3 a general discussion was given of some of the properties of guided electromagnetic waves, but the detailed mathematical consideration of the form of the waves was left until a later chapter. This chapter and the next contain this detailed mathematical analysis of guided wave propagation. The solution of Maxwell's equations are given for propagation along a hollow metal pipe. The dependence of the mode of propagation on the shape and size of the pipe is determined.

The initial assumption is that the bounding walls are a rectangular metal pipe. The three rectangular space coordinates x, y and z are used in the mathematical analysis of the electromagnetic wave propagation inside rectangular waveguide. For convenience, the walls are parallel to two of the three space dimensions, as shown in Fig. 5.1. The pipe is of uniform cross-section in the x–y plane and it extends to infinity in the z-direction, so that the z-direction is the direction of propagation of the wave. To find expressions for the electromagnetic fields inside the waveguide, it is necessary to solve Maxwell's equations with the constraint of these boundary conditions. The solutions will be similar to the one already obtained in Chapter 2 for a plane wave.

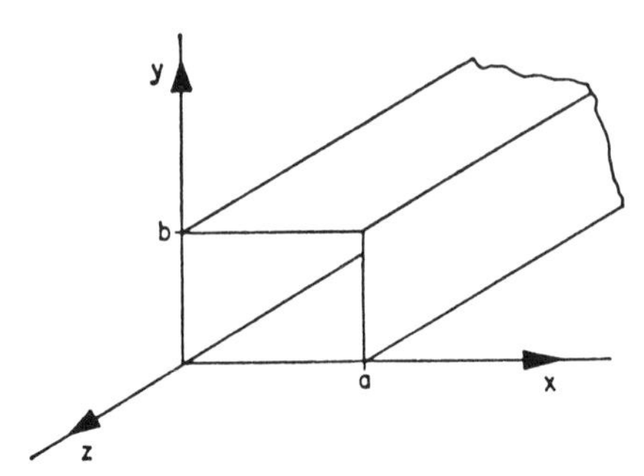

FIG. 5.1. Rectangular waveguide, cross-sectional dimensions $a \times b$, showing its relationship to the axes of the rectangular coordinates.

5.2. Solution of the Wave Equation

The analysis starts by making use of the expressions given in eqns. (2.12) and (2.13). These expressions were obtained as a solution of Maxwell's equations for a non-conducting medium. They are valid for any electromagnetic problem in a non-conducting medium and in particular are used here to analyse electromagnetic wave propagation in hollow metal waveguide. The equations are written here for completeness.

$$\nabla^2 \boldsymbol{E} + \omega^2 \mu\varepsilon \boldsymbol{E} = 0 \tag{2.12}$$

$$\nabla^2 \boldsymbol{H} + \omega^2 \mu\varepsilon \boldsymbol{H} = 0 \tag{2.13}$$

Section 2.3 gives the expansion of the scalar Laplacian and in the rectangular coordinate system the vector Laplacian in the above equations resolves into three similar equations, one for each of the rectangular components of the field. The component equations are of the form

$$\nabla^2 E_x + \omega^2 \mu\varepsilon E_x = 0$$

There are three equations for the components of the electric field and there are another three similar equations for the components of the magnetic field. By use of the relationships given in eqns. (2.1) and (2.2), namely $\boldsymbol{B} = \mu \boldsymbol{H}$ and $\boldsymbol{D} = \varepsilon \boldsymbol{E}$, where μ and ε are constants, it is seen that each component of $\boldsymbol{B}$ and $\boldsymbol{D}$ also satisfies similar wave equations. There are a total of twelve similar equations each one defining a different component of the different fields. It is possible to find solutions to any one of these equations, but the final form of the fields are determined by the boundary conditions. Once a form has been postulated for any one of the twelve components of the fields, it is probable that all the other components of the fields will also be specified since they are related by Maxwell's equations and the material constants, eqns. (2.1) to (2.7). Although we may find possible solutions starting from any one of these twelve equations, experience has shown that the results are easiest to manipulate if solutions are sought for the z-directed components of the electric and magnetic fields. The expressions involving these components are

$$\frac{\partial^2 E_z}{\partial x^2} + \frac{\partial^2 E_z}{\partial y^2} + \frac{\partial^2 E_z}{\partial z^2} = -\omega^2 \mu\varepsilon E_z \tag{5.1}$$

$$\frac{\partial^2 H_z}{\partial x^2} + \frac{\partial^2 H_z}{\partial y^2} + \frac{\partial^2 H_z}{\partial z^2} = -\omega^2 \mu\varepsilon H_z \tag{5.2}$$

Consider first a solution of eqn. (5.1). The solution is obtained using the *separation of variables* technique. The component E_z is generally a function of all three space coordinates. As the space coordinates are independent variables, we assume that E_z is a function of each of these variables independently. Let E_z be in the form of a product of a function of x alone, a function of y alone and a function of z alone. If these functions are denoted by $f_1(x)$, $f_2(y)$ and $f_3(z)$ respectively,

$$E_z = f_1(x) f_2(y) f_3(z) \tag{5.3}$$

Then

$$\frac{\partial^2 E_z}{\partial x^2} = f_1''(x) f_2(y) f_3(z)$$

and two similar relationships for y and z. Substituting these values into eqn. (5.1) gives

$$f_1''(x) f_2(y) f_3(z) + f_1(x) f_2''(y) f_3(z) + f_1(x) f_2(y) f_3''(z)$$
$$= -\omega^2 \mu\varepsilon f_1(x) f_2(y) f_3(z) \tag{5.4}$$

and dividing by E_z from eqn. (5.3)

$$\frac{f_1''(x)}{f_1(x)} + \frac{f_2''(y)}{f_2(y)} + \frac{f_3''(z)}{f_3(z)} = -\omega^2 \mu\varepsilon = -k^2 \quad \text{say} \tag{5.5}$$

It is seen that each term on the left-hand side of eqn. (5.5) is a function of only one of the space coordinates and yet their sum is the constant we have called k^2. This can only be true if each term in the left-hand side of the equation is independently a constant. Therefore we define

$$\frac{f_1''(x)}{f_1(x)} = k_x^2; \qquad \frac{f_2''(y)}{f_2(y)} = k_y^2; \qquad \frac{f_3''(z)}{f_3(z)} = k_z^2; \tag{5.6}$$

where k_x, k_y and k_z are constants which may be complex and

$$k_x^2 + k_y^2 + k_z^2 = -k^2 \tag{5.7}$$

As the z-direction is the direction of propagation, a distinction will be made between k_z and the other two constants. These may be combined into yet another constant which defines the effect of the cross-section of the waveguide. As we shall see, it defines the cut-off conditions of the waveguide and hence is given the subscript c;

$$k_x^2 + k_y^2 = -k_c^2 \tag{5.8}$$

Then substituting into eqn. (5.7)

$$k_z = \pm\sqrt{(k_c^2 - k^2)}$$

The z component of eqn. (5.6) may be written

$$\frac{\partial^2 E_z}{\partial z^2} = k_z^2 E_z$$

This equation has the same form as eqn. (2.16), (2.17) or (2.18) and hence its z-dependence is the same form as that given in eqn. (2.21).

$$E_z = E_0 \exp(j\omega t - \gamma z)$$

It has been assumed that the waveguide is lossless so that there is no attenuation of the propagating wave. Therefore, the phase constant for the waveguide propagation is defined by

$$\gamma^2 = k_z^2 = -\beta^2$$

and

$$\beta = jk_z = \pm j\sqrt{(k_c^2 - k^2)} = \pm\sqrt{(k^2 - k_c^2)} \tag{5.9}$$

Then the solution to eqn. (5.1) takes the form

$$E_z = f_1(x) f_2(y) \exp j(\omega t - \beta z) \tag{5.10}$$

5.3. Cut-off Conditions

If β is real, eqn. (5.10) is an equation for the field of a propagating wave similar to a plane wave, at least the variation of field strength with time and in the direction of propagation is the same as that of a plane wave. If β is imaginary, it is the equation of a lossy wave, and there is an exponential decay of the fields in the z-direction. Even though no power is absorbed by the waveguide lossless propagation cannot occur and the constant k_z becomes the attenuation constant. Here we see mathematically, what has already been shown descriptively in section 3.4, that there is not always a condition for propagation of a bounded wave as there is for a plane wave; a limiting condition is set by the cross-section cut-off constant k_c. The limiting condition for the change from satisfactory propagation to the attenuating condition is

$$\beta^2 = 0$$

then

$$k^2 = k_c^2$$

or

$$k_c = \omega\sqrt{(\mu\varepsilon)}$$

The frequency appropriate to the above limiting condition is the *cut-off frequency*. Hence

$$\omega_c = \frac{k_c}{\sqrt{(\mu\varepsilon)}}$$

$$f_c = \frac{k_c}{2\pi\sqrt{(\mu\varepsilon)}} \tag{5.11}$$

and

$$\lambda_c = \frac{2\pi}{k_c} \tag{5.12}$$

A propagating wave has a wavelength that is different from the characteristic wavelength of a plane wave. It is the *waveguide wavelength*, hence, similar to eqn. (3.7),

$$\lambda_g = \frac{2\pi}{\beta} \tag{5.13}$$

From eqn. (2.30) the plane wave wavelength is given by

$$\lambda = \frac{2\pi}{\omega\sqrt{(\mu\varepsilon)}} = \frac{2\pi}{k} \tag{5.14}$$

Hence substituting from eqns. (5.12) and (5.14) into eqn. (5.9) gives

$$\frac{1}{\lambda_c^2} + \frac{1}{\lambda_g^2} = \frac{1}{\lambda^2} \tag{5.15}$$

If the waveguide is filled with air, it is a very good approximation to a vacuum with constants $\mu = \mu_0$ and $\varepsilon = \varepsilon_0$. Then

$$\lambda = \lambda_0$$

and eqn. (5.15) becomes

$$\frac{1}{\lambda_c^2} + \frac{1}{\lambda_g^2} = \frac{1}{\lambda_0^2} \tag{5.16}$$

Its more familiar form to some microwave engineers is

$$\lambda_g = \frac{\lambda_0}{\sqrt{\left[1 - \left(\frac{\lambda_0}{\lambda_c}\right)^2\right]}} \tag{3.4}$$

5.4. Boundary Conditions

The boundary condition provided by a waveguide is a perfectly conducting surface. The wave propagates in the space inside the boundary and is unaffected by anything outside the boundary. In practice the boundary consists of a high conductivity metal such as copper or silver; sometimes aluminium or brass is also used. Rectangular waveguide consists of a pipe

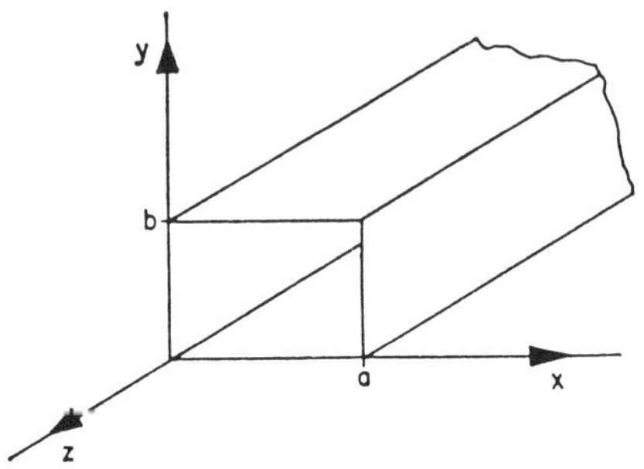

FIG. 5.1. Rectangular waveguide, cross-sectional dimensions $a \times b$, showing its relationship to the axes of the rectangular coordinates.

as shown in Fig. 5.1 whose inside dimensions are a by b. The axes of the rectangular coordinates are chosen to coincide with the waveguide as shown. The diagram only shows the inside surfaces of the waveguide and only the inside dimensions are specified, as anything external to these surfaces does not affect the wave inside the waveguide.

Equation (5.6) shows that the x- and y-dependences of the fields take the same form as the z-dependence, so that it would be possible to postulate an exponential dependence of the fields in the x- and y-dimensions. Alternatively sine and cosine functions may be used and the possible solutions will be given in terms of linear combinations of these trigonometric functions. Therefore let

$$f_1(x) = A \sin(jk_x x) + B \cos(jk_x x) \tag{5.17}$$

$$f_2(y) = C \sin(jk_y y) + D \cos(jk_y y) \tag{5.18}$$

where A, B, C and D are arbitrary constants.

The waveguide consists of a perfectly conducting wall at $x = 0$ and one at $x = a$. At a perfectly conducting wall the electric field parallel to the plane of the wall must vanish and hence the x-dependence of the electric field parallel to the plane of the wall must be such that it is zero at the walls. A suitable solution for eqn. (5.17) is

$$jk_x = \frac{m\pi}{a} \quad \text{or} \quad k_x = -\frac{jm\pi}{a} \tag{5.19}$$

where m is an integer and $B = 0$.

Similarly from eqn. (5.18) for walls at $y = 0$ and $y = b$

$$jk_y = \frac{n\pi}{b} \quad \text{or} \quad k_y = -\frac{jn\pi}{b} \tag{5.20}$$

where n is an integer and $D = 0$.

Hence the full solution for the z-component of the electric field is

$$E_z = E_0 \sin\frac{m\pi x}{a}\sin\frac{n\pi y}{b}\exp j(\omega t - \beta z) \tag{5.21}$$

where the arbitrary constants A and C have been combined in E_0.

By substituting into eqn. (5.8) for k_x and k_y from eqns. (5.19) and (5.20), the cut-off conditions for this mode of propagation in rectangular waveguide are given by

$$k_c = \sqrt{\left[\left(\frac{m\pi}{a}\right)^2 + \left(\frac{n\pi}{b}\right)^2\right]}$$

therefore

$$\lambda_c = \frac{1}{\sqrt{[(m/2a)^2 + (n/2b)^2]}} \tag{5.22}$$

m and n are integers which may take any value. Each different combination of m and n constitutes a separate solution to Maxwell's equations for the given boundary conditions. Mathematically each independent solution is an eigenfunction of this problem and the values for k_x and k_y are the eigenvalues. In electrical terms, each solution gives rise to a different mode of propagation in the waveguide. A discussion of waveguide modes is given in section 5.8 after expressions for the other components of the fields in the waveguide have been derived. We will then be in a position to discuss more fully the properties of the different modes.

5.5. Expressions for the Field Components

To find relationships between the different components of the fields, it is necessary to return to Maxwell's curl equations and consider them in their component form. These equations are given in eqns. (2.24) and (2.25), but we can also substitute the z-dependence of the wave into these equations,

$$\frac{\partial}{\partial z} = -j\beta$$

giving

$$\frac{\partial E_z}{\partial y} + j\beta E_y = -j\omega\mu H_x \tag{5.23}$$

$$-j\beta E_x - \frac{\partial E_z}{\partial x} = -j\omega\mu H_y \tag{5.24}$$

$$\frac{\partial E_y}{\partial x} - \frac{\partial E_x}{\partial y} = -j\omega\mu H_z \tag{5.25}$$

$$\frac{\partial H_z}{\partial y} + j\beta H_y = j\omega\varepsilon E_x \tag{5.26}$$

$$-j\beta H_x - \frac{\partial H_z}{\partial x} = j\omega\varepsilon E_y \tag{5.27}$$

$$\frac{\partial H_y}{\partial x} - \frac{\partial H_x}{\partial y} = j\omega\varepsilon E_z \tag{5.28}$$

Inspection of eqns. (5.23) to (5.28) show that eqns. (5.23), (5.24), (5.26) and (5.27) form two pairs of simultaneous equations in E_y and H_x and E_x and H_y. Hence it is possible to obtain expressions for the x- and y-components of the electric and magnetic fields in terms of the derivatives of the z-directed components of these fields.

Rewriting the above equations in pairs and simplifying gives

$$\beta E_y + \omega\mu H_x = j\frac{\partial E_z}{\partial y} \tag{5.23a}$$

$$\omega\varepsilon E_y + \beta H_x = j\frac{\partial H_z}{\partial x} \tag{5.27a}$$

$$\beta E_x - \omega\mu H_y = j\frac{\partial E_z}{\partial x} \tag{5.24a}$$

$$\omega\varepsilon E_x - \beta H_y = -j\frac{\partial H_z}{\partial y} \tag{5.26a}$$

Hence, using the simplification from eqn. (5.9)

$$\omega^2\mu\varepsilon - \beta^2 = k_c^2$$

the solutions of the simultaneous equations are

$$E_y = \frac{j}{k_c^2}\left(-\beta\frac{\partial E_z}{\partial y} + \omega\mu\frac{\partial H_z}{\partial x}\right) \tag{5.29}$$

$$H_x = \frac{j}{k_c^2}\left(\omega\varepsilon\frac{\partial E_z}{\partial y} - \beta\frac{\partial H_z}{\partial x}\right) \tag{5.30}$$

$$E_x = \frac{-j}{k_c^2}\left(\beta\frac{\partial E_z}{\partial x} + \omega\mu\frac{\partial H_z}{\partial y}\right) \tag{5.31}$$

$$H_y = \frac{-j}{k_c^2}\left(\omega\varepsilon\frac{\partial E_z}{\partial x} + \beta\frac{\partial H_z}{\partial y}\right) \tag{5.32}$$

The components of the fields specified by eqns. (5.29) to (5.32) are functions of E_z and H_z which appear in these equations as independent variables. Maxwell's equations do not provide any other connection between E_z and H_z so that they are independent variables. Hence eqns. (5.29) to (5.32) give two independent sets of field components. One set is a function of E_z and the other set is a function of H_z. It has already been shown that solutions may be obtained to the wave equation, eqns. (5.1) or (5.2), starting with an expression for any of the twelve possible field components. It has now been shown that if possible forms of E_z and H_z are postulated, expressions may be obtained for all the other field components. Just as it has already been shown that there are an infinite number of different possible solutions for Maxwell's equations arising from the expression for E_z and each solution was said to give rise to a different *mode* of propagation in the waveguide, we further postulate that H_z need not exist. That is, we divide the modes of propagation into those modes where E_z exists but $H_z = 0$. Modes where $H_z = 0$ are called transverse magnetic or TM-modes. Similarly there are other possible modes when H_z exists and $E_z = 0$ and they are called transverse electric or TE-modes.

5.6. TM-modes

Substituting $H_z = 0$ and eqn. (5.21) into eqns. (5.29) to (5.32) gives the full expressions for the fields of TM-modes in rectangular waveguide. They are

$$\left.\begin{aligned}
E_x &= -\frac{j\beta E_0}{k_c^2}\frac{m\pi}{a}\cos\frac{m\pi x}{a}\sin\frac{n\pi y}{b}\exp j(\omega t - \beta z)\\
E_y &= -\frac{j\beta E_0}{k_c^2}\frac{n\pi}{b}\sin\frac{m\pi x}{a}\cos\frac{n\pi y}{b}\exp j(\omega t - \beta z)\\
E_z &= E_0\sin\frac{m\pi x}{a}\sin\frac{n\pi y}{b}\exp j(\omega t - \beta z)\\
H_x &= \frac{j\omega\varepsilon E_0}{k_c^2}\frac{n\pi}{b}\sin\frac{m\pi x}{a}\cos\frac{n\pi y}{b}\exp j(\omega t - \beta z)\\
H_y &= -\frac{j\omega\varepsilon E_0}{k_c^2}\frac{m\pi}{a}\cos\frac{m\pi x}{a}\sin\frac{n\pi y}{b}\exp j(\omega t - \beta z)\\
H_z &= 0
\end{aligned}\right\} \tag{5.33}$$

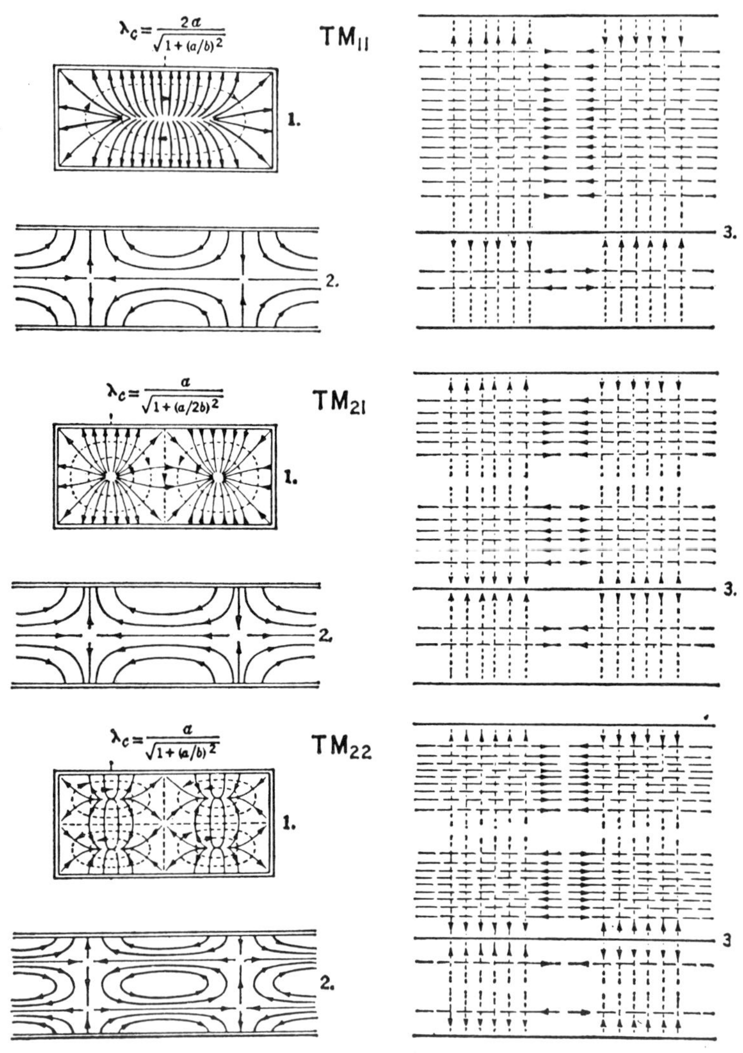

FIG. 5.2. Line representation of the fields of some TM-modes in rectangular waveguide, broad dimension a and narrow dimension b. 1. Cross-sectional view. 2. Longitudinal view. 3. Surface view: – – – electric current; ——— electric field; - - - - magnetic field. (From *Waveguide Handbook*, edited by N. Marcuvitz, McGraw-Hill, 1951. Reproduced by permission of McGraw-Hill Book Co. Inc.)

Line representations of the field distributions of some TM-modes in rectangular waveguide are shown in Fig. 5.2. The rectangular waveguide dimensions, a and b, are such that $a/b = 2.25$ and the frequency is such that $\lambda_g/a = 1.4$. The mode patterns on the left-hand side of the figure depict the electric and magnetic field lines within transverse and longitudinal sections of the waveguide. The patterns on the right-hand side show the magnetic field and current lines on the inner surfaces at the top and side of the waveguide.

5.7. TE-modes

Having obtained expressions for the field components of the modes for which $H_z = 0$, it is now necessary to derive expressions for the field components of the modes for which $E_z = 0$. The boundary condition is that the electric field parallel to the waveguide wall is zero at the wall. As it is hoped to postulate an expression for H_z and then derive all the other field components from it, it is necessary to obtain the boundary conditions in terms of H_z. However, the boundary condition specifies that the electric field is zero at the walls. There is no method of direct specification of the magnetic field at the waveguide walls. For the TE-modes it is necessary to determine the boundary conditions through the electric field components that exist, that is, the transverse components of the electric field. The boundary condition for the rectangular waveguide specifies that E_y is zero at $x = 0$ and $x = a$ and that E_x is zero at $y = 0$ and $y = b$.

For the TE-mode, $E_z = 0$ and eqn. (5.29) shows that, if $E_y = 0$, then

$$\frac{\partial H_z}{\partial x} = 0$$

Hence the boundary condition governing the x-dependence of H_z for the TE-modes is

$$\frac{\partial H_z}{\partial x} = 0 \quad \text{at} \quad x = 0 \quad \text{and} \quad x = a$$

Similarly from eqn. (5.31), the boundary condition governing the y-dependence is

$$\frac{\partial H_z}{\partial y} = 0 \quad \text{at} \quad y = 0 \quad \text{and} \quad y = b$$

Using an argument similar to that already used to obtain the solution to eqn. (5.1), we postulate a general form of expression for H_z as a solution to eqn. (5.2):

$$H_z = [A \sin(jk_x x) + B \cos(jk_x x)][C \sin(jk_y y) + D \cos(jk_y y)] \times \exp j(\omega t - \beta z) \tag{5.34}$$

Differentiating eqn. (5.34) appropriately and applying the boundary conditions, it is found that eqns. (5.19) and (5.20) also apply to the TE-modes and that the constants A and C are both zero. Combining the other two constants into the constant H_0, and substituting from eqns. (5.19) and (5.20) into eqn. (5.34), gives an expression for H_z which is the solution to eqn. (5.2)

$$H_z = H_0 \cos\frac{m\pi x}{a}\cos\frac{n\pi y}{b}\exp j(\omega t - \beta z) \tag{5.35}$$

Substituting this expression for H_z and $E_z = 0$ into eqns. (5.29) to (5.32) gives the full expressions for the fields of TE-modes in rectangular waveguide. They are

$$\left.\begin{aligned}
E_x &= \frac{j\omega\mu H_0}{k_c^2}\frac{n\pi}{b}\cos\frac{m\pi x}{a}\sin\frac{n\pi y}{b}\exp j(\omega t - \beta z)\\
E_y &= -\frac{j\omega\mu H_0}{k_c^2}\frac{m\pi}{a}\sin\frac{m\pi x}{a}\cos\frac{n\pi y}{b}\exp j(\omega t - \beta z)\\
E_z &= 0\\
H_x &= \frac{j\beta H_0}{k_c^2}\frac{m\pi}{a}\sin\frac{m\pi x}{a}\cos\frac{n\pi y}{b}\exp j(\omega t - \beta z)\\
H_y &= \frac{j\beta H_0}{k_c^2}\frac{n\pi}{b}\cos\frac{m\pi x}{a}\sin\frac{n\pi y}{b}\exp j(\omega t - \beta z)\\
H_z &= H_0\cos\frac{m\pi x}{a}\cos\frac{n\pi y}{b}\exp j(\omega t - \beta z)
\end{aligned}\right\} \tag{5.36}$$

Line representations of the field distributions of some TE-modes in rectangular waveguide are shown in Fig. 5.3. As in Fig. 5.2, $a/b = 2.25$ and $\lambda_g/a = 1.4$. The mode patterns on the left depict the field distribution in the waveguide and those on the right depict the current distribution on the inner surface of the waveguide.

5.8. Mode Nomenclature

So far, we have referred to TM-modes and TE-modes without defining any system of mode nomenclature. The broad classification of waveguide modes is in terms of the fields existing in the waveguide when that mode is propagating. Consider first a plane wave propagating in free space, which although it cannot strictly be considered to be a waveguide mode helps in an understanding of mode nomenclature. The plane wave has no field components acting along the direction of propagation. We say that this wave has no *longitudinal* components of its fields. Its fields act entirely in a plane transverse to the direction of propagation; they are said to be entirely *transverse* fields. A wave with only transverse components of both its electric and magnetic fields is said to be a *transverse electric and magnetic* mode or a TEM-mode. A plane wave is an example of a TEM-mode although it does not propagate in a waveguide. The TEM-mode is important because it is the fundamental mode which propagates at any frequency in a two conductor

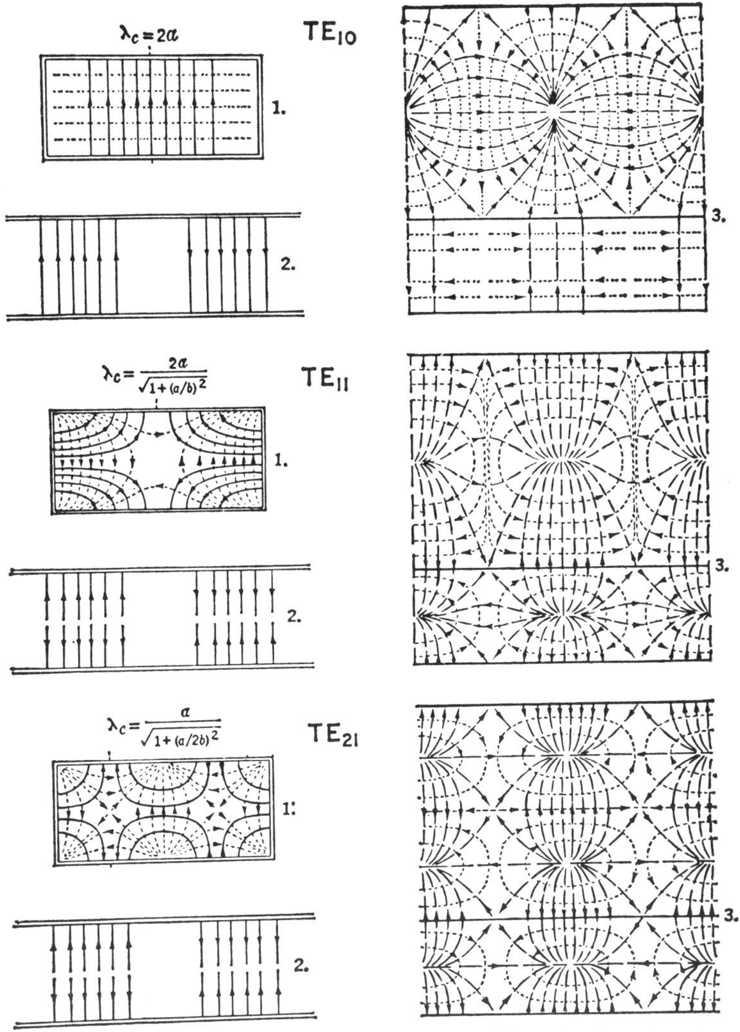

FIG. 5.3. Line representation of the fields of some TE-modes in rectangular waveguide, broad dimension a and narrow dimension b. 1. Cross-sectional view. 2. Longitudinal view. 3. Surface view: — — — electric current; ———— electric field; - - - - magnetic field. (From *Waveguide Handbook*, edited by N. Marcuvitz, McGraw-Hill, 1951. Reproduced by permission of McGraw-Hill Book Co. Inc.)

transmission line. This is the plane wave type of mode which propagates at all frequencies in parallel plate waveguide. It is also the transmission line mode which propagates at all frequencies in any two conductor transmission line such as a coaxial line. The coaxial line is also discussed in more detail in section 6.8.

It has already been shown that there is no mode of propagation in waveguide which has entirely transverse fields. A longitudinal electric or a longitudinal magnetic field is a necessity for any mode which propagates in waveguide. For the waveguide filled with air, or some other homogeneous dielectric material, propagating modes exist with a purely transverse electric field or with a purely transverse magnetic field. There are some situations with inhomogeneous dielectric filling or anisotropic material filling the waveguide where the waveguide modes have both longitudinal electric and magnetic fields but a discussion of these modes and their nomenclature is beyond the scope of this book. The simple waveguide modes are classified as *transverse electric* or *transverse magnetic*, TE- or TM-modes respectively. Sometimes the modes are classified by the field whose longitudinal component exists, so that the TE-modes are called H-modes and the TM-modes are called E-modes. The two major sets of modes are further subdivided by the values taken for the integers m and n in the expressions for the components of the fields. Equations (5.33) give the values for the components of the fields for the TM_{mn}- (or E_{nm}-) mode and eqn. (5.36) give the values for the TE_{mn} (or H_{nm}) mode. Equation (5.22) shows that the different modes have different frequencies below which propagation is cut off and that this cut-off frequency is approximately a function of m and n. Hence the mode with the lowest possible values of m and n is able to propagate at the lowest frequencies and there is a band of frequencies over which this mode and this mode only can propagate. This is called the *dominant mode* and rectangular waveguide is normally used in that frequency range where only the dominant mode can propagate. It is seen from eqn. (5.33) for TM-modes that if either m or n is zero no mode can exist as E_z is zero, and Maxwell's equations are not satisfied. Hence the lowest TM-mode is the TM_{11}-mode. However eqn. (5.36) shows that the TE_{10}- or the TE_{01}-modes are possible and, if it is assumed that a is the broad dimension of the waveguide, TE_{10} is the dominant mode in rectangular waveguide.

There is a desire among most microwave engineers that all standard waveguides should have a 2:1 ratio between the sides, that is $a = 2b$, but unfortunately this is not always true because some of the most popular sizes and presumably some of the first to be chosen as standard sizes have the 2:1 ratio applied to their outside dimensions rather than their inside dimensions. (A list of dimensions of standard rectangular waveguides is given in Table 11.1) However, the assumption is made that all standard rectangular waveguides have internal dimensions which are approximately in the ratio 2:1. Assuming that $a = 2b$ for the TE_{10}-mode $\lambda_c = 2a$ whilst for the TE_{01}- and the TE_{20}-modes $\lambda_c = a$. For the other nearest modes TE_{11}- and the TM_{11}-modes $\lambda_c = 2a/\sqrt{5}$. There is an octave over which the dominant mode is the only mode which can propagate in the waveguide. Because at frequencies near to the cut-off frequency the waveguide wavelength is very long, the recommended frequency band of operation of standard waveguide is approximately $1{:}1\frac{1}{2}$.

5.9. TE_{10}-mode

As in most situations the dominant mode is the only mode that exists in the waveguide, the dominant TE_{10}-mode is discussed in some detail. Substituting the conditions $m = 1$ and $n = 0$ into eqn. (5.36) gives

$$\left.\begin{aligned}
E_x &= 0 \\
E_y &= j\frac{\omega\mu a}{\pi} H_0 \sin\frac{\pi x}{a} \exp j(\omega t - \beta z) \\
E_z &= 0 \\
H_x &= j\frac{\beta a}{\pi} H_0 \sin\frac{\pi x}{a} \exp j(\omega t - \beta z) \\
H_y &= 0 \\
H_z &= H_0 \cos\frac{\pi x}{a} \exp j(\omega t - \beta z)
\end{aligned}\right\} \tag{5.37}$$

This mode is not only the dominant mode but it is also an extremely simple mode as far as the field patterns are concerned. The electric field only acts in the y-direction and the magnetic field acts in the x–z-plane. If we represent these fields by lines, then the electric field consists of straight lines parallel to the y-axis and the magnetic field consists of closed loops in the x–z-plane. These field patterns are shown in Fig. 5.3. For the TE_{10}-mode there is no variation of field strength in the y-direction, and there is a sinusoidal variation in the x-direction. There is also a sinusoidal variation in the z-direction and the whole field pattern appears to be moving in the z-direction with the speed

$$v_p = \frac{\omega}{\beta} = f\lambda_g$$

where v_p is the *phase velocity*. The electric field lines terminate in electric charges in the walls of the waveguide. As the wave travels, the currents in the waveguide walls redistribute these charges so that the electric field is always correctly terminated.

5.10. Waveguide Wall Currents

For the purposes of evaluating the electromagnetic fields inside the waveguide, the waveguide walls have been considered to be a perfect conductor. Hence any currents occurring in the walls have been confined to an infinitely thin surface layer and have been infinitely large. In practice the metal of the waveguide wall possesses finite conductivity so that the currents in the wall are finite. For an exact evaluation of the current density in the walls it would be necessary to perform an analysis similar to that of Chapter 7. No such analysis is made here because we are more interested in the relationships between the different currents in different parts of the waveguide walls, rather than the absolute amplitudes of such currents. The currents may be obtained from eqn. (2.7) which is rewritten here in its component form.

$$J_x = \frac{\partial H_z}{\partial y} - \frac{\partial H_y}{\partial z} - j\omega\varepsilon E_x \tag{5.38}$$

$$J_y = \frac{\partial H_x}{\partial z} - \frac{\partial H_z}{\partial x} - j\omega\varepsilon E_y \tag{5.39}$$

$$J_z = \frac{\partial H_y}{\partial x} - \frac{\partial H_x}{\partial y} - j\omega\varepsilon E_z \tag{5.40}$$

Equations (5.38) to (5.40) are the general relationships that may be applied to any waveguide. The only currents relevant are those tangential to the plane of the wall and the electric field in this plane is zero. Therefore the wall currents are entirely generated by the magnetic field. The space variation of the field that gives rise to these currents is not the normal variation of the field occurring inside the waveguide, but it is the sudden decay of these magnetic fields at the walls. For normal high conductivity metals at microwave frequencies, the rate of decay of the field at the wall is so much faster than the normal variation of the field in the waveguide that the latter variation may be ignored.

As only the currents tangential to the plane of the waveguide wall are relevant for the broad walls it is necessary to find expressions for J_x and J_z at $y = 0$ and $y = b$ and for the narrow walls to find J_y and J_z at $x = 0$ and $x = a$. The electric field tangential to the plane of the wall is zero at the wall and at the broad walls the terms $\partial H_y/\partial z$ and $\partial H_y/\partial x$ are negligibly small. Therefore in the broad walls at $y = 0$ and $y = b$

$$J_x = \frac{\partial H_z}{\partial y} \tag{5.41}$$

$$J_z = -\frac{\partial H_x}{\partial y} \tag{5.42}$$

At the narrow walls the terms $\partial H_x/\partial z$ and $\partial H_x/\partial y$ are negligibly small, therefore in the narrow walls at $x = 0$ and $x = a$,

$$J_y = -\frac{\partial H_z}{\partial x} \tag{5.43}$$

$$J_z = \frac{\partial H_y}{\partial x} \tag{5.44}$$

The rate of decay of the field as it penetrates the metal of the waveguide wall is a function of the conductivity of the metal and the frequency of the wave and it is unaffected by the field pattern inside the waveguide. This means that the differential operations in eqns. (5.41) to (5.44) are all the same, and it is possible to specify a direct proportionality between the currents in the wall and the magnetic field adjacent to the wall. In the broad wall, J_x is proportional to H_z and J_z to H_x, and in the narrow wall, J_y is proportional to H_z and J_z to H_y. For a perfect conductor, these are the fields and currents existing on each side of the boundary. The surface current density, existing in an infinitely thin skin in the metal of the waveguide wall, is proportional to the appropriate magnetic field component, existing inside the waveguide adjacent to the wall. Hence the current streamlines shown in Fig. 5.3 are obtained. Note that in the picture of the field pattern, the current streamlines are not completed by the electric field. The current is transferring charge ready to support the electric field inside the waveguide a half cycle later. There is also a maximum rate of change of electric field at the same point as the node of electric current so that the displacement current in the inside of the waveguide is completing the current streamlines in the waveguide walls.

5.11. Waveguide Attenuation

For microwave propagation along hollow waveguide pipe, the attenuation of the wave is due to losses associated with the finite conductivity of the waveguide walls. The mathematical analysis so far in this chapter assumes that the waveguide walls are perfect conductors which sustain infinite current densities in an infinitely thin skin at the surface of the conductor. Practical materials used for the waveguides are excellent conductors, but they still have a finite conductivity. This section will indicate how allowance may be made for the finite conductivity of the waveguide material and for the fact that there is some penetration of fields into the material of the waveguide walls.

It has already been shown in section 5.10 that the current in the walls is proportional to the tangential magnetic fields at the walls. If it can be assumed that in the process of penetrating the walls, the tangential magnetic field behaves like a plane wave, then the total surface current has the same numerical value as the tangential magnetic field at the surface as shown in section 7.5. Hence at any point on the surface

$$\text{power loss} = \tfrac{1}{2} R_s H_t^2 \tag{5.45}$$

where H_t is, in this instance, the value of the magnetic field *tangential to the wall* of the waveguide and R_s is the equivalent surface resistance of the waveguide wall material at the operating frequency. Looking ahead to the theory in section 7.6, by definition, R_s is given by

$$R_s = \frac{\text{resistivity}}{\text{skin depth}} \tag{5.46}$$

In the Poynting vector expression for power flow, eqn. (2.38), the power flow down the waveguide is proportional to the cross-multiplication of the transverse electric and magnetic fields. As the magnetic field is proportional to the electric field, the power flow is proportional to the square of the electric field, and the power transmitted is given by

$$W = KE_t^2$$

where, in this instance, E_t is the transverse electric field and where K is the constant of proportionality. Hence the power loss is given by

$$\frac{dW}{dz} = 2KE_t\frac{dE_t}{dz}$$

Since the amplitude of the field quantities are all varying with distance as $\exp -\alpha z$,

$$\frac{dE_t}{dz} = -\alpha E_t$$

Therefore,

$$\frac{dW}{dz} = -2K\alpha E_t^2 = -2\alpha W$$

and the attenuation constant is given by

$$\alpha = \frac{1}{2}\left(\frac{dW}{dz}\right)W = \frac{1}{2}\frac{\text{power loss}}{\text{transmitted power}} \tag{5.47}$$

The total power loss is obtained from the integration over the perimeter of the cross-section of the waveguide, of the expression given in eqn. (5.45). The power flow is given by the integration of the Poynting vector across the cross-sectional area of the waveguide. Substituting these expressions into eqn. (5.47) gives the normally quoted form for the attenuation constant of empty waveguide:

$$\alpha = \frac{R_s}{2}\frac{\oint H_t^2\, dl}{\iint \boldsymbol{E} \times \boldsymbol{H} . d\boldsymbol{a}} \tag{5.48}$$

The integral in the numerator of eqn. (5.48) is evaluated over the waveguide perimeter and that in the denominator is evaluated over the cross-section of the waveguide.

Expressions for the attenuation constants of rectangular waveguide modes can be obtained by substituting expressions for the fields into eqn. (5.48). For the TM-modes, the expressions for the fields are given by eqn. (5.33).

The numerator of the expression in eqn. (5.48) is given by

$$\oint H_t^2\, dl = 2\int_0^a H_x^2\, dx + 2\int_0^b H_y^2\, dy$$
$$= \frac{\omega^2\varepsilon^2 E_0^2\pi^2}{ak_c^4}\left(\frac{b}{a}\right)\left[m^2 + n^2\left(\frac{a}{b}\right)^3\right] \tag{5.49}$$

and the denominator of the expression is given by

$$\iint \boldsymbol{E} \times \boldsymbol{H} . d\boldsymbol{a} = \int_0^a \int_0^b (E_x H_y + E_y H_x)\, dx\, dy$$
$$= \frac{\omega\varepsilon\beta E_0^2\pi^2}{4k_c^4}\left(\frac{b}{a}\right)\left[m^2 + n^2\left(\frac{a}{b}\right)^2\right] \tag{5.50}$$

Therefore

$$\alpha = \frac{2R_s\omega\varepsilon}{a\beta}\left[\frac{m^2 + n^2(a/b)^3}{m^2 + n^2(a/b)^2}\right] \tag{5.51}$$

An expression for R_s is derived in section 7.6,

$$R_s = \sqrt{\left(\frac{\omega\mu}{2\sigma}\right)} \tag{5.52}$$

Substituting for R_s and also substituting from eqns. (2.30) and (5.13) to get the result in terms of wavelengths gives

$$\alpha = \frac{2\lambda_g}{a\lambda_0}\left(\frac{\pi}{\lambda_0\eta\sigma}\right)^{1/2}\left[\frac{m^2 + n^2(a/b)^3}{m^2 + n^2(a/b)^2}\right] \text{nepers/m} \tag{5.53}$$

Some results calculated for the first few modes in rectangular waveguide are plotted in Fig. 5.4. These results are scaled in terms of the broad dimension of the waveguide for waveguides having a 2:1 aspect ratio where $a = 2b$. The horizontal axis is λ_0/a, which is dimensionless, or the equivalent frequency is fa in Hz.m because $fa = c/(\lambda_0/a)$. The attenuation then comes out to be $\alpha a^{3/2}$. The attenuation constant is calculated for drawn copper waveguide assuming a conductivity of 4.00×10^7 S/m. This figure is known to be a good approximation at about 10 GHz, and a round figure was chosen for the conductivity of copper to aid scaling. For other waveguide materials, the attenuation constant is scaled according to the square root of the ratio of the conductivities. Using the conversion that 1 neper = 8.686 dB, the vertical axis is calibrated in dB . $\text{m}^{1/2}$. These simple expressions for attenuation constant are only valid when α is a small part of the propagation constant. The expression in eqn. (5.53) (and in eqns. (5.59) to (5.61)) is not valid near to cut-off conditions. However results from eqn. (5.53) (and from eqns. (5.59)

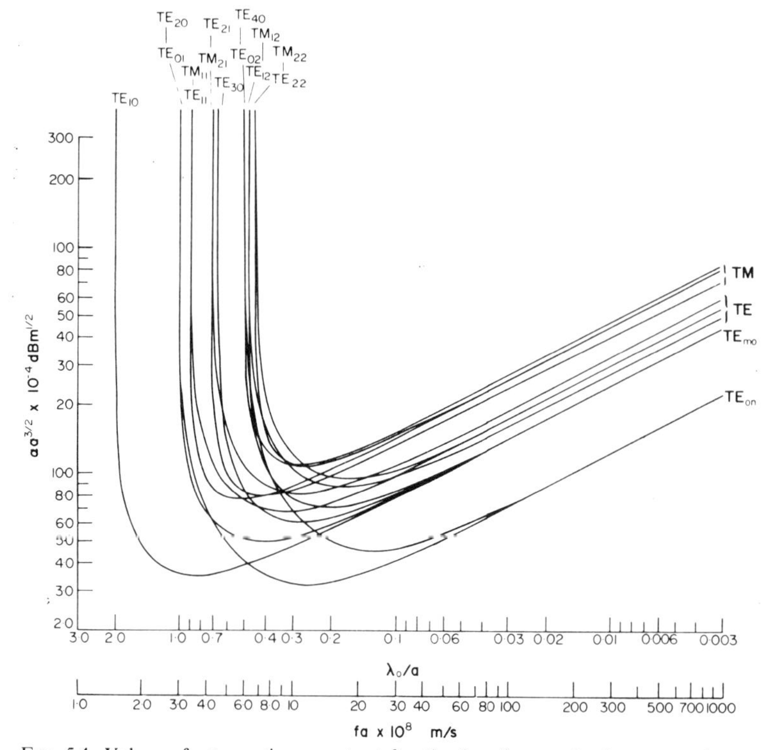

FIG. 5.4. Values of attenuation constant for the first few modes in rectangular waveguide, of 2:1 aspect ratio, width a, plotted against normalized wavelength. These have been plotted to cut-off from the simple formulae even though application of these formulae is not valid near to cut-off.

to (5.61)) have been plotted to cut-off in Fig. 5.4 in order to show simply the cut-off wavelength of each mode.

Similarly expressions for the attenuation constant of the TE-modes are obtained by substituting field values from eqn. (5.36) into eqn. (5.48). The numerator of the expression is given by

$$\oint H_t^2\, dl = 2\int_0^a H_x^2\, dx + 2\int_0^a H_z^2\, dx + 2\int_0^b H_y^2\, dy + 2\int_0^b H_z^2\, dy$$

$$= \frac{\beta^2 H_0^2 \pi^2}{k_c^4}\left(\frac{m^2}{a} + \frac{n^2}{b}\right) + H_0^2(a+b) \qquad (5.54)$$

provided that $m \neq 0$ and $n \neq 0$.

If $n = 0$

$$\oint H_t^2\, dl = \frac{\beta^2 H_0^2 \pi^2 m^2}{k_c^4 a} + H_0^2(a+2b) \qquad (5.55)$$

and if $m = 0$

$$\oint H_t^2\, dl = \frac{\beta^2 H_0^2 \pi^2 n^2}{k_c^4 b} + H_0^2(2a + b) \tag{5.56}$$

The denominator of the expression is given by

$$\iint \boldsymbol{E} \times \boldsymbol{H} . d\boldsymbol{a} = \int_0^a \int_0^b (E_x H_y + E_y H_x)\, dx\, dy$$

$$= \frac{\omega\mu\beta H_0^2 \pi^2}{4k_c^4}\left(\frac{b}{a}\right)\left[m^2 + n^2\left(\frac{a}{b}\right)^2\right] \tag{5.57}$$

provided that $m \neq 0$ and $n \neq 0$.

If $m = 0$ or $n = 0$,

$$\iint \boldsymbol{E} \times \boldsymbol{H} . d\boldsymbol{a} = \frac{\omega\mu\beta H_0^2 \pi^2}{2k_c^4}\left(\frac{b}{a} m^2 \quad \text{or} \quad \frac{a}{b} n^2\right) \tag{5.58}$$

Making substitutions to obtain results in terms of wavelengths, the attenuation constant is given by

$$\alpha = \frac{2\lambda_0}{b\lambda_g}\left(\frac{\pi}{\lambda_0 \eta \sigma}\right)^{1/2}\left[\frac{m^2 + n^2(a/b)}{m^2 + n^2(a/b)^2} + \left(\frac{\lambda_g}{\lambda_c}\right)^2\left(1 + \frac{b}{a}\right)\right] \text{nepers/m} \tag{5.59}$$

provided that $m \neq 0$ and $n \neq 0$.

If $n = 0$,

$$\alpha = \frac{\lambda_0}{b\lambda_g}\left(\frac{\pi}{\lambda_0 \eta \sigma}\right)^{1/2}\left[1 + \left(\frac{\lambda_g}{\lambda_c}\right)^2\left(1 + 2\frac{b}{a}\right)\right] \text{nepers/m} \tag{5.60}$$

and if $m = 0$,

$$\alpha = \frac{\lambda_0}{a\lambda_g}\left(\frac{\pi}{\lambda_0 \eta \sigma}\right)^{1/2}\left[1 + \left(\frac{\lambda_g}{\lambda_c}\right)^2\left(2\frac{a}{b} + 1\right)\right] \text{nepers/m} \tag{5.61}$$

Some values of attenuation constant for the TE-modes are also plotted in Fig. 5.4 for waveguides having a 2:1 aspect ratio, where $a = 2b$. It will be noticed that the dominant TE_{10}-mode is also the mode having the lowest value of attenuation constant but the TE_{01}-mode also has a low value of attenuation constant. If two sizes of waveguide are chosen such that the TE_{10}-mode and the TE_{01}-mode have the same cut-off frequency, then the TE_{01}-mode in the larger size of waveguide has the lower value of attenuation constant at any particular frequency, and this mode of propagation has been used to provide particularly low loss propagation in rectangular waveguide. Some care is needed to prevent excitation of the other modes that can also propagate in this size of waveguide, but these undesired modes can be filtered out by use of suitable slots in the waveguide wall. Study of the field patterns shown in Figs. 5.2 and 5.3 will indicate where slots can be placed in the waveguide wall to be parallel to the current streamlines of the TE_{01}-mode and to cut the current streamlines of all the other modes. Microwave power is radiated out through any slots in the waveguide wall which cut the current streamlines (see problem 5.8).

For a waveguide filled with a low loss material, there is a contribution to the losses and hence a contribution to the attenuation constant from both the losses in the waveguide walls and the losses in the material. As a first approximation, the attenuation constants due to the two effects may be calculated separately and added to give the total attenuation constant. Very often, when there is a material filling the waveguide, the losses due to the material are an order larger than the losses due to the waveguide walls, so that in this case the wall losses may be neglcted. Otherwise, when the waveguide is filled with air, the losses due to the air may be neglected and the losses due to wall currents predominate.

In the process of calculating the attenuation constant we have also calculated expressions for the power flow in the waveguide. The power flow is given by the integration of the Poynting vector across the cross-sectional area of the waveguide. This has been calculated in eqns. (5.50) and (5.57) except that the expressions are in terms of the square of the *peak* value of the electric or magnetic field intensity. So the power flow is given by half the values calculated from eqns. (5.50) and (5.57). Suitable algebraic simplification shows that the power flow is given by

$$P = \left(\frac{\omega\beta ab}{8k_c^2}\right)\begin{Bmatrix} \varepsilon E_0^2 & \text{for TM-modes} \\ \mu H_0^2 & \text{for TE-modes} \end{Bmatrix} \tag{5.62}$$

provided that $m \neq 0$ and $n \neq 0$. If $m = 0$ or $n = 0$,

$$P = \frac{\mu H_0^2 \omega\beta ab}{4k_c^2} = \frac{\eta H_0^2 \lambda_c^2 ab}{4\lambda_0\lambda_g} \tag{5.63}$$

For the dominant mode, $k_c^2 = \pi^2/a^2$, and the expression for the power flow becomes

$$P = \frac{\mu H_0^2 \omega\beta a^3 b}{4\pi^2} = \frac{\eta H_0^2 a^3 b}{\lambda_0\lambda_g} \tag{5.64}$$

5.12. Waveguide Impedance

The wave impedance in waveguide is given by eqn. (3.14):

$$Z_0 = \frac{E_t}{H_t}$$

To find expressions for the transverse components of the field we adopt a new notation. Each vector is split into two components, a component along the direction of propagation—denoted by the subscript z, and a component in the transverse plane—denoted by the subscript t. Then the general vector $\boldsymbol{A}$ has the components A_z and A_t. Now generate a new vector of the same amplitude as $\boldsymbol{A}_t$ but perpendicular to it in the same plane called $\boldsymbol{A}_t^*$. The starring operation rotates the vector through a right angle. Then if

$$\boldsymbol{A}_t = (\boldsymbol{i}\alpha + \boldsymbol{j}\beta)$$

$$\boldsymbol{A}_t^* = (\boldsymbol{i}\beta - \boldsymbol{j}\alpha)$$

where $\boldsymbol{i}$ and $\boldsymbol{j}$ are unit vectors in the x and y directions respectively and

$$(\boldsymbol{A}_t^*)^* = -\boldsymbol{A}_t$$

$$\boldsymbol{A}_t \cdot \boldsymbol{A}_t = \alpha^2 + \beta^2 = \boldsymbol{A}_t^* \cdot \boldsymbol{A}_t^*$$

$$\boldsymbol{A}_t^* \cdot \boldsymbol{B}_t = -\boldsymbol{A}_t \cdot \boldsymbol{B}_t^*$$

Similarly the differential operator ∇ can be divided into the components $\partial/\partial z$ and ∇_t. ∇_t and ∇_t^* have properties similar to those given for $\boldsymbol{A}_t$ and $\boldsymbol{A}_t^*$.

Maxwell's equations, eqns. (2.10) and (2.11), can be denoted in this new notation. Equation (5.28) becomes

$$\frac{\partial H_y}{\partial z} - \frac{\partial H_x}{\partial y} = \nabla_t \cdot \boldsymbol{H}_t^* = j\omega\varepsilon E_z \tag{5.65}$$

Equations (5.26) and (5.27) give

$$\boldsymbol{i}\left(\frac{\partial H_z}{\partial y} + j\beta H_y\right) + \boldsymbol{j}\left(-j\beta H_x - \frac{\partial H_z}{\partial x}\right) = j\omega\varepsilon(\boldsymbol{i}E_x + \boldsymbol{j}E_y)$$

Therefore

$$\nabla_t^* H_z + j\beta \boldsymbol{H}_t^* = j\omega\varepsilon \boldsymbol{E}_t \tag{5.66}$$

Similarly eqns. (5.23) to (5.25) may be written,

$$\nabla_t^* E_z + j\beta \boldsymbol{E}_t^* = -j\omega\mu \boldsymbol{H}_t \tag{5.67}$$

$$\nabla_t \cdot \boldsymbol{E}_t^* = -j\omega\mu H_z \tag{5.68}$$

Applying the starring operation to eqn. (5.67) gives

$$-\nabla_t E_z - j\beta \boldsymbol{E}_t = -j\omega\mu \boldsymbol{H}_t^* \tag{5.69}$$

Equations (5.66) and (5.69) form a pair of simultaneous equations in $\boldsymbol{E}_t$ and $\boldsymbol{H}_t^*$. Rewriting these equations gives

$$j\omega\varepsilon \boldsymbol{E}_t - j\beta \boldsymbol{H}_t^* = \boldsymbol{\nabla}_t^* H_z \qquad (5.66a)$$

$$-j\beta \boldsymbol{E}_t + j\omega\mu \boldsymbol{H}_t^* = \boldsymbol{\nabla}_t E_z \qquad (5.69a)$$

Using the simplification, $\omega^2\mu\varepsilon - \beta^2 = k_c^2$, the solution to these equations is given by

$$k_c^2 \boldsymbol{E}_t = -j\omega\mu \boldsymbol{\nabla}_t^* H_z - j\beta \boldsymbol{\nabla}_t E_z \qquad (5.70)$$

$$k_c^2 \boldsymbol{H}_t^* = j\beta \boldsymbol{\nabla}_t^* H_z - j\omega\varepsilon \boldsymbol{\nabla}_t E_z \qquad (5.71)$$

The wave impedance is a scalar quantity, therefore

$$Z_0 = \frac{|\boldsymbol{E}_t|}{|\boldsymbol{H}_t|} = \frac{|\boldsymbol{E}_t|}{|\boldsymbol{H}_t^*|}$$

Substituting from eqns. (4.70) and (4.71) gives

$$Z_0 = \frac{|\boldsymbol{E}_t|}{|\boldsymbol{H}_t^*|} = \frac{|-j\omega\mu \boldsymbol{\nabla}_t^* H_z - j\beta \boldsymbol{\nabla}_t E_z|}{|j\beta \boldsymbol{\nabla}_t^* H_z - j\omega\varepsilon \boldsymbol{\nabla}_t E_z|} \qquad (5.72)$$

For TM-modes $H_z = 0$, therefore

$$Z_0 = \frac{\beta}{\omega\varepsilon} = \eta \frac{\lambda_0}{\lambda_g} \qquad (5.73)$$

For TE-modes $E_z = 0$, therefore

$$Z_0 = \frac{\omega\mu}{\beta} = \eta \frac{\lambda_g}{\lambda_0} \qquad (5.74)$$

This last result shows that for normal TM- or TE-modes in waveguide, the wave impedance is a constant for any one mode irrespective of the position in the cross-section of the waveguide at which $\boldsymbol{E}_t$ or $\boldsymbol{H}_t$ are taken. Because this derivation is obtained in terms of the transverse components of the field in the waveguide and is independent of the shape of the waveguide, eqns. (5.73) and (5.74) give values for the wave impedance inside any hollow metal waveguide.

From eqns. (5.70) and (5.71), it can also be seen that for normal TM- or TE-modes in waveguide, $\boldsymbol{E}_t$ and $\boldsymbol{H}_t^*$ are parallel vectors. Therefore the electric and magnetic fields inside the waveguide are everywhere perpendicular to one another.

For the particular case of rectangular waveguide, the expressions for the wave impedance can be derived directly from the expressions for the fields in the waveguide. Inspection of eqns. (5.33) and (5.36) shows that the transverse components of the fields occur in pairs having a constant amplitude ratio. Therefore

$$\frac{E_x}{H_y} = -\frac{E_y}{H_x} \tag{5.75}$$

The total transverse field is the vector sum of the two components of the field lying in the transverse plane so that

$$E_t = \sqrt{(E_x^2 + E_y^2)}$$

and

$$Z_0 = \frac{\sqrt{(E_x^2 + E_y^2)}}{\sqrt{(H_x^2 + H_y^2)}} \tag{5.76}$$

and substitution of eqn. (5.75) into eqn. (5.76) shows that

$$Z_0 = \frac{E_t}{H_t} = \frac{E_x}{H_y} = -\frac{E_y}{H_x} \tag{5.77}$$

It is left as an exercise for the reader to substitute values from eqns. (5.33) and (5.36) into eqn. (5.77) to confirm the values for Z_0 given in eqns. (5.73) and (5.74).

The wave impedance is not the only possible definition of characteristic impedance in waveguide. There are also two definitions obtained from voltage or current and power flow in the waveguide,

$$Z_0 = \frac{V^2}{P} = \frac{P}{I^2} \tag{5.78}$$

For the dominant mode it is comparatively easy to determine the voltage form of this expression, and it is useful for determining the relative impedance of items added in shunt across the waveguide. Then the voltage is the integral across the height of the waveguide of the electric field intensity at the centre of the broad face of the waveguide, i.e. the maximum value of E_y. Substituting from eqn. (5.37), the voltage is given by

$$V = \int_0^b E_y \, dy = -\frac{j\omega\mu a H_0}{\pi\sqrt{2}} b \tag{5.79}$$

Therefore the characteristic impedance for shunt components is given by

$$Z_0 = \frac{V^2}{P} = \frac{2\omega\mu b}{\beta a} = (\text{wave impedance})\frac{2b}{a} \tag{5.80}$$

which is equal to the wave impedance when $a = 2b$.

It is seen that the wave impedance equals the intrinsic impedance for TEM-modes where $\lambda_g = \lambda_0$.

5.13. Resonant Cavity

Consider a microwave resonator which consists of a closed rectangular box. As with waveguide, only the inside shape and size are of importance. Because wall losses lower the Q-factor of the resonator, it is important that microwave resonators are made with walls of high conductivity metal. The hollow metal enclosure used to make a microwave resonator is often called

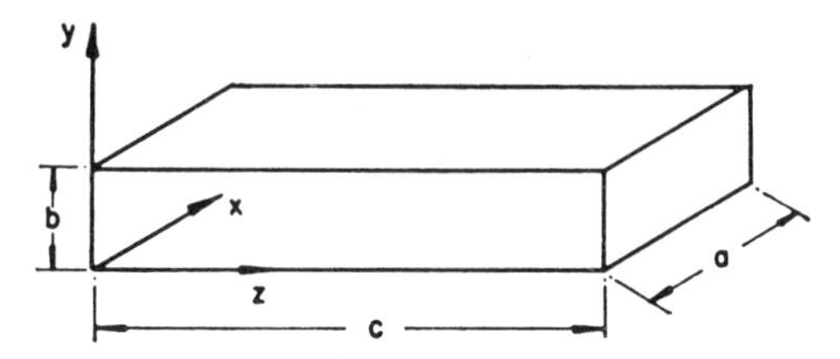

FIG. 5.5. Rectangular resonant cavity showing its dimensions and its relationship to the rectangular coordinate system.

a cavity. A rectangular cavity is shown diagrammatically in Fig. 5.5 together with its relationship to a rectangular coordinate system. Since in the cavity there is no propagation in any direction, mathematically the system is symmetrical and eqn. (5.7) may be written

$$\omega^2\mu\varepsilon = k^2 = -k_x^2 - k_y^2 - k_z^2$$

and the resonant frequency of the cavity is given by

$$2\pi f_0 = \sqrt{\left(\frac{-k_x^2 - k_y^2 - k_z^2}{\mu\varepsilon}\right)} \tag{5.81}$$

Starting with eqn. (5.81) and using an argument similar to that used in section 5.3 we can arrive at an expression for the resonant frequency of a rectangular cavity of dimensions a, b and c as shown in Fig. 5.5. l, m and n are the integers providing the mode numbers for the field patterns inside the cavity. Then

$$f_0 = \tfrac{1}{2}\sqrt{\left[\frac{\left(\frac{l}{a}\right)^2 + \left(\frac{m}{b}\right)^2 + \left(\frac{n}{c}\right)^2}{\mu\varepsilon}\right]} \tag{5.82}$$

and the mode in the cavity is specified as TE_{lmn} or TM_{lmn}.

The resonant cavity can also be considered as a length of uniform waveguide with a short circuit at each end. There is a sinusoidal distribution of field between the two ends of the waveguide so that we can write the field distribution in the form

$$E = f(x, y)\sin\frac{n\pi z}{c}$$

This can be considered to be the standing wave due to two equal and opposite waves since

$$\sin\frac{n\pi z}{c} = \tfrac{1}{2}\exp j\frac{n\pi z}{c} - \tfrac{1}{2}\exp -j\frac{n\pi z}{c}$$

where the phase constant of these waves is

$$\beta = \frac{n\pi}{c}$$

therefore

$$\lambda_g = \frac{2c}{n}$$

so that the resonant cavity can be considered to be a number of half-wavelength sections of waveguide with a complete standing wave inside it. It might also be considered that the two waves travelling in opposite directions are constantly being perfectly reflected from the end walls of the cavity.

5.14. Summary

5.1. The boundary is a ***rectangular metal pipe***. The walls are perfectly conducting and the electromagnetic wave propagates inside the pipe.

5.2. The solution to Maxwell's equations in rectangular coordinates gives

$$\frac{\partial^2 E_z}{\partial x^2} + \frac{\partial^2 E_z}{\partial y^2} + \frac{\partial^2 E_z}{\partial z^2} = -\omega^2\mu\varepsilon E_z \tag{5.1}$$

$$\frac{\partial^2 H_z}{\partial x^2} + \frac{\partial^2 H_z}{\partial y^2} + \frac{\partial^2 H_z}{\partial z^2} = -\omega^2\mu\varepsilon H_z \tag{5.2}$$

$$k^2 = \omega^2\mu\varepsilon \tag{5.5}$$

$$\beta = \pm\sqrt{(k^2 - k_c^2)} \tag{5.9}$$

5.3. ***Cut-off condition*** is $k_c = \omega\sqrt{(\mu\varepsilon)}$

Waveguide wavelength is given by

$$\frac{1}{\lambda_g^2} + \frac{1}{\lambda_c^2} = \frac{1}{\lambda_0^2} \tag{5.16}$$

5.4. Rectangular waveguide boundary conditions are such that

$$\lambda_c = \frac{1}{\sqrt{[(m/2a)^2 + (n/2b)^2]}} \tag{5.22}$$

where $a \times b$ are the cross-section dimensions of the waveguide and m and n are integers.

The z-component of the electric field is

$$E_z = E_0 \sin\frac{m\pi x}{a} \sin\frac{n\pi y}{b} \exp j(\omega t - \beta z) \tag{5.21}$$

5.5. The other components of the electric and magnetic fields can all be expressed in terms of the space derivatives of the two longitudinal components.

5.6. The field components of the TM_{mn}-mode are given in eqn. (5.33) and line representations of the field pattern are given in Fig. 5.2.

5.7. The z-component of the magnetic field is

$$H_z = H_0 \cos\frac{m\pi x}{a} \cos\frac{n\pi y}{b} \exp j(\omega t - \beta z) \tag{5.35}$$

The field components of the TE_{mn}-mode are given in eqn. (5.36) and line representations of the field pattern are given in Fig. 5.3.

5.8. The modes are split into:

Transverse electric and magnetic, TEM-modes (for plane waves and two-conductor transmission lines)

Transverse magnetic, TM-modes

Transverse electric, TE-modes

The ***dominant mode*** in rectangular waveguide is the TE_{10}-mode.

5.9. The TE_{10}-mode field components are

$$\left.\begin{aligned}
E_y &= -j\frac{\omega\mu a}{\pi} H_0 \sin\frac{\pi x}{a} \exp j(\omega t - \beta z) \\
H_x &= j\frac{\beta a}{\pi} H_0 \sin\frac{\pi x}{a} \exp j(\omega t - \beta z) \\
H_z &= H_0 \cos\frac{\pi x}{a} \exp j(\omega t - \beta z) \\
E_x &= E_z = H_y = 0
\end{aligned}\right\} \tag{5.37}$$

5.10. The ***waveguide wall currents*** are functions of the magnetic fields inside the waveguide adjacent to the wall.

The wall currents in the broad walls are given by

$$J_x \propto H_z; \qquad J_z \propto -H_x$$

and in the narrow walls by

$$J_y \propto H_z; \qquad J_z \propto H_y$$

where the constant of proportionality is the same in each case.

5.11. ***Waveguide attenuation*** $\alpha = \frac{1}{2}\dfrac{\text{power loss}}{\text{transmitted power}}$ (5.47)

For empty waveguide with an equivalent surface resistivity R_s

$$\alpha = \frac{R_s}{2} \frac{\oint H_t^2 \, dl}{\iint \boldsymbol{E} \times \boldsymbol{H} . d\boldsymbol{a}} \tag{5.48}$$

Attenuation constants for some modes in rectangular waveguide are shown in Fig. 5.4.

Power flow

$$P = \left(\frac{\omega\beta ab}{8k_c^2}\right)\begin{Bmatrix}\varepsilon E_0^2 & \text{for TM-modes} \\ \mu H_0^2 & \text{for TE-modes}\end{Bmatrix} \tag{5.62}$$

5.12. ***Waveguide impedance***
Wave impedance

$$Z_0 = \eta \frac{\lambda_0}{\lambda_g} \quad \text{for TM-modes} \tag{5.73}$$

$$Z_0 = \eta \frac{\lambda_g}{\lambda_0} \quad \text{for TE-modes} \tag{5.74}$$

5.13. For a rectangular resonator of size $a \times b \times c$,

$$f_0 = \tfrac{1}{2} \sqrt{\left[\frac{\left(\frac{l}{a}\right)^2 + \left(\frac{m}{b}\right)^2 + \left(\frac{n}{c}\right)^2}{\mu\varepsilon}\right]} \tag{5.82}$$

for the TE_{imn}- or TM_{lmn}-modes.

Problems

5.1. (a) Calculate the cut-off frequency of the following modes in rectangular waveguide whose inside dimensions are 2 cm by 1 cm: TE_{10}-mode, TM_{11}-mode, TE_{01}-mode, TM_{21}-mode, TE_{22}-mode. [7.5, 16.7, 15.0, 21.2, 33.5 GHz]
(b) Repeat the calculations for a square waveguide 4 cm by 4 cm inside dimensions. [3.75, 5.3, 3.75, 8.4, 10.6 GHz]

5.2. Calculate a few points and plot a graph of frequency against waveguide wavelength for the dominant mode in rectangular waveguide.

5.3. A waveguide for use at 10 GHz has the inside dimensions 0.90 in. by 0.40 in. Calculate the frequency range over which the dominant mode and it alone can propagate and, with reference to the graph of problem 5.2, explain why the recommended range of operation is 8.2–12.5 GHz.

5.4. Discuss whether eqns. (5.29) to (5.32) are universally true or whether they only apply to the conditions inside rectangular waveguide.

5.5. A parallel plate waveguide (see Fig. 3.1) may be considered as a rectangular waveguide with an infinite broad dimension. Derive expressions for the cut-off conditions and components of the fields inside parallel plate waveguide.

5.6. Perform the substitutions and confirm the accuracy of eqns. (5.33) and (5.36).

5.7. (a) Sketch graphs of the amplitude of the different components of the field of the TE_{10}-mode against position inside the waveguide.

(b) Sketch graphs of the amplitude of the components of the wall current of the TE_{10}-mode against position in the waveguide.

5.8. Which of the sections of slotted waveguide shown in Fig. 5.6 radiate for the dominant mode, on the principle that slots which do not cut lines of current flow do not radiate? In particular, which section preferentially accepts the TE_{01}-mode without radiating, whilst radiating for other modes? [*a, b, d, e*; *a*]

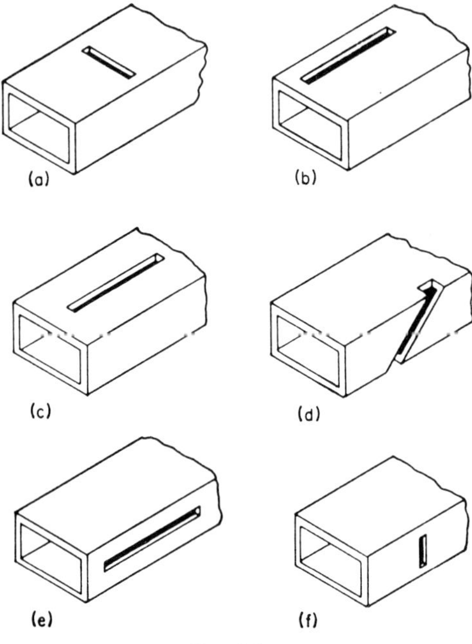

FIG. 5.6

5.9. Calculate the size of the quarter wavelength section of waveguide needed to match a junction between rectangular waveguides 2 × 1 cm and 4 × 1 cm at 10 GHz. At what frequency are higher order modes troublesome? [2.42 × 1 cm; 7.5 GHz]

5.10. Calculate a few of the resonant frequencies of a cavity made from a section of rectangular waveguide 2 cm by 1 cm by 3 cm long. Identify the mode of oscillation in the cavity appropriate to each frequency. [9.0, 12.5, 15.8, 17.5 GHz]

6

Circular Waveguides

6.1. Circular Pipe

In the previous chapter expressions were obtained for the electromagnetic field components in rectangular waveguide in terms of rectangular coordinates. In this chapter expressions are derived for the field components inside circular waveguide, that is, inside a length of uniform circular metal pipe. Although circular waveguide appears to be simple, there is an indeterminacy of most modes of propagation which makes rectangular waveguide the preferred shape for most applications. Circular waveguide is sometimes used, however, so that it is necessary to study the electromagnetic fields inside a circular boundary. Because the boundary is circular in one plane, cylindrical polar coordinates, r, θ and z, are used in this part of the analysis. The axis of the coordinates is taken to be coincident with the axis of the circular waveguide with propagation occurring in the z-direction as shown in Fig. 6.1.

6.2. Wave Equation in Cylindrical Polar Coordinates

As has already been shown in Chapter 2, Maxwell's equations can be reduced to two, eqns. (2.12) and (2.13). For rectangular waveguide, all the

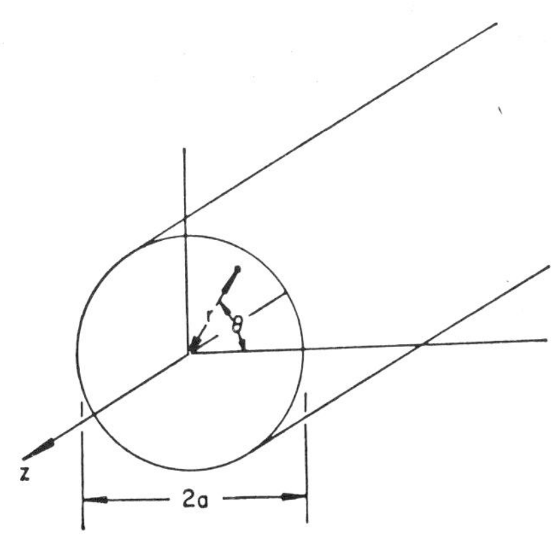

FIG. 6.1. Circular waveguide of radius a showing its relationship to the axes of the cylindrical polar coordinates.

field components may be expressed in terms of the z-directed component of the magnetic and electric fields. It will be shown that the same is true when the fields are described by components parallel to the cylindrical polar coordinates. In this case, it cannot be said that the components are all parallel to the axes. The z-directed component is the component parallel to the z-axis. The r-directed component is that component pointing radially away from the z-axis which is perpendicular to that axis and the θ-directed component is perpendicular to the other two components at that point. In distinction to the rectangular coordinate system, the r and θ components respectively are not all parallel to one another except in a plane of constant θ. The components of a field at a typical point in space together with the coordinates of that point are shown in Fig. 6.2.

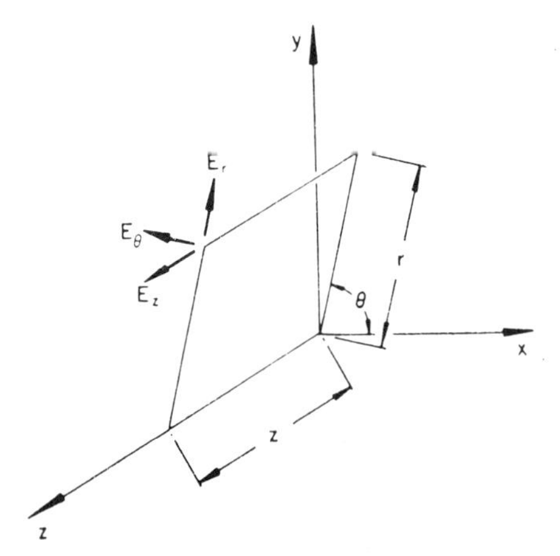

FIG. 6.2. The electric field components at some point (r, θ, z) in cylindrical polar coordinates showing their relationship to the axes of the rectangular coordinate system.

In the vector wave equation (eqns. (2.12) and (2.13)) in cylindrical polar coordinates, the Laplacian of the vector may be resolved into an equation in terms of the Laplacian of the longitudinal component of the vector and an equation involving the transverse components of the vector. There is no simple separation of the transverse components of the vector as occurs with a rectangular coordinate system. However the longitudinal, or z-directed, component of the vector field in eqn. (2.12) may be separated giving

$$\nabla^2 E_z = -\omega^2 \mu\varepsilon E_z$$

and similarly from eqn. (2.13)

$$\nabla^2 H_z = -\omega^2 \mu\varepsilon H_z$$

Expanding the scalar Laplacian from the expression given in section 2.3, these equations become

$$\frac{\partial^2 E_z}{\partial r^2} + \frac{1}{r}\frac{\partial E_z}{\partial r} + \frac{1}{r^2}\frac{\partial^2 E_z}{\partial \theta^2} + \frac{\partial^2 E_z}{\partial z^2} = -\omega^2 \mu\varepsilon E_z \tag{6.1}$$

$$\frac{\partial^2 H_z}{\partial r^2} + \frac{1}{r}\frac{\partial H_z}{\partial r} + \frac{1}{r^2}\frac{\partial^2 H_z}{\partial \theta^2} + \frac{\partial^2 H_z}{\partial z^2} = -\omega^2 \mu\varepsilon H_z \tag{6.2}$$

We have already assumed that propagation is in the z-direction because the z-axis is coincident with the axis of the waveguide. If β is the phase constant of the wave, the z-dependence of the fields will be $\exp -j\beta z$. As the waveguide is empty, the fields inside are continuous and the field pattern must repeat itself on turning through an angle $\theta = 2\pi$, so that there will probably be a sinusoidal distribution of fields in the circumferential direction. As with the time dependence, a sinusoidal variation is depicted mathematically by an exponential dependence, so that the θ-dependence of the fields is assumed to be $\exp -jn\theta$, where n is a positive or negative integer. Hence the differential operations with regard to θ and z become

$$\frac{\partial^2}{\partial \theta^2} = -n^2 \quad \text{and} \quad \frac{\partial^2}{\partial z^2} = -\beta^2$$

Substitution of these values into eqn. (6.1) gives

$$\frac{\partial^2 E_z}{\partial r^2} + \frac{1}{r}\frac{\partial E_z}{\partial r} - \frac{n^2}{r^2}E_z - \beta^2 E_z = -\omega^2 \mu\varepsilon E_z \tag{6.3}$$

If k_c is defined similar to eqn. (5.9)

$$k_c^2 = k^2 - \beta^2$$

or substituting $k^2 = \omega^2\mu\varepsilon$ from eqn. (5.5)

$$k_c^2 = \omega^2\mu\varepsilon - \beta^2$$

and eqn. (6.3) becomes

$$\frac{\partial^2 E_z}{\partial r^2} + \frac{1}{r}\frac{\partial E_z}{\partial r} + \left(k_c^2 - \frac{n^2}{r^2}\right)E_z = 0 \tag{6.4}$$

Equation (6.4) does not have an analytic solution, but it is of a form that occurs frequently in analysis of scientific problems and its solutions have been tabulated. It is one form of what is called Bessel's equation and the solution is

$$E_z = AJ_n(k_c r) + BY_n(k_c r) \tag{6.6}$$

where A and B are arbitrary constants of integration whose values are determined by the boundary conditions, and J_n and Y_n are the *Bessel functions* of order n of the first and second kind respectively. Some of the properties

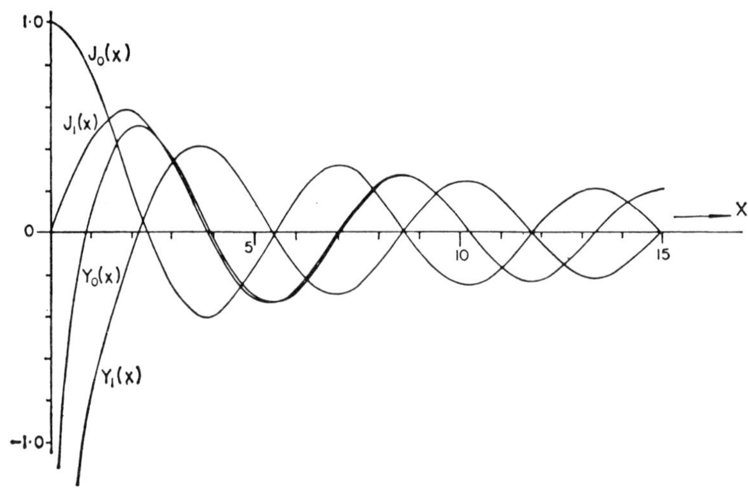

FIG. 6.3. Bessel functions of the first and second kinds of order 0 and 1.

of the various Bessel functions are given in Table 6.1. There are Bessel functions other than those of the first and second kinds, but they are not required in the analysis of electromagnetic propagation inside air-filled round pipe. As we are mainly interested in the lower order modes with $n = 0$ or 1, $J_0(x)$, $Y_0(x)$, $J_1(x)$ and $Y_1(x)$ are plotted in Fig. 6.3. The Bessel functions of the first and second kinds have distinct properties when the argument is zero. They are

$$J_0(0) = 1$$

$$J_n(0) = 0 \quad n > 0 \quad (n = 1, 2, 3 \text{ etc.})$$

$$Y_n(0) = -\infty \quad n \geqslant 0 \quad (n = 0, 1, 2 \text{ etc.})$$

TABLE 6.1
Bessel Equations and Their Solutions

Equation	Solution
$x^2 \dfrac{d^2y}{dx^2} + x\dfrac{dy}{dx} + (x^2 - n^2)y = 0$	$J_n(x)$, $Y_n(x)$
$x^2 \dfrac{d^2y}{dx^2} + x\dfrac{dy}{dx} - (x^2 + n^2)y = 0$	$I_n(x), K_n(x). [I_n(x) = j^{-n}J_n(jx)]$
$x^2 \dfrac{d^2y}{dx^2} + x\dfrac{dy}{dx} - jx^2y = 0$	$J_0(j^{3/2}x)$, $Y_0(j^{3/2}x)$. $[J_0(j^{3/2}x) = \text{ber}\, x + j\, \text{bei}\, x]$
$x^2 \dfrac{d^2y}{dx^2} + x(1 - 2a)\dfrac{dy}{dx} + [(pqx^q)^2 + a^2 - n^2q^2]y = 0$	$x^a J_n(px^q)$, $x^a Y_n(px^q)$

For larger arguments, the values of these Bessel functions oscillate rather like a decaying sine wave and, apart from $Y_n(x)$ near to zero argument, their value is less than one.

Equation (6.5) is a solution of eqn. (6.4) and eqn. (6.4) is a simplified version of eqn. (6.1) so that the full solution to eqn. (6.1) is

$$E_z = [AJ_n(k_c r) + BY_n(k_c r)]\exp j(\omega t - n\theta - \beta z) \tag{6.6}$$

The effect of cut off is the same as discussed in the previous chapter in consideration of rectangular waveguide. The actual value of the constant k_c is governed by the dimensions of the waveguide and determines the cut-off frequency for that waveguide.

6.3. Boundary Conditions

Figure 6.1 shows the waveguide and its relationship to the axes of the coordinate system being used for this part of the analysis. The boundary condition is seen to be that the waveguide wall is at $r = a$ everywhere. The boundary is a perfectly conducting wall so that the component of the electric field tangential to the wall is zero at the boundary. The θ-component is always directed tangentially to the circular cross-section of the waveguide and the longitudinal component is also tangential to the wall so that E_θ and E_z are both zero at $r = a$. For the modes with a longitudinal electric field, eqn. (6.6) gives as the boundary condition

$$AJ_n(k_c a) + BY_n(k_c a) = 0 \tag{6.7}$$

The electric field must be continuous and finite at the centre of the waveguide so that no term in $Y_n(x)$ could be allowed to exist in eqn. (6.6). Hence $B = 0$ and the boundary condition becomes

$$J_n(k_c a) = 0 \tag{6.8}$$

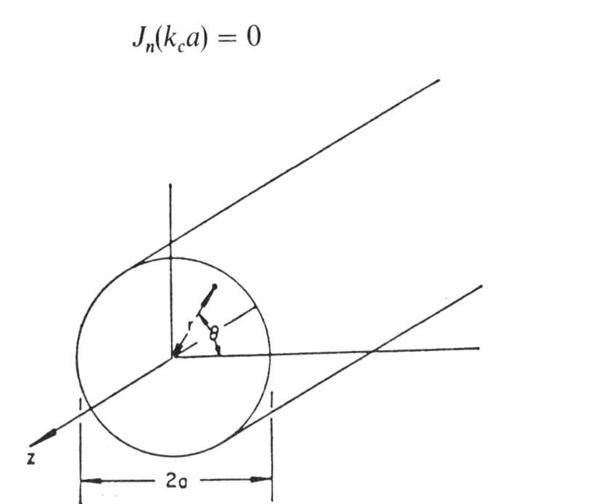

FIG. 6.1. Circular waveguide of radius a showing its relationship to the axes of the cylindrical polar coordinates.

As seen from Fig. 6.3 there are an infinite number of zeros of $J_n(x)$. These zeros are numbered starting with 1 for the zero having the smallest value of argument. Each different solution of eqn. (6.8) is equivalent to a different mode of propagation in the waveguide. The modes are separated into two classes of modes, the TM- (or E-) modes and the TE- (or H-) modes. The mode having a longitudinal electric field is a TM-mode. The mode is named by means of number subscripts similar to the rectangular waveguide modes. The subscripts are n and m. n is the integer denoted by n in the expressions for the electric field. It is the order of the Bessel functions in eqn. (6.6) and is a measure of the circumferential variation in the field pattern. m is the number of the zero of the Bessel function of order n and hence it is a measure of the radial variation of the field pattern. Hence eqn. (6.6) is the expression for the field of the TM_{nm}- (or E_{nm}-) mode and eqn. (6.8) provides the necessary value for k_c. A discussion of the boundary conditions for the TE-modes will be deferred until after expressions have been found for the different components of the field.

6.4. Expressions for the Field Components

In order to find the relationships between the different components of the fields inside the waveguide it is necessary to go to Maxwell's curl equations. Writing these in their component form in cylindrical polar coordinates, eqn. (2.11) becomes

$$\frac{1}{r}\frac{\partial H_z}{\partial \theta} - \frac{\partial H_\theta}{\partial z} = j\omega\varepsilon E_r \tag{6.9}$$

$$\frac{\partial H_r}{\partial z} - \frac{\partial H_z}{\partial r} = j\omega\varepsilon E_\theta \tag{6.10}$$

$$\frac{1}{r} H_\theta + \frac{\partial H_\theta}{\partial r} - \frac{1}{r}\frac{\partial H_r}{\partial \theta} = j\omega\varepsilon E_z \tag{6.11}$$

and eqn. (2.10) becomes

$$\frac{1}{r}\frac{\partial E_z}{\partial \theta} - \frac{\partial E_\theta}{\partial z} = -j\omega\mu H_r \tag{6.12}$$

$$\frac{\partial E_r}{\partial z} - \frac{\partial E_z}{\partial r} = -j\omega\mu H_\theta \tag{6.13}$$

$$\frac{1}{r} E_\theta + \frac{\partial E_\theta}{\partial r} - \frac{1}{r}\frac{\partial E_r}{\partial \theta} = -j\omega\mu H_z \tag{6.14}$$

In deriving eqn. (6.3) a z, θ and t dependence has already been assumed. The time dependence has already been substituted into eqns. (2.10) and (2.11).

The z and θ dependence of $\exp j(-n\theta - \beta z)$ gives

$$\frac{\partial}{\partial z} = -j\beta; \qquad \frac{\partial}{\partial \theta} = -jn$$

and these are substituted into eqns. (6.9) to (6.14) giving

$$-\frac{jn}{r}H_z + j\beta H_\theta = j\omega\varepsilon E_r \tag{6.15}$$

$$-j\beta H_r - \frac{\partial H_z}{\partial r} = j\omega\varepsilon E_\theta \tag{6.16}$$

$$\frac{1}{r}H_\theta + \frac{\partial H_\theta}{\partial r} + \frac{jn}{r}H_r = j\omega\varepsilon E_z \tag{6.17}$$

$$-\frac{jn}{r}E_z + j\beta E_\theta = -j\omega\mu H_r \tag{6.18}$$

$$-j\beta E_r - \frac{\partial E_z}{\partial r} = -j\omega\mu H_\theta \tag{6.19}$$

$$\frac{1}{r}E_\theta + \frac{\partial E_\theta}{\partial r} + \frac{jn}{r}E_r = -j\omega\mu H_z \tag{6.20}$$

Inspection of eqns. (6.15) to (6.20) shows that eqns. (6.15), (6.16), (6.18) and (6.19) form two pairs of simultaneous equations in E_r and H_θ and in E_θ and H_r. As might be expected, this is a relationship similar to that obtained in rectangular coordinates. The solutions are

$$E_r = \frac{1}{k_c^2}\left(-\frac{\omega\mu n}{r}H_z - j\beta\frac{\partial E_z}{\partial r}\right) \tag{6.21}$$

$$H_\theta = \frac{1}{k_c^2}\left(-\frac{\beta n}{r}H_z - j\omega\varepsilon\frac{\partial E_z}{\partial r}\right) \tag{6.22}$$

$$E_\theta = \frac{1}{k_c^2}\left(j\omega\mu\frac{\partial H_z}{\partial r} - \frac{\beta n}{r}E_z\right) \tag{6.23}$$

$$H_r = \frac{1}{k_c^2}\left(-j\beta\frac{\partial H_z}{\partial r} + \frac{\omega\varepsilon n}{r}E_z\right) \tag{6.24}$$

It is seen from these equations that a system of fields of a propagating mode may be obtained in terms of either a longitudinal electric field or a longitudinal magnetic field. Hence, as has been already stated, there are TM- (or E-) modes and TE- (or H-) modes. Equations (6.21) to (6.24) show that some components of the fields are derived in terms of the differentiation of the longitudinal component with respect to r. As it is most likely that the longitudinal component occurs with a Bessel function dependence with r, the differentiation of the Bessel functions are required. The notation is defined

$$J_n'(x) = \frac{dJ_n(x)}{dx}$$

which may be evaluated from the recurrence relationships:

$$J_n'(kr) = J_{n-1}(kr) - \frac{n}{kr} J_n(kr)$$

$$J_n'(kr) = \frac{n}{kr} J_n(kr) - J_{n+1}(kr)$$

$$J_{n+1}(kr) = \frac{2n}{kr} J_n(kr) - J_{n-1}(kr)$$

6.5. TM-modes

Equation (6.6) gives an expression for the E_z component of the field. For the TM-modes, $H_z = 0$ and the arbitrary constant in eqn. (6.6), $B = 0$. Let the arbitrary constant A be replaced by a constant specifying the amplitude of the fields, $A = E_0$. Substituting these values into eqns. (6.21) to (6.24) gives the components of the fields for the TM_{nm}-mode in circular waveguide. They are

$$\left.\begin{aligned}
E_r &= -\frac{j\beta}{k_c} E_0 J_n'(k_c r) \exp j(\omega t - n\theta - \beta z) \\
E_\theta &= -\frac{\beta n}{k_c^2} E_0 \frac{1}{r} J_n(k_c r) \exp j(\omega t - n\theta - \beta z) \\
E_z &= E_0 J_n(k_c r) \exp j(\omega t - n\theta - \beta z) \\
H_r &= \frac{\omega \varepsilon n}{k_c^2} E_0 \frac{1}{r} J_n(k_c r) \exp j(\omega t - n\theta - \beta z) \\
H_\theta &= -\frac{j\omega\varepsilon}{k_c} E_0 J_n'(k_c r) \exp j(\omega t - n\theta - \beta z) \\
H_z &= 0
\end{aligned}\right\} \qquad (6.25)$$

Line representations of the field distributions of some TM-modes in circular waveguide are shown in Fig. 6.4. The frequency is such that $\lambda_g/a = 4.2$. The mode patterns on the left-hand side of the figure depict the electric and magnetic field lines on transverse and longitudinal planes in which the radial electric field is a maximum. The patterns on the right-hand side show a development of the magnetic field and current lines on the inner surface of half the waveguide circumference.

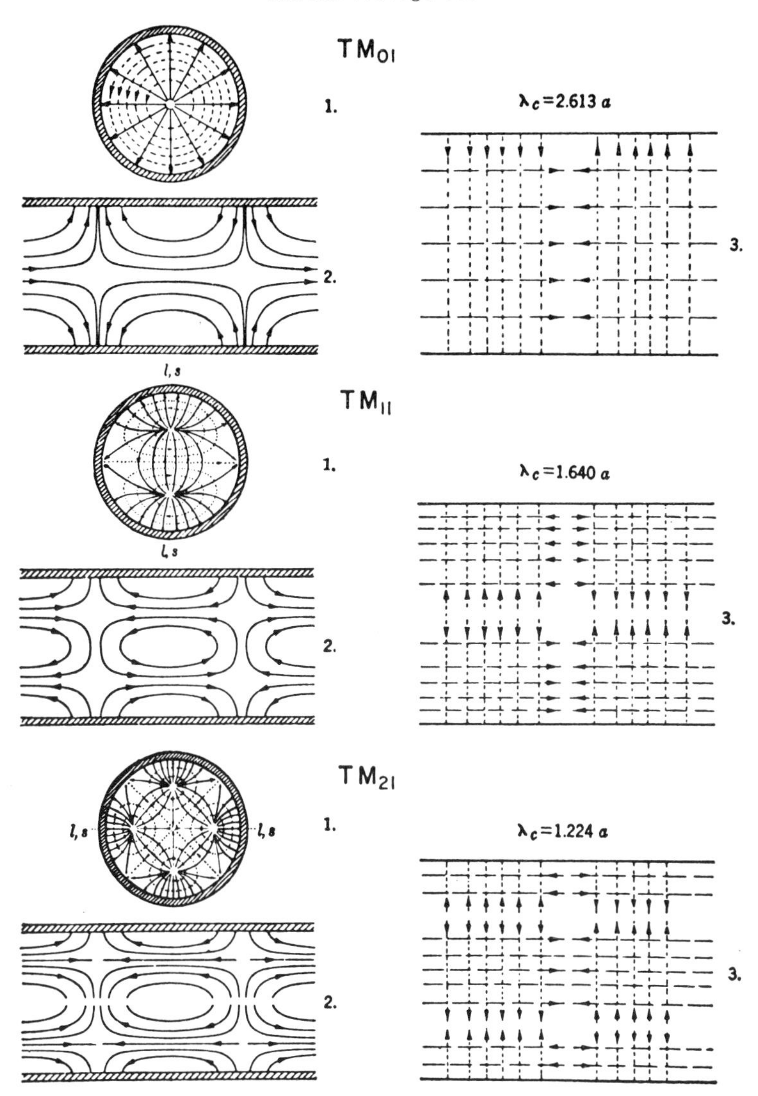

FIG. 6.4. Line representation of the fields of some TM-modes in circular waveguide, radius a. 1. Cross-sectional view. 2. Longitudinal view through plane l–l. 3. Surface view from s–s: — — —electric current; ——— electric field; – – – – magnetic field. (From *Waveguide Handbook*, edited by N. Marcuvitz, McGraw-Hill, 1951. Reproduced by permission of McGraw-Hill Book Co. Inc.)

6.6. TE-modes

Having derived the components of the fields for the TM-modes in circular waveguide, it is now necessary to consider the boundary conditions for the TE-modes in circular waveguide. As considered in connection with the TM-modes, the tangential components of the electric field are zero at the waveguide wall. E_θ and E_z are both zero at $r = a$. For the TE-modes, E_z is zero everywhere throughout the inside of the waveguide. The boundary condition is given by $E_\theta = 0$ at $r = a$. As was considered in section 5.7, it is impossible to go directly in the application of the boundary conditions to the longitudinal component of the magnetic field, H_z. It is necessary to use the condition $E_\theta = 0$. Hence from eqn. (6.23) the boundary condition is

$$\frac{\partial H_z}{\partial r} = 0$$

An expression for the longitudinal component of the magnetic field is given by a solution to eqn. (6.2). This solution may be obtained by a process similar to that used to obtain eqn. (6.6). The solution is

$$H_z = [AJ_n(k_c r) + BY_n(k_c r)] \exp j(\omega t - n\theta - \beta z) \tag{6.26}$$

As the fields must be finite at the centre of the waveguide where $r = 0$, then $B = 0$. Hence the boundary condition is

$$J_n'(k_c a) = 0$$

The differential of the Bessel function is also an oscillatory function having an infinite number of zeros. The mode nomenclature of the TE-modes is similar to that of the TM-modes in that the second subscript is the number of the zero of the function $J_n'(x)$.

Substituting $E_z = 0$ and H_z from eqn. (6.26), with $A = H_0$, into eqns. (6.21) to (6.24) gives expressions for the components of the fields for the TE_{nm}-mode in circular waveguide. They are

$$\left.\begin{aligned}
E_r &= -\frac{\omega\mu n}{k_c^2} H_0 \frac{1}{r} J_n(k_c r) \exp j(\omega t - n\theta - \beta z) \\
E_\theta &= \frac{j\omega\mu}{k_c} H_0 J_n'(k_c r) \exp j(\omega t - n\theta - \beta z) \\
E_z &= 0 \\
H_r &= -\frac{j\beta}{k_c} H_0 J_n'(k_c r) \exp j(\omega t - n\theta - \beta z) \\
H_\theta &= -\frac{\beta n}{k_c^2} H_0 \frac{1}{r} J_n(k_c r) \exp j(\omega t - n\theta - \beta z) \\
H_z &= H_0 J_n(k_c r) \exp j(\omega t - n\theta - \beta z)
\end{aligned}\right\} \tag{6.27}$$

Line representations of the field patterns of some TE-modes in circular waveguide are shown in Fig. 6.5. As with Fig. 6.4, the frequency is such that $\lambda_g/a = 4.2$. The transverse and longitudinal views are in the plane of the maximum electric field. The patterns on the right-hand side show a

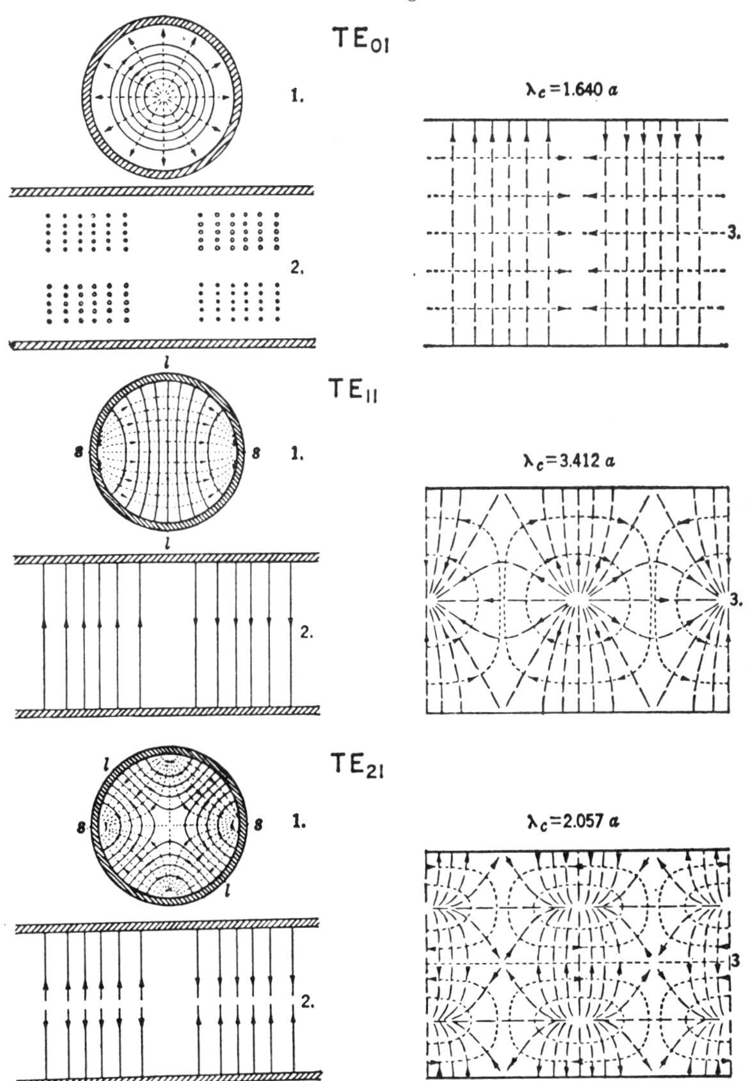

FIG. 6.5. Line representation of the fields of some TE-modes in circular waveguide, radius a. 1. Cross-sectional view. 2. Longitudinal view through plane l–l. 3. Surface view from s–s: — — — electric current; ———— electric field; - - - - magnetic field. (From *Waveguide Handbook*, edited by N. Marcuvitz, McGraw-Hill, 1951. Reproduced by permission of McGraw-Hill Book Co. Inc.)

development of the magnetic field and current distribution on half the waveguide circumference. It is of interest to note in comparison with Fig. 5.3 that the TE_{11}-mode in circular waveguide has a similar field pattern to the TE_{10}-mode in rectangular waveguide and that they are both the dominant mode.

TABLE 6.2

Zeros of Bessel Functions, $J_n(x)$ and $J_n'(x)$

Mode	*m*th zero	of	Value of x
TE_{11}	1	$J_1'(x)$	1.84
TM_{01}	1	$J_0(x)$	2.40
TE_{21}	1	$J_2'(x)$	3.05
TM_{11}	1	$J_1(x)$	3.83
TE_{01}	1	$J_0'(x)$	3.83
TE_{31}	1	$J_3'(x)$	4.20
TM_{21}	1	$J_2(x)$	5.14
TE_{41}	1	$J_4'(x)$	5.32
TE_{12}	2	$J_1'(x)$	5.33
TM_{02}	2	$J_0(x)$	5.52
TM_{31}	1	$J_3(x)$	6.38
TE_{51}	1	$J_5'(x)$	6.42
TE_{22}	2	$J_2'(x)$	6.71
TM_{12}	2	$J_1(x)$	7.02
TE_{02}	2	$J_0'(x)$	7.02

The values of the argument giving fifteen zeros of the Bessel functions appropriate to the first fifteen modes in circular waveguide are given in Table 6.2. If the value of the argument is denoted by x and the waveguide radius is a, the cut-off condition is given by

$$k_c = x/a$$

and by substitution into eqn. (5.12)

$$\lambda_c = \frac{2\pi a}{x}$$

It is seen that there is a much smaller separation between the cut-off frequency of the dominant mode and that of the next mode than in standard rectangular waveguide, so that circular waveguide operates satisfactorily in a single mode over a much smaller bandwidth than rectangular waveguide.

6.7. Polarization

The angular dependence of the circular waveguide modes has been specified as $\exp -jn\theta$ where n is any integer. This means that as the wave propagates forward in space, the axis of the whole field pattern rotates about the z-axis. Consider the dominant TE_{11}-mode. The direction of maximum electric field appears to trace out a helix in space at any instant of time. The helix is shown figuratively in Fig. 6.6. It can be called a *helical* wave although it is commonly called a *circularly polarized wave.*

If it were possible to use a directional detector so that only the component of the electric field in one direction can be observed, the field appears to have the normal sinusoidal variation in amplitude as it propagates. The

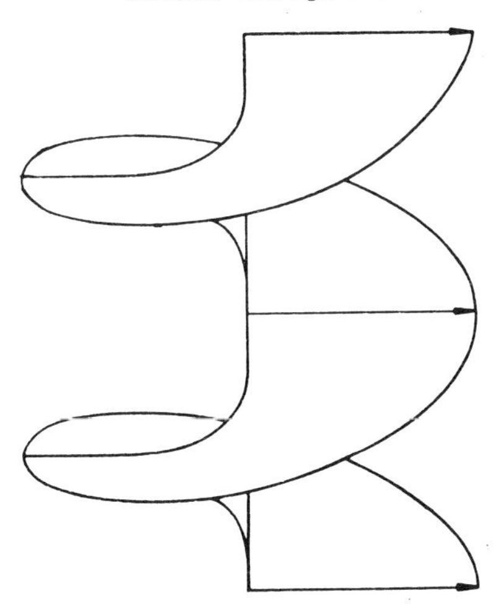

FIG. 6.6. Locus of the electric field vector of a helical wave, two wavelengths.

exponential dependence in angle can be resolved into two components in phase quadrature:

$$\exp - jn\theta = \cos n\theta - j\sin n\theta$$

For the dominant mode, $n = 1$ and the two components are perpendicular to one another. The circularly polarized wave can be resolved into an x-directed field component and an equal amplitude y-directed field component. In the mathematics there is no reason why n must be a positive integer. If n is a negative integer, the circularly polarized wave is rotating in the opposite direction to one in which n is positive. Hence the definition of *positive* and *negative circularly polarized waves* is obvious. This definition is tied to the particular convention adopted for the cylindrical polar coordinates and is perfectly suitable for any discussion which is completely self-contained, but if communication with others is required a more rigorous system of identifying circular polarization is required. A *right-hand circularly polarized* wave has a positive n in a right-handed cylindrical coordinate system. A *left-hand circularly polarized* wave has a negative value for n in a right-handed cylindrical coordinate system and a positive value for n in a left-handed coordinate system. The coordinate system as defined in Fig. 6.1 is a right-handed system. Because eqn. (6.4) only contains n^2, the order of the Bessel functions is $|n|$, always positive, but signed values for n would be used elsewhere in eqns. (6.25) and (6.27).

From a descriptive point of view the definition of a right-hand circularly polarized wave means that if time is frozen, the locus of the maximum of the electric field traces a helix in space, the direction of rotation is counter-clockwise. Alternatively if the observer remains in one place so that z remains constant, the maximum of the electric field rotates clockwise.

The plane wave described in section 2.8 has its electric field always directed in one direction throughout space. Such a wave is called a *linearly polarized* wave and the direction in which the maximum of the electric field lies is called the *plane of polarization* of the wave. Reference to the cylindrical polar components of the electric field shown in Fig. 6.2 show that the rectangular components may be resolved into the cylindrical polar components using the relationships

$$E_r = E_x \cos\theta + E_y \sin\theta$$

$$E_\theta = E_y \cos\theta - E_x \sin\theta$$

The plane wave defined by eqns. (2.20) and (2.28) may be resolved into cylindrical polar coordinates, giving

$$\left.\begin{aligned} E_r &= E_0 \cos\theta \exp j(\omega t - \beta z) \\ E_\theta &= E_0 \sin\theta \exp j(\omega t - \beta z) \\ \eta H_r &= E_0 \sin\theta \exp j(\omega t - \beta z) \\ \eta H_\theta &= E_0 \cos\theta \exp j(\omega t - \beta z) \end{aligned}\right\} \quad (6.28)$$

In eqn. (6.28) the angular dependence may be extracted to show one of the properties of linear polarization

$$E_0 \cos\theta = \tfrac{1}{2}E_0 \exp j\theta + \tfrac{1}{2}E_0 \exp - j\theta$$

A linearly polarized wave may be resolved into two equal circularly polarized waves of opposite hand. Similarly, we have already shown that a circularly polarized wave may be resolved into two equal linearly polarized waves perpendicular to one another and 90° out of phase. The mathematics of this chapter have shown that mathematically there is no advantage when considering electromagnetic fields in cylindrical polar coordinates, in using either circular or linear polarization to describe any particular wave. The preference depends on the physical system that is being investigated at any time. A linearly polarized plane wave may be described either in terms of rectangular coordinates or in terms of linearly polarized cylindrical coordinates, as in eqn. (6.28).

The *plane of polarization* of any linearly polarized plane wave is the plane containing the direction of propagation and the direction of maximum electric field strength. In circular waveguide modes, the plane of polarization is the plane containing the maximum electric field strength on the axis of the waveguide. This description only holds strictly for the dominant mode which approximates to a plane wave at the centre of the waveguide. For the dominant mode in circular waveguide which has been launched directly from the dominant mode in rectangular waveguide, the wave is linearly polarized and the linear polarization description of the fields inside the waveguide ought to be used. In Chapter 8 a system is investigated where the circularly polarized description of the field quantities is the only one which is mathematically acceptable and so the same description has also been used in eqns. (6.25) and (6.27).

6.8. TEM-modes in Cylindrical Coordinates

A plane wave in an unbounded medium had zero longitudinal components of both its electric and magnetic fields. This section considers mathematically the conditions under which such a wave can be described in cylindrical polar coordinates. Assume the same angle, time and distance dependence as the field components given in eqn. (6.6). Substituting $H_z = 0$ and $E_z = 0$ into eqns. (6.15) to (6.20) and simplifying gives

$$\beta H_\theta = \omega\varepsilon E_r \tag{6.29}$$

$$\beta H_r = -\omega\varepsilon E_\theta \tag{6.30}$$

$$\frac{1}{r}H_\theta + \frac{\partial H_\theta}{\partial r} + j\frac{n}{r}H_r = 0 \tag{6.31}$$

$$\beta E_\theta = -\omega\mu H_r \tag{6.32}$$

$$\beta E_r = \omega\mu H_\theta \tag{6.33}$$

$$\frac{1}{r}E_\theta + \frac{\partial E_\theta}{\partial r} + j\frac{n}{r}E_r = 0 \tag{6.34}$$

It is seen that eqns. (6.29) and (6.33) and eqns. (6.30) and (6.32) form independent pairs of equations. Solution of these pairs gives

$$\beta^2 = \omega^2\mu\varepsilon \tag{6.35}$$

which shows that any wave which is characterized by eqns. (6.29) to (6.34) has the same propagation constant as a plane wave. This is the TEM-mode. If it is remembered that jn is equivalent to turning through 90° in space, it changes $\cos\theta$ into $\sin\theta$ and vice versa, so that eqn. (6.28) satisfies the above equations and one solution is a plane wave.

$$\left.\begin{aligned} \mathrm{E}_r &= E_0 \cos\theta \exp j(\omega t - \beta z) \\ \mathrm{E}_\theta &= E_0 \sin\theta \exp j(\omega t - \beta z) \\ \eta \mathrm{H}_r &= E_0 \sin\theta \exp j(\omega t - \beta z) \\ \eta \mathrm{H}_\theta &= E_0 \cos\theta \exp j(\omega t - \beta z) \end{aligned}\right\} \tag{6.28}$$

For a plane wave $n = 1$ and there is no variation with respect to r so that eqns. (6.31) and (6.34) become

$$\left.\begin{aligned} \frac{1}{r}H_\theta &= -j\frac{1}{r}H_r \\ \frac{1}{r}E_\theta &= -j\frac{1}{r}E_r \end{aligned}\right\} \tag{6.36}$$

If it is remembered again that in these equations, the j operator denotes a rotation of 90° in space rather than a difference in phase, it is found that eqn. (6.28) also satisfies eqn. (6.36).

There is also a solution to eqns. (6.29) to (6.34) which allows for the boundary conditions inside circular waveguide. It is still desired to specify a TEM-mode so that the propagating conditions are given by the solution to eqns. (6.29), (6.30), (6.32) and (6.33). However if $n = 0$, and allowing some radial variation of the fields, eqns. (6.31) and (6.34) become

$$\left.\begin{aligned} \frac{1}{r}H_\theta + \frac{\partial H_\theta}{\partial r} &= 0 \\ \frac{1}{r}E_\theta + \frac{\partial E_\theta}{\partial r} &= 0 \end{aligned}\right\} \tag{6.37}$$

The solution to eqn. (6.37) is

$$\left.\begin{aligned} H_\theta &= \frac{1}{r}H_0 \exp j(\omega t - \beta z) \\ E_\theta &= \frac{1}{r}E_0 \exp j(\omega t - \beta z) \end{aligned}\right\} \tag{6.38}$$

The fields in eqn. (6.38) are infinite at the origin so that they cannot be allowed to exist there. Hence one boundary condition is that there must be a conductor at the origin where $r = 0$. These are the fields that surround a wire along which an electromagnetic wave is propagating. The circular conductor also provides the boundary condition that $E_\theta = 0$ everywhere on its surface. Therefore $E_0 = 0$ and the only possible fields surrounding the conductor are

$$H_\theta = \frac{1}{r}H_0 \exp j(\omega t - \beta z) \tag{6.39}$$

$$E_r = \eta\frac{1}{r}H_0 \exp j(\omega t - \beta z) \tag{6.40}$$

$$E_\theta = E_z = H_r = H_z = 0$$

This field must be enclosed with an outer conductor concentric with the inner circular conductor, as shown in Fig. 6.7. This is a coaxial transmission line which is approximated to by the coaxial cable. There are no cut-off conditions inherent in eqns. (6.39) and (6.40) because the propagation constant, given by eqn. (6.35), is that of the plane wave, and the mode propagates at all frequencies down to d.c. It is noticed that the field dependence is the same as that due to a direct current in the wire and an electrostatic potential between the conductors. The only difference is that the time-varying fields are related. The TEM-mode is the dominant mode in a coaxial waveguide, but it is not the only possible mode. There are also

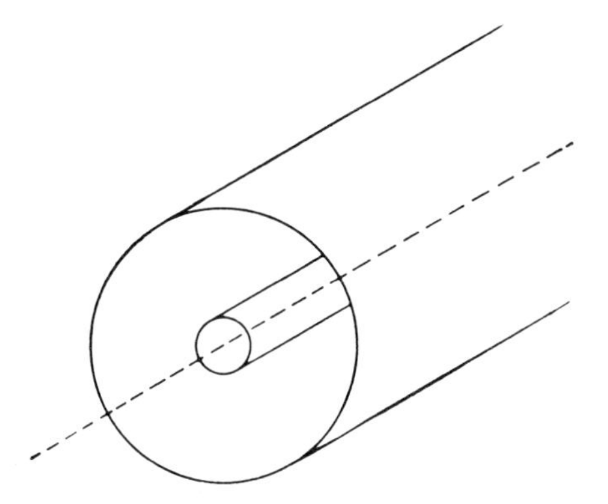

FIG. 6.7. Coaxial transmission line.

higher modes which exhibit cut off and are called waveguide modes.

6.9. Waveguide Modes in Coaxial Line

Assume that there is at least one field with a longitudinal component. For a TM-mode, the wave equation is eqn. (6.1) which has eqn. (6.6) as a general solution. If the waveguide consists of a circular metal pipe of inner radius a with inside it a concentric circular metal rod of radius b, the boundary conditions are that $E_z = 0$ at $r = a$ and $r = b$ if the metal walls are perfectly conducting. There are now two boundary equations:

$$AJ_n(k_c a) + BY_n(k_c a) = 0 \tag{6.41}$$

$$AJ_n(k_c b) + BY_n(k_c b) = 0 \tag{6.42}$$

The constant B is not necessarily zero because the fields do not exist at the origin of the coordinate system which is the axis of the waveguide. Equations (6.41) and (6.42) may be combined into the matrix form

$$\begin{vmatrix} A \\ B \end{vmatrix} \begin{vmatrix} J_n(k_c a) & Y_n(k_c a) \\ J_n(k_c b) & Y_n(k_c b) \end{vmatrix} = 0$$

As $\begin{vmatrix} A \\ B \end{vmatrix} = 0$ is a trivial solution to this equation, the useful solution is

$$\begin{vmatrix} J_n(k_c a) & Y_n(k_c a) \\ J_n(k_c b) & Y_n(k_c b) \end{vmatrix} = 0 \tag{6.43}$$

where the determinant of the matrix is zero. Equation (6.43) is called the characteristic equation for this waveguide and gives a relationship between k_c, a and b. The solutions of eqn. (6.43) have been calculated and are available.

Similarly for the TE-modes, eqn. (6.26) is a general solution to the wave equation. The boundary condition is

$$\frac{\partial H_z}{\partial r} = 0$$

at $r = a$ and $r = b$, giving the boundary equations

$$AJ_n'(k_c a) + BY_n'(k_c a) = 0$$

$$AJ_n'(k_c b) + BY_n'(k_c b) = 0$$

which give the characteristic equation

$$\begin{vmatrix} J_n'(k_c a) & Y_n'(k_c a) \\ J_n'(k_c b) & Y(k_c b) \end{vmatrix} = 0 \tag{6.44}$$

Equation (6.43) is the characteristic equation for the TM-modes and eqn. (6.44) is the characteristic equation for the TE modes.

Having found the characteristic value for k_c the expressions for the components of the fields are similar to eqns. (6.25) and (6.27). They are not given here, but it is left as an exercise for the reader to write down these expressions for the fields. Good approximations for the cut-off wavelengths of the waveguide modes in coaxial waveguide are

$$\lambda_c \approx \frac{2(a-b)}{m} \quad \text{for the TM}_{nm}\text{-modes}$$

$$\lambda_c \approx \frac{\pi(a+b)}{n} \quad \text{for the TE}_{n1}\text{-modes}$$

$$\lambda_c \approx \frac{2(a-b)}{(m-1)} \quad \text{for the TE}_{nm}\text{-modes} \quad (m \geqslant 2)$$

Line representations of the field distribution of some TM-modes in coaxial waveguide are shown in Fig. 6.8 and those of some TE-modes are shown in Fig. 6.9. The waveguide dimensions are such that $a/b = 3$ and the frequency is such that $\lambda_g/a = 4.24$. The mode patterns on the left-hand side of the figure depict the electric and magnetic field lines in the transverse and longitudinal planes on which the radial electric field is a maximum. The patterns on the right-hand side show a development of the magnetic field and current lines on the inner surface of half the circumference of the outer conductor.

6.10. Waveguide Impedance

It has already been shown in section 5.12 that the wave impedance inside any hollow metal waveguide which will support pure TM- or TE-modes is given by eqns. (5.73) and (5.74). Alternatively, the expressions for the wave

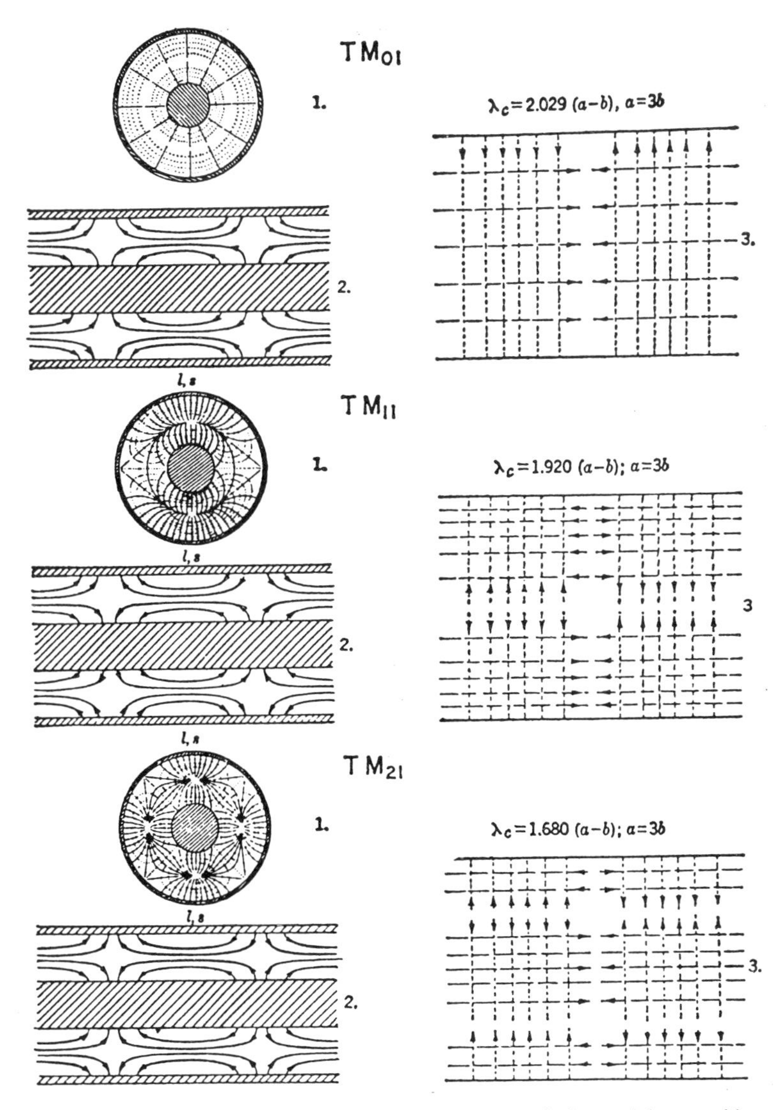

FIG. 6.8. Line representation of the fields of some TM-modes in coaxial waveguide, outer conductor radius a and inner conductor radius b. 1. Cross-sectional view. 2. Longitudinal view through plane l–l. 3. Surface view from s–s: — — — electric current; ———— electric field; - - - - magnetic field. (From *Waveguide Handbook*, edited by N. Marcuvitz, McGraw-Hill, 1951. Reproduced by permission of McGraw-Hill Book Co. Inc.)

impedance can be derived directly from the expressions for the fields in the waveguide. Inspection of eqns. (6.21) to (6.24) show that the transverse components of the fields occur in pairs having a constant amplitude ratio.

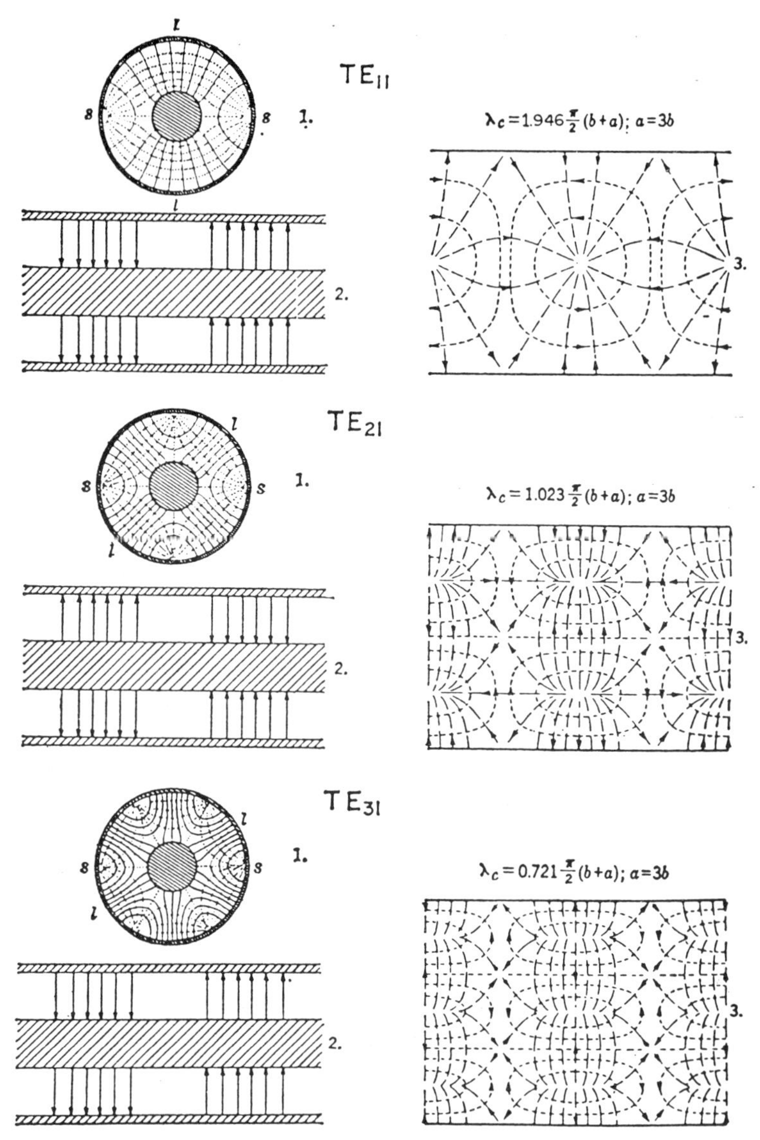

FIG. 6.9. Line representation of the fields of some TE-modes in coaxial waveguide outer conductor radius a and inner conductor radius b. 1. Cross-sectional view. 2. Longitudinal view through plane l–l. 3. Surface view from s–s: — — — electric current; ———— electric field; - - - - magnetic field. (From *Waveguide Handbook*, edited by N. Marcuvitz, McGraw-Hill, 1951. Reproduced by permission of McGraw-Hill Book Co. Inc.)

Therefore we can obtain an expression for the wave impedance similar to eqn. (5.77):

$$Z_0 = \frac{E_t}{H_t} = \frac{E_r}{H_\theta} = -\frac{E_\theta}{H_r} \tag{6.45}$$

Substituting values into eqn. (6.45) from eqns. (6.21) to (6.24) shows that eqns. (5.73) and (5.74) are true for waveguide modes in circular waveguide and coaxial waveguide.

$$Z_0 = \begin{cases} \dfrac{\beta}{\omega\varepsilon} = \eta\dfrac{\lambda_0}{\lambda_g} & \text{for TM-modes} \\[2ex] \dfrac{\omega\mu}{\beta} = \eta\dfrac{\lambda_g}{\lambda_0} & \text{for TE-modes} \end{cases}$$

6.11. Waveguide Attenuation

General expressions for the attenuation constant of a propagating mode in waveguide have been given in eqn. (5.48). Expressions for the field values of the TM- and TE-modes in circular waveguide may be substituted into this equation to derive expressions for the attenuation constant of the various modes in circular waveguide. Integration of the Bessel functions is aided by using an expression for the power flow in terms of the longitudinal components of the microwave field in the waveguide. It can be shown* that, for a propagating waveguide mode, the integral of the Poynting vector across the cross-sectional area of the waveguide can be expressed in terms of the integral across the same cross-sectional area of the longitudinal component of the field. The power transmitted is given by

$$P = \iint_{\text{area}} \mathbf{E} \times \mathbf{H} \,.\, d\mathbf{a} = \frac{\omega\varepsilon\beta}{k_c^2} \iint_{\text{area}} E_z^2 \, da \tag{6.46}$$

for TM-modes and

$$P = \iint_{\text{area}} \mathbf{E} \times \mathbf{H} \,.\, d\mathbf{a} = \frac{\omega\mu\beta}{k_c^2} \iint_{\text{area}} H_z^2 \, da \tag{6.47}$$

for TE-modes.

Expressions for the attenuation constant of the TM-modes are obtained by substituting field values from eqn. (6.25) into eqn. (5.48). The numerator of the expression is given by

$$\oint H_t^2 \, dl = \int_0^{2\pi} H_\theta^2 a \, d\theta = \frac{\pi a \omega^2 \varepsilon^2 E_0^2}{k_c^2} [J_n'(k_c a)]^2 \tag{6.48}$$

where a is the radius of the waveguide.

* See for example: L. Lewin, *Theory of Waveguides*. Newnes-Butterworths. 1975. pp. 32–34.

Using an expression for the integral of the Bessel function,

$$\int_0^a J_n^2(kr)r\,dr = \tfrac{1}{2}a^2[J_n^2(ka) - J_{n-1}(ka)J_{n+1}(ka)]$$

the denominator of the expression is given by

$$\iint_{\text{area}} \boldsymbol{E} \times \boldsymbol{H}.\boldsymbol{da} = \frac{\omega\varepsilon\beta}{k_c^2}\iint_{\text{area}} E_z^2\,da$$

$$= \frac{\pi\omega\varepsilon\beta E_0^2 a^2}{2k_c^2}[J_n^2(k_c a) - J_{n-1}(k_c a)J_{n+1}(k_c a)] \tag{6.49}$$

However, for the TM-modes $J_n(k_c a) = 0$ so that from the recurrence relationships given on p. 130, we obtain

$$J_n'(ka) = -J_{n+1}(ka) \quad \text{and} \quad J_n'(ka) = J_{n-1}(ka)$$

and the attenuation constant for TM-modes in circular waveguide is given by

$$\alpha = R_s\frac{\omega\varepsilon}{\beta a} = \frac{\lambda_g}{a\lambda_0}\sqrt{\left(\frac{\pi}{\lambda_0\eta\sigma}\right)} \tag{6.50}$$

Expressions for the attenuation constant of the TE-modes are obtained by substituting field values from eqn. (6.27) into eqn. (5.48). The numerator of the expression is given by

$$\oint H_t^2\,dl = \int_0^{2\pi} H_\theta^2 a\,d\theta + \int_0^{2\pi} H_z^2 a\,d\theta = \pi a H_0^2\left(\frac{\beta^2 n^2}{k_c^4 a^2} + 1\right)J_n^2(k_c a) \tag{6.51}$$

and the denominator is given by

$$\iint_{\text{area}} \boldsymbol{E} \times \boldsymbol{H}.\boldsymbol{da} = \frac{\omega\mu\beta}{k_c^2}\iint_{\text{area}} H_z^2\,da$$

$$= \frac{\pi\omega\mu\beta H_0^2 a^2}{2k_c^2}[J_n^2(k_c a) - J_{n-1}(k_c a)J_{n+1}(k_c a)] \tag{6.52}$$

For the TE-modes, $J_n'(k_c a) = 0$ so that from the recurrence relationships are obtained

$$kaJ_{n+1}(ka) = nJ_n(ka) \quad \text{and} \quad kaJ_{n-1}(ka) = nJ_n(ka)$$

Therefore eqn. (6.52) becomes

$$\iint \boldsymbol{E} \times \boldsymbol{H}.\boldsymbol{da} = \frac{\pi\omega\mu\beta H_0^2 a^2}{2k_c^2}\left(1 - \frac{n^2}{k_c^2 a^2}\right)J_n^2(k_c a) \tag{6.53}$$

and the attenuation constant for TE-modes in circular waveguide is given by

$$\alpha = \frac{R_s k_c^2}{a\omega\mu\beta}\left(1 + \frac{\beta^2 n^2}{k_c^4 a^2}\right) \Big/ \left(1 - \frac{n^2}{k_c^2 a^2}\right)$$

$$= \frac{\lambda_g \lambda_0}{a\lambda_c^2}\left(\frac{\pi}{\lambda_0 \eta \sigma}\right)^{1/2}\left[1 + \left(\frac{\lambda_c}{\lambda_0}\right)^2 \frac{n^2}{x^2 - n^2}\right] \tag{6.54}$$

where $x = k_c a = 2\pi a/\lambda_c$ is the number given in Table 6.2. If $n = 0$, the expression in eqn. (6.54) simplifies to

$$\alpha = \frac{\lambda_g \lambda_0}{a\lambda_c^2}\sqrt{\left(\frac{\pi}{\lambda_0 \eta \sigma}\right)} \tag{6.55}$$

which shows that for the TE_{0m}-modes, the attenuation continuously decreases with increasing frequency. Attenuation constant values, calculated from eqns. (6.50) and (6.54), have been plotted in Fig. 6.10 for the modes listed in Table 6.2. Similarly to Fig. 5.4, the results have been normalized so that they are applicable to any size of circular waveguide. The horizontal axis is λ_0/d which is dimensionless and the attenuation comes out to be $\alpha d^{3/2}$. As in Fig. 5.4, the vertical axis is calibrated in $\mathrm{dB\,.\,m^{1/2}}$, and a round figure for the conductivity of the copper of drawn waveguide is taken to be 4.00×10^7 S/m. Also the results are plotted to cut off although the mathematical expressions for α are only valid when α is a small part of the propagation constant.

The curves in Fig. 6.10 clearly show the low attenuation of some of the higher order TE-modes in large size waveguide. These low attenuations help the designer to choose the best mode of operation when designing high Q cavities as described in section 6.13. Also the TE_{01}-mode can provide a very low loss microwave transmission medium. The waveguide wall currents of the TE_{01}-mode are circumferential, so circumferential slots in the waveguide wall filter out all other possible modes in the waveguide except other TE_{0m}-modes. Filter sections are constructed where the waveguide wall consists of a close-fitting spiral of copper wire insulated between each turn.

6.12. Elliptical Waveguide

Hollow metal pipe having an elliptical cross-section is also used as a waveguide. The shape of the waveguide and the field patterns of the dominant mode are shown in Fig. 6.11. It can be seen that the mode is similar to the dominant mode in circular waveguide except that there are now two different linearly polarized waves, called the even and odd modes, which have different propagation constants. These modes can probably be more easily understood by considering them similar to either the TE_{11}-mode in circular waveguide

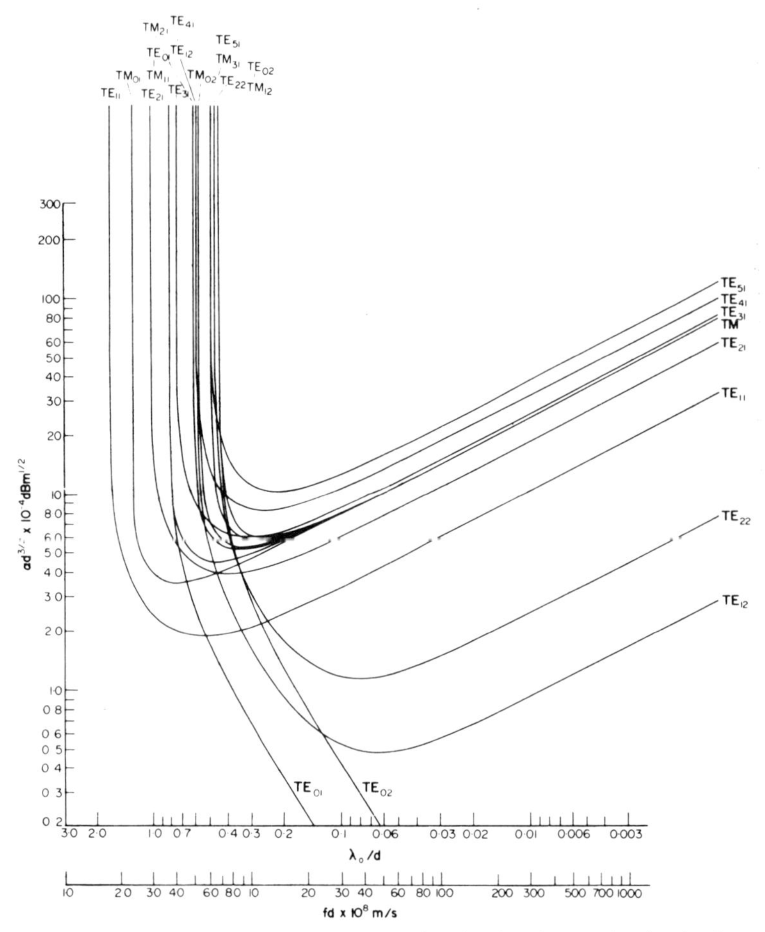

FIG. 6.10. Values of attenuation constant for the first few modes in circular waveguide, of diameter *d*, plotted against normalized wavelength. These have been plotted to cut off from the simple formulae even though application of these formulae is not valid near to cut off.

or the TE_{10}-mode or TE_{01}-mode in rectangular waveguide. One important consideration is that even for a very small amount of ellipticity, any mode in circular waveguide is resolved into two perpendicular linearly polarized modes, having slightly different phase constants.

Artificially induced ellipticity in circular waveguide can be used to convert a circularly polarized wave into a linearly polarized wave and vice versa. One method of introducing the perpendicular mode separation in circular waveguide is to introduce a vane of dielectric material diametrically across the waveguide as shown in Fig. 6.12. Then the mode with its plane of polarization having the electric field parallel to the dielectric vane has a shorter wavelength than the perpendicularly polarized mode. If the length

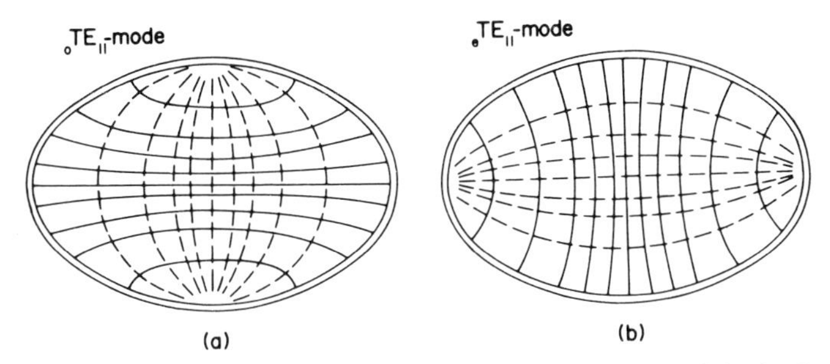

FIG. 6.11. Distribution of the transverse components of the field of the dominant TE_{11}-mode in elliptical waveguide. (a) Odd mode; (b) even mode. ——— electric field; - - - - - magnetic field.

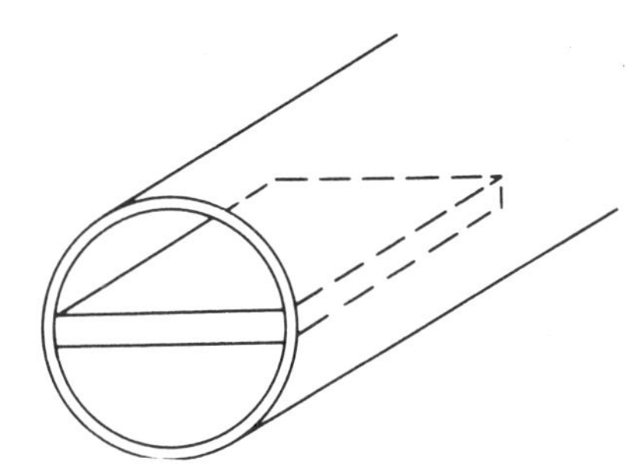

FIG. 6.12. Dielectric vane mounted diametrically across a circular waveguide.

of the dielectric vane in a device is such that there is a quarter wavelength phase difference between the two perpendicular polarizations, the device is called a quarter-wave plate. Similarly a longer vane giving a half wavelength phase difference is called a half-wave plate.

A quarter-wave plate can be used as a circular polarizer, to convert linear polarization into circular polarization. If the vane is mounted at 45° to the plane of polarization of the input linearly polarized wave as shown in Fig. 6.13, it is resolved into two equal waves having their planes of polarization parallel to and perpendicular to the vane. After passage through the quarter-wave plate, the wave having its electric field parallel to the vane is retarded in phase by 90° compared with the perpendicular wave. The two output waves are also shown in Fig. 6.13 and by reference to section 6.7 it can be seen that the output wave with a 90° phase difference between the two perpendicular waves is a circularly polarized wave. If the dielectric vane is rotated through 90° so that it is on the opposite diagonal, the output is a circularly polarized wave of opposite hand. Similarly, if the input to the quarter-wave plate is a circularly polarized wave, the output is a linearly polarized wave. The plane of polarization of the output linearly polarized wave depends on the hand of rotation of the circularly polarized wave and the angular position of the dielectric vane.

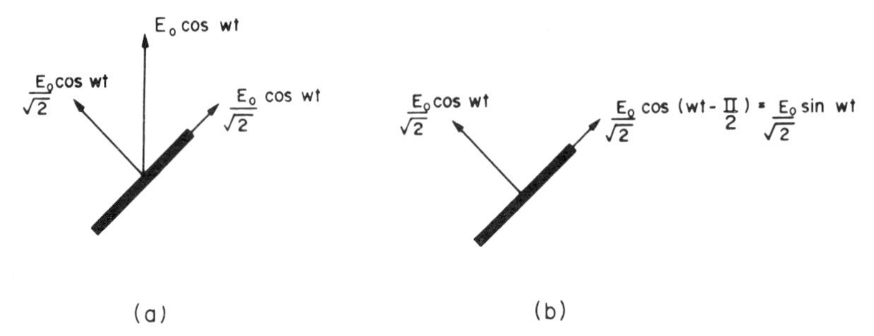

FIG. 6.13. A quarter-wave plate showing (a) the input linearly polarized wave and (b) the output circularly polarized wave.

Similarly, it can be seen from Fig. 6.14, that a half-wave plate rotates the input linearly polarized wave through a right angle. It is left as an exercise for the reader to prove that, if there is some angle θ between the vane and the plane of polarization of the input linearly polarized wave, the output from the half-wave plate is also a linearly polarized wave whose plane of polarization has been rotated through an angle 2θ compared with the plane of polarization of the input wave.

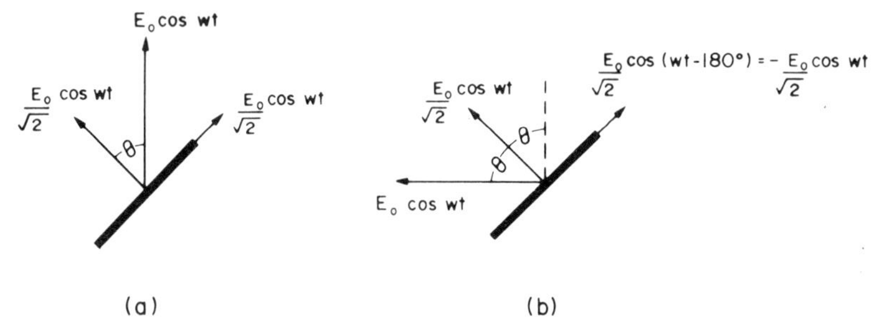

FIG. 6.14. A half-wave plate showing (a) the input linearly polarized wave and (b) the output linearly polarized wave rotated through 90°.

6.13. Resonant Cavity

A microwave resonator may be made out of a length of circular waveguide or coaxial waveguide closed at each end with a short circuit in the same way that a rectangular box makes a rectangular cavity. The cylindrical cavity is not symmetrical in the coordinate system in the same way as a rectangular cavity, so that it is not possible to produce a simple formula similar to eqn. (5.82) for the resonant frequency of a cylindrical cavity. The resonant frequency of a circular or coaxial cavity must be calculated from the cut-off conditions of the equivalent circular or coaxial waveguide modes. In order to simplify the calculation of the resonant frequency of a cylindrical cavity and to help in the design, a mode chart has been constructed connecting the resonant frequency of a cylindrical cavity with its dimensions. The cavity is shown diagrammatically in Fig. 6.15 and the mode chart in Fig. 6.16. The third number in the mode nomenclature denotes the length in half-

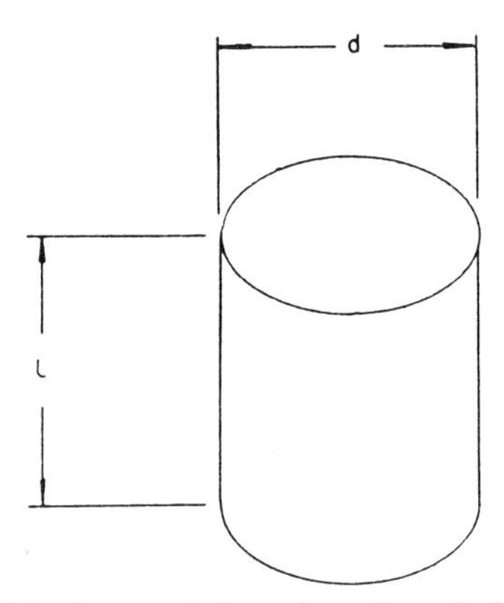

FIG. 6.15. The shape and dimensions of a cylindrical cavity.

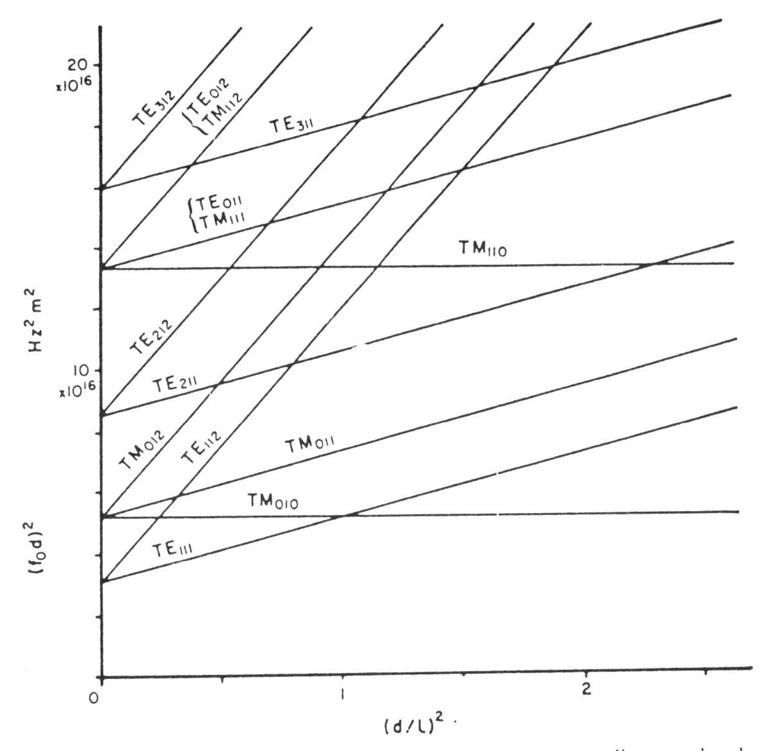

FIG. 6.16. Mode chart for a right circular cylindrical resonator, diameter d and length l.

wavelengths of the equivalent circular waveguide. Using the same notation as has been used for waveguide modes, the cylindrical cavity mode is TM_{nml} or TE_{nml}. It is interesting to note that there are TM-modes with no variation of the fields along the length of the cavity.

Further design of resonant cavities is concerned with achieving high Q-factors and precise control of the resonant frequency. The losses in a cavity which does not have any material filling are entirely controlled by the resistivity of the cavity walls. Since the losses are proportional to the area of the walls but the stored power is proportional to the volume of the cavity, large cavities have a large Q-factor. Large cavities, however, are able to support a number of waveguide modes at different frequencies so it is necessary to determine all the possible resonant frequencies of any cavity. Circular cavities are popular since they are easy to make to the precise dimensions required. Hence the mode chart of Fig. 6.16 is useful in determining the shape and size of any cavity so that it supports only one mode over the proposed frequency band of operation.

6.14. Summary

6.1. The boundary is a ***circular metal pipe*** with perfectly conducting walls.

6.2. The solution to Maxwell's equations in cylindrical polar coordinates gives

$$\frac{\partial^2 E_z}{\partial r^2} + \frac{1}{r}\frac{\partial E_z}{\partial r} + \frac{1}{r^2}\frac{\partial^2 E_z}{\partial \theta^2} + \frac{\partial^2 E_z}{\partial z^2} = -\omega^2 \mu\varepsilon E_z \tag{6.1}$$

$$\frac{\partial^2 H_z}{\partial r^2} + \frac{1}{r}\frac{\partial H_z}{\partial r} + \frac{1}{r^2}\frac{\partial^2 H_z}{\partial \theta^2} + \frac{\partial^2 H_z}{\partial z^2} = -\omega^2 \mu\varepsilon H_z \tag{6.2}$$

The longitudinal variation is

$$\exp -j\beta z$$

The angular variation is

$$\exp -jn\theta$$

and

$$k_c^2 = \omega^2 \mu\varepsilon - \beta^2$$

which leads to ***Bessel's equation***

$$\frac{\partial^2 E_z}{\partial r^2} + \frac{1}{r}\frac{\partial E_z}{\partial r} + \left(k_c^2 - \frac{n^2}{r^2}\right) E_z = 0 \tag{6.4}$$

with the solution

$$E_z = [AJ_n(k_c r) + BY_n(k_c r)]\exp j(\omega t - n\theta - \beta z) \tag{6.6}$$

6.3. Circular waveguide boundary conditions are given by the solution to the equations

$$J_n(k_c a) = 0 \quad \text{for the TM-modes} \tag{6.8}$$

6.6.

$$J_n'(k_c a) = 0 \quad \text{for the TE-modes}$$

6.4. The other components of the electric and magnetic fields can all be expressed in terms of the derivatives of the two longitudinal components.

6.5. The field components of the TM_{nm}-mode are given in eqn. (6.25).

6.6. The z component of the magnetic field is

$$H_z = [AJ_n(k_c r) + BY_n(k_c r)]\exp j(\omega t - n\theta - \beta z) \tag{6.26}$$

The field components of the TE_{nm}-mode are given in eqn. (6.27).

6.7. ***Circularly polarized wave***—the direction of the maximum of the electric field appears to trace out a helix in space.

Linearly polarized wave—the maximum of the electric field always lies in one plane—the plane of polarization.

A linearly polarized wave may be resolved into two equal amplitude circularly polarized waves of opposite hand.

A circularly polarized wave may be resolved into two equal amplitude linearly polarized waves in phase quadrature whose planes of polarization are perpendicular.

6.8. A TEM-mode propagates in coaxial line. Its field components are

$$H_\theta = \frac{1}{r} H_0 \exp j(\omega t - \beta z) \tag{6.39}$$

$$E_r = \eta \frac{1}{r} H_0 \exp j(\omega t - \beta z) \tag{6.40}$$

6.9. The characteristic equations for waveguide modes in coaxial line are

$$\begin{vmatrix} J_n(k_c a) & Y_n(k_c a) \\ J_n(k_c b) & Y_n(k_c b) \end{vmatrix} = 0 \quad \text{for the TM-modes} \tag{6.43}$$

$$\begin{vmatrix} J_n'(k_c a) & Y_n'(k_c a) \\ J_n'(k_c b) & Y_n'(k_c b) \end{vmatrix} = 0 \quad \text{for the TE-modes} \tag{6.44}$$

6.10. ***Waveguide impedance***

$$Z_0 = \begin{cases} \eta \dfrac{\lambda_0}{\lambda_g} & \text{for TM-modes} \\[2ex] \eta \dfrac{\lambda_g}{\lambda_0} & \text{for TE-modes} \end{cases}$$

6.11. ***Attenuation constants*** for some modes in circular waveguide are shown in Fig. 6.10.

6.12. The ***quarter-wave plate*** can be used to change a linearly polarized wave into a circularly polarized wave and vice versa.

6.13. The resonant frequency of a cylindrical resonator is given by the mode chart in Fig. 6.16.

Problems

6.1. Calculate the cut-off frequencies of the following modes in circular waveguide whose inside diameter is 2 cm: TM_{01}-mode, TE_{01}-mode, TE_{11}-mode, TM_{11}-mode, TM_{12}-mode.
[11.4, 18.3, 8.8, 18.3, 33.6 GHz]

6.2. Discuss whether eqns. (6.21) to (6.24) are universally true or whether they only apply to the conditions inside circular waveguide. Compare eqns. (6.21) to (6.24) with eqns. (5.29) to (5.32).

6.3. Using the recurrence relationships given in section 6.4 obtain expressions for: $J_1'(x)$, $J_0'(x)$, $J_3'(x)$ in terms of $J_0(x)$ and $J_1(x)$.

6.4. Perform the substitutions and confirm the accuracy of eqns. (6.25) and (6.27).

6.5. A circular polarizer can be made from material with directional dielectric properties. In the plane of the material, for an electromagnetic plane wave with its electric field parallel to the plane of the material, the permittivity of the material appears to be $\varepsilon_0\varepsilon_r$. Perpendicular to the plane of the material, for an electromagnetic plane wave with its electric field perpendicular to the plane of the material, the permittivity appears to be ε_0. Write down an expression for the field components of the plane wave which entered the material as a linearly polarized plane wave with its electric field at an angle of 45° to the plane of the material. What length of material is required to give an output which is a circularly polarized plane wave? What is the output if twice this length of material is used? In the latter condition, what is the effect of altering the angle between the plane of polarization of the incident wave and the plane of the material?

6.6. Starting with eqns. (6.39) and (6.40) and the basic field relationships, obtain expressions for the current and potential difference in the coaxial transmission line.

6.7. By using the asymptotic expressions for large argument,

$$J_n(x) = \sqrt{\left(\frac{2}{\pi x}\right)}\cos\left(x - \tfrac{1}{2}n\pi - \tfrac{1}{4}\pi\right)$$

$$Y_n(x) = \sqrt{\left(\frac{2}{\pi x}\right)}\sin\left(x - \tfrac{1}{2}n\pi - \tfrac{1}{4}\pi\right)$$

show that in the limit of large diameter but with a fixed radial distance between the conductors, the cut-off conditions of coaxial waveguide are the same as those of parallel plate waveguide with the same separation between the conductors.

6.8. The first three roots of the equation

$$J_1(x)Y_1(10x) - Y_1(x)J_1(10x) = 0$$

are $x = 0.394, 0.733, 1.075$.

Calculate the cut-off frequencies of the modes appropriate to these roots in coaxial line of dimensions: inner conductor, O.D. 1 mm; outer conductor, I.D. 1 cm, and label the modes.
[37.6, 70.2, 102.8 GHz]

6.9. Write out the expressions for the components of the fields of the waveguide modes in coaxial waveguide.

6.10. It is desired to design a cylindrical cavity to be resonant to two frequencies, one twice the other. Suggest approximate values for d/l and fd and identify the modes for a system to satisfy this requirement.

7

Conducting Media

7.1. Conducting Media

The conductivity of a perfect conductor is assumed to be infinite and any electromagnetic radiation is perfectly reflected from its surface. In many microwave problems, the metals may be considered to be perfect conductors, but there are situations where the finite conductivity of the conducting medium must be taken into account. In this chapter, the propagation of electromagnetic radiation through conducting media is considered. Although some of the situations included in this chapter are more appropriate to frequencies lower than microwaves, they are included here for completeness in the treatment of electromagnetic radiation.

In a medium of finite conductivity, eqn. (2.3) gives

$$\boldsymbol{J} = \sigma \boldsymbol{E}$$

and eqn. (2.7) becomes

$$\nabla \times \boldsymbol{H} = \sigma \boldsymbol{E} + j\omega\varepsilon \boldsymbol{E} = (\sigma + j\omega\varepsilon)\boldsymbol{E} \tag{7.1}$$

In eqn. (7.1) it is seen that the term in parentheses may be considered as a single constant. An effective permittivity may be defined and a direct solution to Maxwell's equations can be found in terms of this permittivity. Let us define the effective permittivity,

$$\varepsilon_{\text{eff}} = \varepsilon - j\frac{\sigma}{\omega} \tag{7.2}$$

which is seen to have real and imaginary parts. The real part is the permittivity of the equivalent non-conducting or lossless material and the imaginary part is a permittivity effect of the conduction of the material. As a conduction current serves to transfer power from the electromagnetic wave to heat in the material, the imaginary part of the effective permittivity is a measure of the lossiness of the material. Hence, as we have already seen in section 3.10, any material which causes attenuation of an electromagnetic wave may be described in terms of a complex permittivity. The imaginary part of the complex permittivity is a measure of the power lost by the wave to the material. In general terms, a material of complex permittivity

$$\varepsilon = \varepsilon' - j\varepsilon''$$

has some losses which cannot be attributed to its conductivity and the effective permittivity becomes

$$\varepsilon_{\text{eff}} = \varepsilon'_{\text{eff}} - j\varepsilon''_{\text{eff}} = \varepsilon' - j\left(\varepsilon'' + \frac{\sigma}{\omega}\right) \tag{7.3}$$

It is seen that for all intents and purposes, the total microwave losses due to any material can be combined into either an apparent conductivity or into the imaginary part of the complex permittivity. There is no way of differentiating between different loss mechanisms to the electromagnetic wave. In many tables of published results of measurements of the microwave properties of different materials, the imaginary part of the complex permittivity is quoted in the form of a loss tangent, defined by

$$\tan \delta_e = \frac{\varepsilon''}{\varepsilon'}$$

7.2. Plane Wave

If it is assumed that there is a plane wave propagating through an infinite conducting medium, Maxwell's equations are similar to eqns. (2.8) to (2.11) except that ε is replaced by ε_{eff}. The solution to Maxwell's equations is of the form

$$H_y = H_0 \exp(j\omega t - \gamma z)$$

where

$$\gamma^2 = -\omega^2\mu\left(\varepsilon - j\frac{\sigma}{\omega}\right) \tag{7.4}$$

If the propagation constant is split into its component parts

$$\gamma = \alpha + j\beta$$

substitution into eqn. (7.4) gives

$$\alpha^2 = \tfrac{1}{2}\omega^2\mu\varepsilon\left(-1 + \sqrt{\left[1 + \left(\frac{\sigma}{\omega\varepsilon}\right)^2\right]}\right) \tag{7.5}$$

$$\beta^2 = \tfrac{1}{2}\omega^2\mu\varepsilon\left(1 + \sqrt{\left[1 + \left(\frac{\sigma}{\omega\varepsilon}\right)^2\right]}\right) \tag{7.6}$$

It is seen that the propagation constant now has both real and imaginary parts. This means that the wave propagates through the material with a wavelength that is modified by the finite conductivity, and the amplitude of the wave experiences an exponential decay. This is a general condition for any electromagnetic wave which is propagating with loss to its surroundings. It will be noticed that for lossless propagation along waveguides, there are two possible types of propagation. Either the wave is propagated without loss in an oscillatory mode if the frequency is above the cut-off frequency, or the wave experiences an exponential decay with distance if it is below cut off. In a conducting medium, it is a propagating mode which is experiencing an exponential decay in amplitude.

Applying the conditions for a plane wave in a conducting medium to Maxwell's equations gives results similar to eqns. (2.26) and (2.27). Hence

$$\gamma E_x = j\omega\mu H_y \tag{7.7}$$

$$\gamma H_y = j\omega\varepsilon_{\text{eff}} E_x \tag{7.8}$$

A form for H_y has already been given, but it is rewritten here for completeness together with an expression for the other field component obtained by substituting into eqn. (7.7) and simplifying:

$$H_y = H_0 \exp(j\omega t - \alpha z - j\beta z) \tag{7.9}$$

$$E_x = \frac{j(\alpha - j\beta)}{\omega\varepsilon\sqrt{[1 + (\sigma/\omega\varepsilon)^2]}} H_0 \exp(j\omega t - \alpha z - j\beta z) \tag{7.10}$$

where α and β are given by eqns. (7.5) and (7.6). It is seen that this plane wave has many of the characteristics of a plane wave propagating through a non-conducting medium. The wave still consists of only two field components which are both perpendicular to the direction of propagation and perpendicular to one another, but the two fields are not in phase with one another; there is some phase angle between them.

7.3. Plane Surface

It is obviously unrealistic to consider plane wave propagation in an unbounded conducting medium, as either the losses due to conduction must be so small that they may be neglected or the amplitude of the wave will be so small as to be useless. There are no situations which approximate to a plane wave propagating in an unbounded conducting medium, but the plane wave results may be used to investigate the current flowing near the surface of a semi-infinite block of conducting material. The coordinate system and the directions of the field components are shown in Fig. 7.1. The electric field is always at the same phase throughout any plane parallel to the surface and this electric field gives rise to a current

$$J_x = \sigma E_x$$

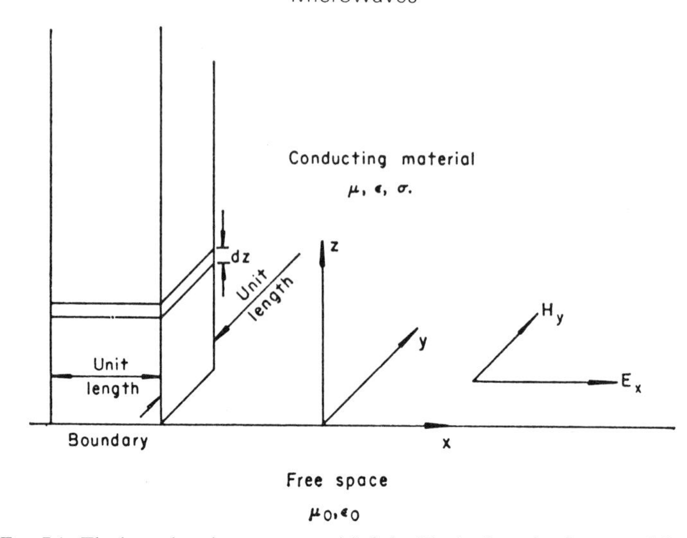

FIG. 7.1 The boundary between a semi-infinite block of conducting material and free space, showing diagrammatically its relationship to the coordinate system and the shape of the unit element for the calculation of current flow and power dissipation.

Therefore the current may be considered to be a current sheet flowing parallel to the surface. It is generated either by a relatively low-frequency current flowing in a high conductivity material or by a plane wave incident normally to the surface.

The components of the electromagnetic fields inside the block at any depth z are given by eqns. (7.9) and (7.10). The current flowing through any element dz of unit width is

$$dI = \sigma E_x dz \tag{7.11}$$

and the total current flowing per unit length of surface is

$$I = \int_0^\infty \sigma E_x \, dz$$

$$= \int_0^\infty \frac{j\sigma(\alpha - j\beta)}{\omega\varepsilon\sqrt{[1 + (\sigma/\omega\varepsilon)^2]}} H_0 \, e^{-(\alpha + j\beta)z} \, dz \exp j\omega t$$

$$I = \frac{\sigma(\sigma - j\omega\varepsilon)}{(\sigma^2 + \omega^2\varepsilon^2)} H_0 \exp j\omega t \tag{7.12}$$

7.4. High Conductivity Material

In most conductors, even at high frequencies, it is found that

$$\sigma \gg \omega\varepsilon$$

so that eqns. (7.5) to (7.12) may be simplified. It is only in semiconductors at microwave frequencies that the above condition does not apply. The condition means in physical terms that the conduction current term in eqn. (2.7) is dominant and that the displacement current term may be neglected. Therefore the simplified expressions are

$$\alpha = \beta = \sqrt{\left(\frac{\omega\mu\sigma}{2}\right)} \tag{7.13}$$

and

$$E_x = (1 + j)\sqrt{\left(\frac{\omega\mu}{2\sigma}\right)} H_0 \exp\{j\omega t - \alpha(1 + j)z\} \tag{7.14}$$

If I_0 is the peak value of the current I such that

$$I = I_0 \exp j\omega t$$

then from eqn. (7.12)

$$I_0 = H_0 \tag{7.15}$$

H is the magnetic field strength in amperes/metre and I is the surface current density also in amperes/metre.

7.5. Power Loss

As the electromagnetic fields penetrate into the conducting block of material, the amplitude of the fields decreases and power is lost in heating the material. The power lost is due to the flow of current in the material. Hence the power loss density is

$$\begin{aligned} p &= \tfrac{1}{2}E_x J_x^* = \tfrac{1}{2}\sigma E_x E_x^* \\ &= \tfrac{1}{2}\sigma(1 + j)\sqrt{\left(\frac{\omega\mu}{2\sigma}\right)} H_0\, e^{-\alpha(1+j)z} . (1 - j)\sqrt{\left(\frac{\omega\mu}{2\sigma}\right)} H_0\, e^{-\alpha(1-j)z} \\ &= \tfrac{1}{2}\omega\mu H_0^2 e^{-2\alpha z} \end{aligned}$$

The power loss in a unit element thin slice, as shown in Fig. 7.1, is given by

$$dP = \tfrac{1}{2}\omega\mu H_0^2 e^{-2\alpha z}\, dz$$

and the total power loss in the unit element of infinite length is

$$P = \tfrac{1}{2}\int_0^\infty \omega\mu H_0^2 e^{-2\alpha z}\, dz$$

$$= \tfrac{1}{2}\sqrt{\left(\frac{\omega\mu}{2\sigma}\right)} H_0^2 \tag{7.16}$$

Substituting from eqn. (7.15) into eqn. (7.16) gives

$$P = \tfrac{1}{2}\sqrt{\left(\frac{\omega\mu}{2\sigma}\right)}I_0^2 \tag{7.17}$$

7.6. Skin Depth

If the power loss is put in the form $\frac{1}{2}RI_0^2$, it is found that eqn. (7.17) can be put into the form

$$P = \frac{1}{2z_0\sigma}I_0^2 \tag{7.18}$$

where z_0 is called the *skin depth*. It is found that the power loss is the same as if the same total current were flowing equally distributed in the depth z_0. It is also obvious from the expression for the fields that the skin depth is the distance from the surface at which the field strength has fallen to $1/e$ of its strength at the surface. That is, it is the depth beyond which for many purposes the field strength becomes negligible:

$$z_0 = \frac{1}{\alpha} = \sqrt{\left(\frac{2}{\omega\mu\sigma}\right)} \tag{7.19}$$

The concept of skin depth is useful at many different frequencies. At low frequencies it may be used to calculate the thickness required for the laminations of transformer cores, so that the laminations are thin compared with the skin depth to give good penetration of the magnetic field. It may be used to calculate the a.c. resistance of conductors since due to skin effect the a.c. resistance is larger than the d.c. resistance of the same conductor. So far we have only considered an infinite plane surface, and it is doubtful whether any conductor at low frequencies will approximate to our model. In the rest of this chapter expressions for the alternating current distribution in a circular wire are developed. At higher frequencies, the concept of skin depth is useful in determining the necessary thickness of waveguide wall. There are some applications where, for the achievement of light weight or minimum thermal conductivity, a thin-wall waveguide is necessary. Also in some ferrite control devices such as the variable attenuator described in section 13.8, it is necessary to obtain a relatively low frequency magnetic field inside the waveguide. Under these conditions, the waveguide wall must be thinner than the skin depth at the frequency of the low-frequency magnetic field but it must also be thicker than the skin depth at the microwave frequency.

The wavelength of propagation into the surface of the conductor is given by

$$\lambda = 2\pi z_0 \tag{7.20}$$

and it is seen that this wavelength is very much smaller than the characteristic wavelength for the same frequency. Consequently, it is perfectly justifiable to use the plane wave approximation in the calculation of currents in the waveguide walls. Hence it is seen that an equivalent surface resistivity of the waveguide wall, R_s, may be postulated as already discussed in section 5.11. From eqn. (7.18), R_s may be defined as

$$R_s = \frac{1}{z_0 \sigma} \tag{7.21}$$

which is another way of writing eqn. (5.46). Substituting into eqn. (7.16) gives

$$P = \tfrac{1}{2} R_s H_0^2$$

which is another way of writing eqn. (5.45).

7.7. Cylindrical Polar Coordinates

The problem of conduction at a high frequency along a long straight wire will be solved by the solution of Maxwell's equations using cylindrical polar coordinates. It is assumed that the axis of the coordinate system is the same as the axis of the wire, as shown in Fig. 7.2. The solution of Maxwell's equations is the wave equation in cylindrical polar coordinates, eqn. (6.4), which is rewritten here

$$\frac{\partial^2 E_z}{\partial r^2} + \frac{1}{r}\frac{\partial E_z}{\partial r} + \left(k_c^2 - \frac{n^2}{r^2}\right) E_z = 0 \tag{7.22}$$

but k_c now has a different value

$$k_c^2 = \omega^2 \mu \varepsilon - j\omega\mu\sigma - \beta^2 \tag{7.23}$$

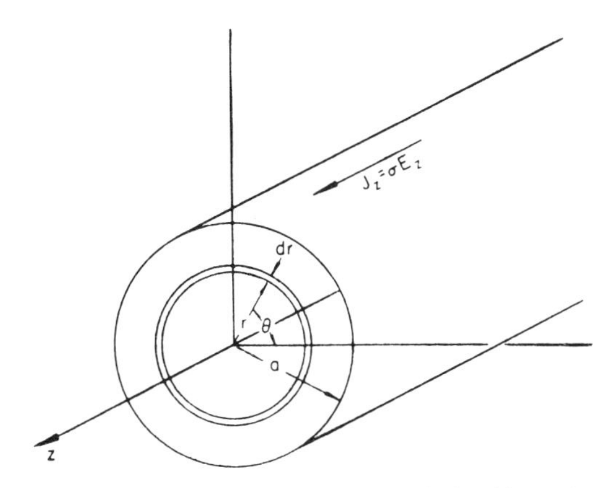

FIG. 7.2. A circular wire of radius a showing its relationship to the axes of the cylindrical coordinate system.

In this equation k_c^2 is complex and so k_c can be expressed in terms of real and imaginary parts. The solution of eqn. (7.22) is still

$$E_z = AJ_n(k_c r) + BY_n(k_c r) \tag{7.24}$$

except that in this case all the terms in eqn. (7.24) are complex. $J_n(x)$ and $Y_n(x)$ can each be expressed in terms of a series so that complex Bessel functions in terms of complex arguments can be evaluated. However the problem may be further simplified by applying some of the obvious conditions applicable to a circular wire which lead to results in terms of some complex Bessel functions that have been tabulated.

7.8. Circular Symmetry

For relatively low-frequency conduction along a metal wire, it may be assumed that there is no variation in current or fields along the wire. For higher frequency currents, it has already been shown that the rate of variation of the fields into a plane surface is so much larger than any variation parallel to the surface that the latter may be ignored. Similarly, for a circular wire the radial variation of the fields is so much larger than any variation parallel to the surface that the latter may be ignored; hence it is assumed that there is no variation of the fields in the z-direction and $\beta = 0$.

It is assumed that the wire is a good conductor so that

$$\omega\varepsilon \ll \sigma$$

hence simplifying eqn. (7.23),

$$k_c^2 = -j\omega\mu\sigma = j^3\omega\mu\sigma$$

Let

$$m = \sqrt{(\omega\mu\sigma)}$$

then

$$k_c = j^{3/2}m$$

Since a symmetrical circular wire is being considered, the fields are symmetrical around the wire and there is no variation of the fields with change of θ. Hence $n = 0$. Also the fields have to be finite or zero at $r = 0$, therefore $B = 0$ and the expression for the electric field becomes

$$E_z = AJ_0(j^{3/2}mr) \tag{7.25}$$

The real and imaginary parts of the Bessel function in eqn. (7.25) are called *Kelvin functions*. They have been tabulated and are termed *bessel real*, ber, and *besel imaginary*, bei, hence we have by definition

$$J_0(j^{3/2}x) = \text{ber}(x) + j\,\text{bei}(x)$$

and eqn. (7.25) becomes

$$E_z = A[\text{ber}(mr) + j\,\text{bei}(mr)] \tag{7.26}$$

where A is the amplitude of the field and is determined by the boundary conditions. E_z is the only component of the electric field which exists. Comparison with eqns. (6.21) to (6.24) shows that if $n = 0$ and $\beta = 0$ and provided $H_z = 0$, then $E_r = E_\theta = H_r = 0$ and

$$H_\theta = -\frac{\sigma}{k_c^2}\frac{\partial E_z}{\partial r} \tag{7.27}$$

Therefore

$$H_\theta = -\frac{j\sigma A}{m}[\text{ber}'(mr) + j\,\text{bei}'(mr)]$$

or

$$H_\theta = \frac{\sigma A}{m}[\text{bei}'(mr) - j\,\text{ber}'(mr)] \tag{7.28}$$

7.9. Current Distribution in a Circular Wire

The current density at any radius inside the wire is given by

$$J_z = \sigma E_z$$

hence the total current in a wire of radius a is

$$I_0 = \int_0^a \sigma E_z \,.\, 2\pi r\, dr$$

$$= 2\pi A\sigma \int_0^a rJ_0(j^{3/2}\,mr)\,dr$$

The integration of a Bessel function is given by

$$\int_0^z xJ_0(x)\,dx = zJ_1(z)$$

and the other relationships are

$$J_1(z) = -J_0'(z)$$

and

$$j^{3/2}J_0'(j^{3/2}z) = \text{ber}'\,z + j\,\text{bei}'\,z$$

hence

$$I_0 = \frac{2\pi a A\sigma}{m}[\text{bei}'(ma) - j\,\text{ber}'(ma)] \tag{7.29}$$

If H_0 is the amplitude of the magnetic field at the surface of the wire, that is, the value of H_θ at $r = a$, then comparison with eqn. (7.28) shows that

$$I_0 = 2\pi a H_0$$

Substituting the value of A from eqn. (7.29) into eqn. (7.26) gives an expression for the current density at any radius

$$J_z = \frac{I_0 m}{2\pi a}\left[\frac{\text{ber}(mr) + j\,\text{bei}(mr)}{\text{bei}'(ma) - j\,\text{ber}'(ma)}\right] \tag{7.30}$$

7.10. Summary

7.1. Consideration of finite conductivity $\boldsymbol{J} - \sigma\boldsymbol{E}$ leads to an ***effective permittivity***

$$\varepsilon_{\text{eff}} = \varepsilon - j\frac{\sigma}{\omega} \tag{7.2}$$

7.2. A plane wave has the propagation constant

$$\gamma^2 = -\omega^2\mu\left(\varepsilon - j\frac{\sigma}{\omega}\right) \tag{7.4}$$

The plane wave has only two field components perpendicular to one another but there is a phase difference between them.

7.4. For a good conductor $\sigma \gg \omega\varepsilon$ and for a plane wave

$$H_y = H_0 \exp\{j\omega t - \alpha(1 + j)z\} \tag{7.9}$$

$$E_x = (1 + j)\frac{\alpha}{\sigma}H_0 \exp\{j\omega t - \alpha(1 + j)z\} \tag{7.14}$$

where

$$\alpha = \sqrt{\left(\frac{\omega\mu\sigma}{2}\right)} \tag{7.13}$$

If H_0 is the amplitude of the magnetic field at the plane surface of a conductor, the total current flowing parallel to the surface,

$$I_0 = H_0 \tag{7.15}$$

7.6. Power loss in the conductor $= \dfrac{1}{2z_0\sigma}I_0^2$ (7.18)

The ***skin depth*** $z_0 = \frac{1}{\alpha} = \sqrt{\left(\frac{2}{\omega\mu\sigma}\right)}$ (7.19)

The wavelength of propagation into the conductor is given by

$$\lambda = 2\pi z_0 \tag{7.20}$$

The equivalent surface resistivity of the conductor is given by

$$R_s = \frac{1}{z_0\sigma} \tag{7.21}$$

7.8. For a circular wire, the results occur in terms of the real and imaginary parts of a complex Bessel function:

$$E_z = AJ_0(j^{3/2}\,mr) = A[\mathrm{ber}\,(mr) + j\,\mathrm{bei}\,(mr)] \tag{7.26}$$

where

$$m = \sqrt{(\omega\mu\sigma)}$$

$$H_\theta = \frac{\sigma A}{m}[\mathrm{bei}'\,(mr) - j\,\mathrm{ber}'\,(mr)] \tag{7.28}$$

7.9. The current distribution in a wire of radius a is

$$J_z = \frac{I_0 m}{2\pi a}\left[\frac{\mathrm{ber}\,(mr) + j\,\mathrm{bei}\,(mr)}{\mathrm{bei}'\,(ma) - j\,\mathrm{ber}'\,(ma)}\right] \tag{7.30}$$

Problems

For copper the following material constants may be assumed: $\mu = \mu_0$, $\varepsilon = \varepsilon_0$, $\sigma = 5 \times 10^7$ S/m.

7.1. Calculate a few points and plot a graph of the two components, α and β, of the propagation constant of a plane wave in copper against a wide range of frequencies and hence show at what frequencies the assumptions of section 7.4 are not valid.

7.2. Starting from first principles (Maxwell's equations) derive eqns. (7.9) and (7.10).

7.3. Calculate the ratio of the plane wave wavelength in copper to the characteristic wavelength of an electromagnetic wave at: 1 kHz, 1 MHz, 1 GHz, 10 GHz.
[2.12×10^7, 6.7×10^5, 2.12×10^4, 6.7×10^3]

7.4. Calculate the skin depth in copper at the frequencies: 1 kHz, 1 MHz, 1 GHz, 10 GHz.
[2.25 mm, 71 μm, 2.25 μm, 0.71 μm]

7.5. A ferrite variable attenuator has a solenoid wound on the outside of the waveguide to magnetize the ferrite inside the waveguide. Is it possible to vary the magnetic field in the ferrite at a frequency of up to 1 MHz when the microwave signal in the waveguide is 10 GHz? Unless it is thin, the waveguide wall provides a shorted turn to the solenoid and effectively shields the ferrite inside the waveguide from the externally applied magnetic field. The waveguide will be made of copper.

7.6. Calculate the VSWR in the air space for a plane wave in air normally incident onto the plane surface of a medium of conductivity σ.

7.7. Discuss in terms of skin depth and calculate approximate sizes, guessing values for material parameters where appropriate, for the following:

laminated transformer cores for use at mains supply frequency;
laminated transformer cores for use at high frequencies;
copper-plated steel wire for use at high frequencies;
microstrip;
thin wall waveguide;
copper-plated waveguide.

7.8. Derive eqn. (7.22) from first principles (Maxwell's equations).

7.9. Calculate the lowest frequency at which a plane surface approximation would be valid for a copper wire of diameter, (a) 1 mm and (b) 1 μm. [10 MHz, 10^{14} Hz]

7.10. Calculate a few points and plot the amplitude of the current distribution in a copper wire for the two conditions $ma = 1$ and $ma = 5$. Some values of the Kelvin function are given in the table; assume any reasonable diameter for the wire, and hence quote the frequency of operation.

TABLE 7.1
Kelvin Functions

x	ber x	bei x
0.0	1.00	0.00
0.2	1.00	0.01
0.4	1.00	0.04
0.6	0.99	0.09
0.8	0.99	0.16
1.0	0.98	0.25
2.0	0.75	0.97
3.0	−0.22	1.93
4.0	−2.56	2.29
5.0	−6.23	0.12
x	ber′ x	bei′ x
1.0	−0.06	0.50
5.0	−3.84	−4.35

8

Ferrite Media

8.1. Magnetic Materials

Electromagnetic wave propagation normally takes place through media which would commonly be termed non-magnetic. The common magnetic materials are those metals of the iron family and their compounds which are ferromagnetic and which have a relative permeability of the order of a thousand. Because of their good conductivity, there is little interaction between these magnetic materials and an electromagnetic wave. However, *ferrites* and *magnetic garnets* have strong magnetic properties and are also insulators. They have enabled certain properties of ferromagnetism to be used at microwave frequencies.

First a summary of the properties of magnetic materials is given. All electrons behave as if they are spinning magnetic tops. The rotation of the electric charge of the electron due to the spin gives rise to a magnetic moment associated with each electron. The direction of the magnetic moment of the electron is parallel to the axis of spin and dependent on the direction of rotation of the spin. The spin axes of the electrons in any atom are aligned but they usually occur in antiparallel pairs so that the total effect external to the atom is zero. In some elements, however, there are a number of unpaired electron spins in the atom and each atom has some magnetic moment. For example, the ferrous ion Fe^{2+} has a spin magnetic moment of 5, the ferric ion Fe^{3+} has one of 6 and nickel Ni^{2+} has one of 3. In these materials in the solid there is a very strong coupling between the different atoms to align these spin magnetic moments, so that the total magnetic effect is large. In ferromagnetic substances, such as iron, the spin magnetic moments of all the atoms act together giving the maximum possible magnetic effect. The effect is shown in Fig. 8.1. In ferrites and garnets, however, the effect of the coupling is to divide the magnetic atoms into two groups having oppositely oriented spins. If the spin magnetic moments in each group are unequal, there still is some external magnetic field but it is smaller than that of ferromagnetic substances. These are called ferrimagnetic materials. It may be said that, in ferrimagnetic materials, the coupling aligns the electron spins antiparallel in unequal quantities and there is some external magnetic field. In some substances, the spin magnetic moments in each group are equal and

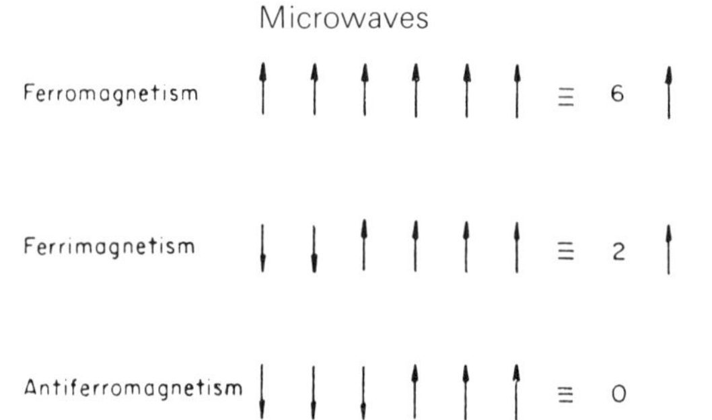

FIG. 8.1. Types of spin coupled magnetization.

there is no external magnetic field. These are called antiferromagnetic materials where the coupling aligns the electron spins in equal quantities; the magnetic moment cancels out inside the material and there is no magnetic effect. All these types of ferromagnetism are illustrated in Fig. 8.1.

There is another type of magnetic effect which is also due to unbalanced electron spin in the atom, which is paramagnetism. The magnetic effect is slight because in paramagnetic substances the coupling between the spins of individual atoms is so small that it may be neglected. The spin magnetic moment of each individual atom aligns itself individually with any external magnetic field giving a very weak internal magnetization.

Ferrites and some garnets are ferrimagnetic and they are also insulators. They provide a medium in which there can be some interaction between microwave electromagnetic fields and ferromagnetic electron spin. The interaction to be described in the rest of this chapter is true for all ferromagnetic materials because it is only dependent on the properties of a spinning magnetic top. All ferromagnetic materials possess these magnetic properties, but normally it is only in the ferrite and garnet materials that the required interaction with electromagnetic waves can be obtained.

8.2. Elementary Properties of Magnetic Materials

A classical description of magnetism will be used to explain some of the properties of ferromagnetic materials. The electron behaves as if it were a negatively charged sphere which is spinning about its own axis with a fixed angular momentum. The rotation of charge gives the electron a magnetic moment which is a function of its charge, angular velocity and size, so that the electron behaves as if it were a spinning magnetic top whose magnetic moment lies along its axis of rotation. It is similar to a spinning gyroscope suspended at a point other than its centre of gravity; the difference is that the electron, acting as a gyroscope, moves due to the influence of magnetic forces whereas a gyroscope is under the influence of gravitational forces. The forces acting on the electron, being magnetic in origin, are coincident with any applied magnetic field.

When the electron is acted upon by a magnetic field, it lines up with the field for minimum potential energy. If the electron is disturbed from this equilibrium position, it does not return to the position of minimum energy but precesses about the axis of the magnetic field, as illustrated in Fig. 8.2 where the spinning gyroscope makes an angle θ with the direction of the magnetic field. The equilibrium motion, if there are no losses, is a precessional motion about the vertical axis with a velocity ω.

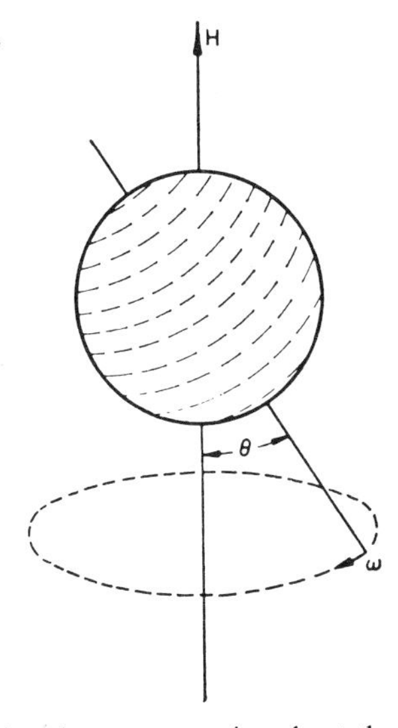

FIG. 8.2. Spinning electron precessing about the magnetic field H.

This classical description of a magnetism can be used to describe the motion of the electrons in a ferrite. The ferrite is magnetically saturated by a field H. If an alternating magnetic field acting in a plane perpendicular to H is superimposed onto the field H, the resultant field alternates between the two directions A and B shown in Fig. 8.3. Initially consider the gyroscope pointing vertically under the influence of the force H. If the direction of the force H is suddenly altered to the position A, the gyroscope precesses about the axis A along the circular path a–b. If, when the gyroscope has reached the position b, the direction of the force H changes to the position B, the gyroscope precesses along the new circular path b–c. If the force then moves back to the position A, the gyroscope continues in the circle c–d. It is seen that if the alternating motion of the force H continues in step with the motion of the gyroscope, the orbit of the gyroscope continues increasing indefinitely. However, in a material there are forces other than the magnetic field acting on the movement of the electron spin axes and these forces tend to oppose the precessional motion of the electrons. In practice, any gyroscope set into motion and then left to precess will slowly spiral to an equilibrium position. The loss of precessional energy is due to friction and other losses

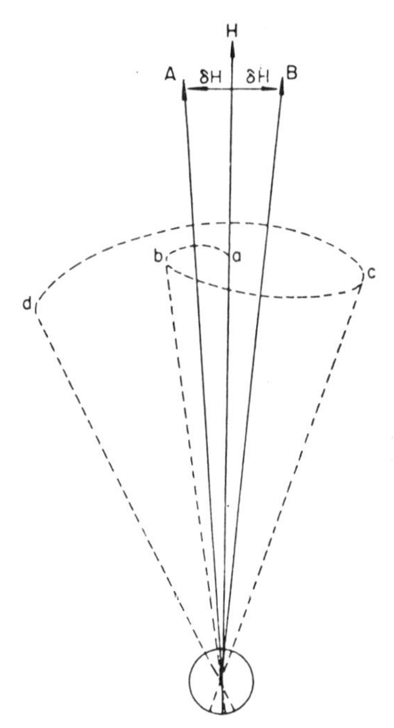

FIG. 8.3. Precessional motion of a spinning electron in a magnetic field which oscillates between the directions A and B.

in the system. In a similar way, there are frictional and other damping mechanisms in a ferromagnetic material to limit the precessional motion of the axis of the spinning electron. Obviously the orbit of the gyroscope in Fig. 8.3 cannot continue increasing indefinitely and it will reach some equilibrium position where the losses in the material exactly offset the driving effect of the alternating force.

8.3. Resonance Absorption

If a ferrite material is initially saturated by a steady magnetic field, the electrons come to rest with their magnetic moments parallel to the field H. If an additional alternating magnetic field at the correct frequency is applied perpendicular to the static field H, the electrons begin to precess in larger and larger circles until they finally reach some equilibrium precession orbit under the influence of the magnetic fields and the internal friction damping. There is a transfer of power from the alternating magnetic field to the precessing electrons in the ferrite. The precessing electrons dissipate the power in internal friction which appears as heat in the material. The ferrite absorbs power from the disturbing magnetic field.

As the transfer of power from the alternating magnetic field to heat in the ferrite material only occurs if the frequency of the alternating magnetic field coincides with the precession frequency of the electrons in the ferrite, the phenomenon is called *resonance absorption*. If the disturbing alternating magnetic field is provided by the magnetic field from an electromagnetic wave, the ferrite absorbs power from the electromagnetic wave. The relationship between power absorbed and the frequency of the alternating magnetic field is a normal resonance curve similar to that shown in Fig. 3.13.

In the description of the generation of precession shown in Fig. 8.3, it is assumed that the disturbing magnetic field jumps between the two positions A and B and that δH takes the form of a square wave. In a linearly polarized wave, the disturbing magnetic field is in the form of a sine wave and the effect of the disturbance is similar to that already described. If the disturbing magnetic field is circularly polarized, however, there is an even greater interaction between the field and the precessing electron. Instead of the field acting to increase the precession orbit just twice in each cycle, the circularly polarized field is acting to increase the orbit all the time, provided that the direction of rotation of the circularly polarized field is the same as the direction of rotation of the precession orbit. Unless the frequency of the alternating magnetic field is the same as the precession frequency of the electrons in the ferrite, and the direction of the rotation of the field coincides with the precessing direction of the electrons, there is very little coupling between the electromagnetic wave and the ferrite. As the precession frequency of the electrons depends on the strength of the static magnetic field, the relationship between power absorbed and variation of static magnetic field for a fixed frequency of alternating magnetic field is the same as that between power absorbed and frequency for a fixed static magnetic field.

8.4. Magnetization Equation

We shall now obtain some quantitative relationships to describe the interaction between a magnetic material and an electromagnetic wave. Initially we consider the magnetic effect and a mathematical description of the precessing electrons. Any ferromagnetic substance has an internal intensity of magnetization $\boldsymbol{M}$ which need not necessarily be parallel to the applied magnetic field $\boldsymbol{H}$. The total magnetic field is given by

$$\boldsymbol{B} = \mu_0 \boldsymbol{H} + \boldsymbol{M}$$

Each minute element in the material may be considered to be a magnetic top, in which the magnetic moment and the angular momentum are parallel vectors. Their ratio is a constant called the *gyromagnetic ratio*, for which the symbol γ is used. In this chapter γ is used to denote the gyromagnetic ratio although in the rest of this book it denotes the propagation constant. Similarly in this chapter, $\boldsymbol{J}$ is used to denote the angular momentum of the spinning electron although it is used to denote current density in the rest of this book.

The magnetization is the volume integration of the magnetic moment of each element so that the magnetic moment is proportional to the magnetization, where K is the constant of proportionality. Hence

$$\gamma = \frac{K\boldsymbol{M}}{\boldsymbol{J}} \tag{8.1}$$

There are two forces acting on the atomic tops, the external magnetic field and the exchange forces within the substance tending to align the magnetic moments of all the tops. It is the action of these exchange forces which allow us to consider the magnetization in the previous equation rather than the magnetic moment of an individual top.

Consider a ferromagnetic body of arbitrary shape and size. The equilibrium direction of each atomic top is such that there is no torque on any top. Under the influence of high frequency magnetic fields, there will be small deviations from the equilibrium position. The equation of motion of any gyroscope is given by

$$\text{torque} = \frac{d\boldsymbol{J}}{dt} \tag{8.2}$$

The torque exerted on any top is given by the cross-multiplication of the magnetic field and the magnetic moment, in this case represented by the magnetization. Hence

$$\text{torque} = K\boldsymbol{M} \times \left(\boldsymbol{H} + \frac{\boldsymbol{M}}{\mu_0}\right) \tag{8.3}$$

and substituting from eqns. (8.1) and (8.2) and remembering that $\boldsymbol{M} \times \boldsymbol{M} = 0$ gives

$$\frac{d\boldsymbol{M}}{dt} = \gamma \boldsymbol{M} \times \boldsymbol{H} \tag{8.4}$$

8.5. Permeability Tensor

To consider the electromagnetic effect of the magnetization equation, eqn. (8.4), a relationship between $\boldsymbol{B}$ and $\boldsymbol{H}$ for the magnetic material is derived which is the microwave permeability of the material. It is seen that this permeability is frequency dependent, which might be expected from the form of eqn. (8.4). Consider an orthogonal system of axes in rectangular coordinates x, y and z. Let the ferromagnetic body be magnetized to saturation by a static magnetic field, H_0, in the z-direction which generates a static magnetization M_0 in the material. It is magnetized to saturation because then all the atomic magnets in the material are aligned with the static magnetic field and eqn. (8.4) applies to the bulk of the material.

Let there also be a time-varying magnetic field, of time dependence $\exp j\omega t$, that is small compared with the saturation magnetic field. This magnetic field $\boldsymbol{H}$ gives rise to a magnetization $\boldsymbol{M}$ in the material. The total magnetic field and magnetization in rectangular components is:

$$\left.\begin{array}{l} H_x \\ H_y \\ H_z + H_0 \end{array}\right\} \text{gives rise to} \left\{\begin{array}{l} M_x \\ M_y \\ M_z + M_0 \end{array}\right.$$

where H_x, H_y and H_z and M_x, M_y and M_z are the time-varying components of the fields and H_0 and M_0 are the static fields. These fields are substituted into eqn. (8.4), giving

$$\begin{aligned} j\omega M_x &= \gamma M_y(H_0 + H_z) - \gamma(M_0 + M_z)H_y \\ j\omega M_y &= \gamma(M_0 + M_z)H_x - \gamma M_x(H_0 + H_z) \\ j\omega M_z &= \gamma M_x H_y - \gamma M_y H_x \end{aligned}$$

If it is assumed that the microwave magnetic fields are so much smaller than the static field that they may be neglected compared with the static field, then all the product terms between components of the time-varying magnetic field in the right-hand side of the above equations may be neglected. These simplified equations are

$$\left.\begin{aligned} j\omega M_x - \gamma H_0 M_y &= -\gamma M_0 H_y \\ \gamma H_0 M_x + j\omega M_y &= \gamma M_0 H_x \\ j\omega M_z &= 0 \end{aligned}\right\} \tag{8.5}$$

It is seen that the first two of the equations (8.5) are simultaneous equations in M_x and M_y. They may be solved and if the abbreviation is adopted

$$\kappa = -\frac{\omega\gamma M_0}{\gamma^2 H_0^2 - \omega^2} \tag{8.6}$$

$$\chi = \frac{\gamma^2 H_0 M_0}{\gamma^2 H_0^2 - \omega^2} \tag{8.7}$$

then the solution of eqn. (8.5) is

$$\left.\begin{aligned} M_x &= \chi H_x + j\kappa H_y \\ M_y &= -j\kappa H_x + \chi H_y \\ M_z &= 0 \end{aligned}\right\} \tag{8.8}$$

The total microwave field is $\boldsymbol{B} = (\mu_0 \boldsymbol{H} + \boldsymbol{M})$ and the field components are

$$\begin{aligned} B_x &= (\mu_0 + \chi)H_x + j\kappa H_y \\ B_y &= -j\kappa H_x + (\mu_0 + \chi)H_y \\ B_z &= \mu_0 H_z \end{aligned}$$

Define

$$\mu = (\mu_0 + \chi) = \frac{\gamma^2 H_0(\mu_0 H_0 + M_0) - \omega^2 \mu_0}{\gamma^2 H_0^2 - \omega^2} = \frac{\gamma^2 H_0 B_0 - \omega^2 \mu_0}{\gamma^2 H_0^2 - \omega^2} \tag{8.9}$$

and the relationship between $\boldsymbol{B}$ and $\boldsymbol{H}$ is

$$\left.\begin{aligned} B_x &= \mu H_x + j\kappa H_y \\ B_y &= -j\kappa H_x + \mu H_y \\ B_z &= \mu_0 H_z \end{aligned}\right\} \tag{8.10}$$

or written in vector form

$$\boldsymbol{B} = \begin{vmatrix} \mu & j\kappa & 0 \\ -j\kappa & \mu & 0 \\ 0 & 0 & \mu_0 \end{vmatrix} \boldsymbol{H} \tag{8.11}$$

where the permeability is a tensor,

$$\boldsymbol{\mu} = \begin{vmatrix} \mu & j\kappa & 0 \\ -j\kappa & \mu & 0 \\ 0 & 0 & \mu_0 \end{vmatrix}$$

The permeability tensor is a matrix representation of the relationship between two vectors. It means that there is a two-dimensional relationship between $\boldsymbol{B}$ and $\boldsymbol{H}$. A magnetic field in one direction gives rise to a resultant magnetic flux in a direction perpendicular to it as well as one parallel to it.

The permeability tensor has been derived from a simple classical model of a ferromagnetic material magnetized to saturation. However, the relationship in eqns. (8.10) and (8.11) has no restrictions except for a certain symmetry, provided that no reference is made to a special model. The relations are generally applicable to any isotropic substance since the only condition to be satisfied by the permeability tensor is the rotational symmetry about the axis of the static magnetization. As long as no reference is made to a special model, μ and κ may be arbitrary quantities which are constant only under conditions of constant frequency and constant static magnetic fields.

For most ferrite materials, γ is 1.76×10^{11} rad/s.T (or 28 GHz/T) and values of μ and κ can be calculated for the ferrite when magnetized to saturation. It is noticed that, as here quoted, the dimensions of γ are incorrect. The values quoted are for γ/μ_0, because it leads to simplicity in calculations. In most practical systems, H_0 is an applied magnetic flux density rather than a magnetic field strength and is measured in units of Tesla rather than A/m. It is noticed that eqns. (8.6) and (8.9) contain a number of angular frequency terms. A value for γ is used in calculations so that frequencies in Hertz may be inserted into the equations.

The resonant frequency is the frequency at which the resonance absorption, described in section 8.3, occurs. It is also the frequency at which the elements of eqn. (8.11) become infinite. This is when the denominator in the expressions in eqns. (8.6) and (8.7) is zero. The resonance frequency is denoted by ω_0 and is given by

$$\omega_0 = \gamma H_0$$

Define

$$\omega_m = \frac{\gamma M_0}{\mu_0}$$

then the elements of eqn. (8.11) become

$$\mu = \mu_0\left(1 + \frac{\omega_0 \omega_m}{\omega_0^2 - \omega^2}\right) \tag{8.9a}$$

$$\kappa = -\mu_0 \frac{\omega \omega_m}{\omega_0^2 - \omega^2} \tag{8.6a}$$

8.6. Plane Wave

Consider the propagation of a plane wave through a statically magnetized ferrite material. The simplest solution is obtained if it is assumed that the direction of any static magnetic field is the same as the direction of propagation of the wave and the analysis will be confined to a consideration of this case. Then the permeability of the ferrite material to the plane wave is given by eqn. (8.11). Otherwise the ferrite material is a normal insulator and can be considered as a non-conducting medium with a permittivity ε and peculiar magnetic properties given by eqn. (8.10). Let the direction of the static magnetic field and the direction of propagation of the wave be coincident with the z-direction of the rectangular coordinate system as shown in Fig. 8.4.

Then the relation between $\boldsymbol{B}$ and $\boldsymbol{H}$ fields of the electromagnetic wave is given by eqn. (8.10) which is rewritten here for convenience:

$$\left.\begin{aligned} B_x &= \mu H_x + j\kappa H_y \\ B_y &= -j\kappa H_x + \mu H_y \\ B_z &= \mu_0 H_z \end{aligned}\right\} \tag{8.10}$$

It is necessary to substitute the relationship of eqn. (8.10) into Maxwell's curl

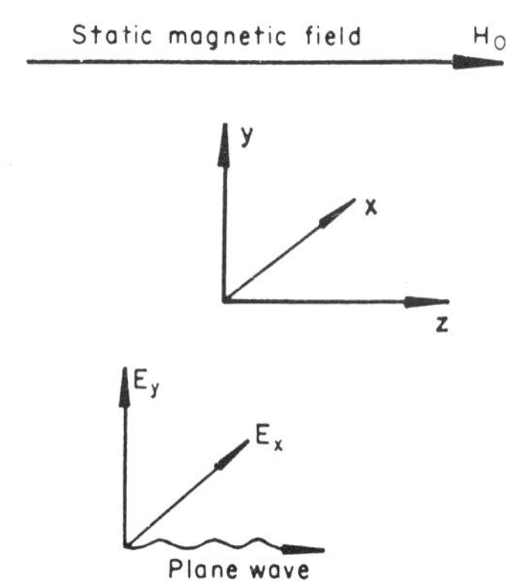

FIG. 8.4. Showing the relationship between the static magnetization in the ferrite material and both the rectangular coordinate system and the direction of propagation of a plane wave. The electric field components of the plane wave are shown.

equations and to solve them as has been done previously. Equation (2.6) becomes

$$\left.\begin{aligned} \frac{\partial E_z}{\partial y} - \frac{\partial E_y}{\partial z} &= -j\omega\mu H_x + \omega\kappa H_y \\ \frac{\partial E_x}{\partial z} - \frac{\partial E_z}{\partial x} &= -\omega\kappa H_x - j\omega\mu H_y \\ \frac{\partial E_y}{\partial x} - \frac{\partial E_x}{\partial y} &= -j\omega\mu_0 H_z \end{aligned}\right\} \tag{8.12}$$

and the other curl equation is given by eqn. (2.25). The conditions for a plane wave lead to the assumptions

$$\frac{\partial}{\partial x} = \frac{\partial}{\partial y} = 0 \quad \text{and} \quad \frac{\partial}{\partial z} = -j\beta$$

Substituting these conditions into eqn. (8.12) gives

$$\left.\begin{aligned} \beta E_y &= -\omega\mu H_x - j\omega\kappa H_y \\ \beta E_x &= -j\omega\kappa H_x + \omega\mu H_y \\ 0 &= H_z \end{aligned}\right\} \tag{8.13}$$

and into eqn. (2.25) gives

$$\left.\begin{aligned} \beta H_y &= \omega\epsilon E_x \\ \beta H_x &= -\omega\epsilon E_y \\ 0 &= E_z \end{aligned}\right\} \tag{8.14}$$

Elimination of the magnetic field components between eqns. (8.13) and (8.14) gives

$$\left.\begin{aligned}(\beta^2 - \omega^2\mu\varepsilon)E_y &= -j\omega^2\varepsilon\kappa E_x\\ (\beta^2 - \omega^2\mu\varepsilon)E_x &= j\omega^2\varepsilon\kappa E_y\end{aligned}\right\} \quad (8.15)$$

Hence the propagation conditions for a plane wave are

$$(\beta^2 - \omega^2\mu\varepsilon)^2 = \omega^4\varepsilon^2\kappa^2 \quad (8.16)$$

Then

$$\beta^2 - \omega^2\mu\varepsilon = \pm\omega^2\varepsilon\kappa$$

or

$$\beta^2 = \omega^2\varepsilon(\mu \pm \kappa) \quad (8.17)$$

There are now two solutions, showing that two modes of propagation are possible having all their components in the transverse plane. Let us define

$$\beta^+ = \omega\sqrt{[\varepsilon(\mu + \kappa)]} \quad (8.18)$$

$$\beta^- = \omega\sqrt{[\varepsilon(\mu - \kappa)]} \quad (8.19)$$

Now there has arisen for propagation through statically magnetized ferrite material a condition where there are two possible solutions to the propagation equation. This means that there are two possible modes of plane wave propagation through a magnetized material. The modes have been labelled as positive and negative modes as shown by the notation used in eqns. (8.18) and (8.19). What is more, these modes do not show the independence between the two transverse components of the fields that is shown in eqn. (2.28). Both components of the electric field and the magnetic field appear to exist in the transverse plane. Substitution for β into eqn. (8.15) gives the relationship between the components of the electric field.

$$E_x = \frac{(\beta^2 - \omega^2\mu\varepsilon)E_y}{-j\omega^2\varepsilon\kappa} = \frac{\omega^2\varepsilon(\mu \pm \kappa - \mu)E_y}{-j\omega^2\varepsilon\kappa}$$

Therefore

$$E_x = \pm jE_y$$

or continuing the notation of eqns. (8.18) and (8.19) for the components of the fields of the two modes

$$E_x^+ = jE_y^+ \quad (8.20)$$

$$E_x^- = -jE_y^- \quad (8.21)$$

The relationship given in eqns. (8.20) and (8.21) is that of a circularly polarized wave. It will be remembered that a circularly polarized wave can be considered to be the sum of two linearly polarized waves perpendicular to one another and in phase quadrature. This is just the relationship given in eqns. (8.20) and (8.21). The two modes propagating in the ferrite material are two circularly polarized plane waves with opposite hands of rotation.

From eqn. (8.14) the components of the magnetic field can be obtained, giving

$$H_y = \frac{\omega\varepsilon}{\beta} E_x \tag{8.22}$$

$$H_x = -\frac{\omega\varepsilon}{\beta} E_y \tag{8.23}$$

and the constant of proportionality can be shown to be effectively the free space impedance of the ferrite material:

$$\frac{\beta}{\omega\varepsilon} = \frac{\sqrt{[\omega^2\varepsilon(\mu \pm \kappa)]}}{\omega\varepsilon} = \sqrt{\left[\frac{(\mu \pm \kappa)}{\varepsilon}\right]} = \eta^{\pm} \tag{8.24}$$

8.7. Effective Permeability

For a circularly polarized plane wave propagating through a magnetized ferrite material, the permeability of the material appears as a single constant in the expressions for the phase constant, eqns. (8.18) and (8.19), and the free space impedance, eqn. (8.24). It is called the *effective permeability* of the ferrite material and is given by

$$\left.\begin{aligned} \mu^{+} &= \mu + \kappa \\ \mu^{-} &= \mu - \kappa \end{aligned}\right\} \tag{8.25}$$

The value of this effective permeability is dependent on the frequency of operation as well as being a function of different materials. It is also different for the two hands of circular polarization and is different for a different relationship between the direction of static magnetization and the direction of propagation of the plane wave. The variation of effective permeability with change of static magnetic field at a fixed frequency is given in Fig. 8.5. Reference to eqns. (8.6) and (8.9) aids understanding some of the characteristics of the diagram. To a first approximation for low fields, μ is constant and κ is proportional to the static magnetic field. Hence the approximately linear relationship below saturation. Above saturation there is a much slower variation, due to the fact that M_0 is now constant, until the resonance condition is reached. At resonance both μ and κ theoretically become infinite and the negative effective permeability ought to become infinite also. In practice the permeability follows a relationship similar to that shown in Fig. 8.5, but at the resonance condition any wave is so heavily attenuated that it is very difficult to measure the phase constant. Below resonance, it is seen

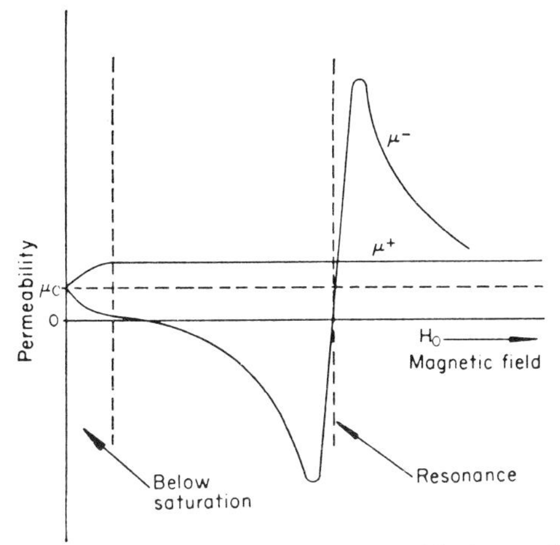

FIG. 8.5. The variation of microwave permeability with change of magnetic field of an infinite ferrite medium for two opposite hands of circularly polarized plane waves at a fixed frequency.

from eqn. (8.6) that κ is positive and above resonance κ is negative. The linear variation of permeability below saturation means that numerical values can be used for μ and κ even when the ferrite is not magnetized to saturation.

8.8. Cylindrical Coordinates

As it has been shown that the plane waves propagating in magnetized ferrite are circularly polarized, a solution will be obtained in terms of cylindrical polar coordinates. The notation of Chapter 6 is used. The permeability tensor relationship of eqn. (8.11) is circularly symmetric about the axis of the static magnetization. If the direction of the static magnetic field and the direction of propagation of the wave are taken to be the z-direction in the cylindrical polar coordinates, eqn. (8.11) may be written

$$\begin{aligned} B_r &= \mu H_r + j\kappa H_\theta \\ B_\theta &= -j\kappa H_r + \mu H_\theta \\ B_z &= \mu_0 H_z \end{aligned} \tag{8.26}$$

Assuming a plane wave, its properties give

$$E_z = H_z = 0$$

and

$$\frac{\partial}{\partial r} = 0, \qquad \frac{\partial}{\partial \theta} = -jn, \qquad \frac{\partial}{\partial z} = -j\beta, \qquad \frac{\partial}{\partial t} = j\omega$$

If these conditions together with the relationship given in eqn. (8.26) are substituted into eqns. (2.6) and (2.7), we obtain by comparison with eqns. (6.9) to (6.14)

$$\left.\begin{aligned} j\beta E_\theta &= -j\omega\mu H_r + \omega\kappa H_\theta \\ j\beta E_r &= \omega\kappa H_r + j\omega\mu H_\theta \end{aligned}\right\} \tag{8.27}$$

$$\frac{1}{r}E_\theta + \frac{jn}{r}E_r = 0 \tag{8.28}$$

$$\left.\begin{aligned} \beta H_\theta &= \omega\varepsilon E_r \\ \beta H_r &= -\omega\varepsilon E_\theta \end{aligned}\right\} \tag{8.29}$$

$$\frac{1}{r}H_\theta + \frac{jn}{r}H_r = 0 \tag{8.30}$$

As before, the solution to eqns. (8.27) and (8.29) is

$$\beta^2 = \omega^2\varepsilon(\mu \pm \kappa) \tag{8.17}$$

together with the relationship

$$E_r = \pm jE_\theta$$

whence

$$E_r^+ = jE_\theta^+ \tag{8.31}$$

$$E_r^- = -jE_\theta^- \tag{8.32}$$

Substitution of the results given in eqns. (8.31) and (8.32) into eqn. (8.28) show that the positive wave having the propagation constant β^+ is associated with a value, $n = 1$, and the negative wave is associated with the value, $n = -1$. Reference to Chapter 6, shows that β^+ is the propagation constant of a positive circularly polarized wave and β^- is the propagation constant of a negative circularly polarized wave. For propagation through ferrite material, it is seen that circularly polarized modes are the fundamental modes of propagation and that a linearly polarized mode must be constructed as the sum of two circularly polarized modes.

8.9. Faraday Rotation

Because the propagation constants of the two hands of circular polarization are different for propagation through magnetized ferrite, one hand of circular polarization rotates further than the other in a fixed length of ferrite material. Linear polarization can be considered as the sum of two equal circularly polarized waves of opposite hand. Because it appears that circular polarization is the fundamental mode of propagation in ferrite, any linearly polarized wave is separated into its circularly polarized components while in the ferrite material. If at any spot it is desired to detect the resultant linearly polarized wave, it is taken as the sum of the two circularly polarized waves. If the two hands of circular polarization have rotated through different angles since being generated from an incident linearly polarized wave, the plane of polarization of the detected linearly polarized wave is rotated compared with the incident wave.

Rotation may be understood simply by reference to the model of the precessing electrons used earlier. One hand of rotation of the circularly polarized wave couples slightly with the precessing electrons and is accelerated, while the other is rotating contrary to the precessing electrons and is retarded. It is seen that the direction of precession of the electrons is determined by the direction of the static magnetic field and hence the direction of rotation of the electromagnetic wave is determined by the direction of the static magnetic field and not by the direction of propagation of the electromagnetic wave. This property leads to one of the most important properties of ferrite materials which is nonreciprocity. A *nonreciprocal* device is a device where a wave travelling in the forward direction is affected differently from a wave travelling in the reverse direction so that forward and reverse waves may be separated. For example, if in traversing a finite length of ferrite material a linearly polarized wave is rotated through 45° and then reflected, it will be rotated a further 45° in the same direction as before and will arrive at the beginning with the plane of polarization at 90° to where it started.

To find a mathematical expression for rotation, the wavelength of the two circularly polarized modes is given by

$$\lambda^+ = \frac{2\pi}{\beta^+} \quad \text{and} \quad \lambda^- = \frac{2\pi}{\beta^-}$$

where the phase constants of the two modes are already defined by eqns. (8.18) and (8.19). The wavelength of the equivalent linearly polarized wave is the mean of the wavelengths of the two circularly polarized waves,

$$\lambda = \tfrac{1}{2}(\lambda^+ + \lambda^-) = \pi\left[\frac{1}{\beta^+} + \frac{1}{\beta^-}\right] \tag{8.33}$$

In the distance of one wavelength of the linearly polarized wave, the phase change of the positive wave is

$$\phi^+ = \beta^+\lambda = \pi\left(1 + \frac{\beta^+}{\beta^-}\right)$$

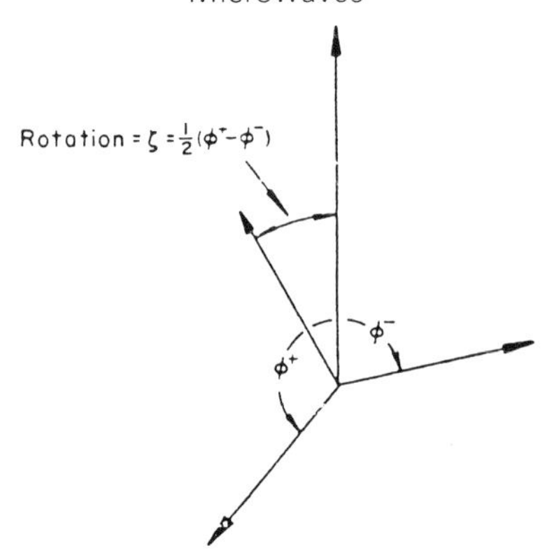

FIG. 8.6. The angle of rotation of the positive and negative circularly polarized waves and the consequent angle of rotation of the linearly polarized wave.

and that of the negative wave is

$$\phi^- = \beta^-\lambda = \pi\left(1 + \frac{\beta^-}{\beta^+}\right)$$

The rotation of the plane of polarization of the linearly polarized wave in one wavelength, λ, is seen by reference to Fig. 8.6 to be

$$\zeta = \tfrac{1}{2}(\phi^+ - \phi^-) = \tfrac{1}{2}(\beta^+ - \beta^-)\lambda = \frac{\pi}{2}\left(\frac{\beta^+}{\beta^-} - \frac{\beta^-}{\beta^+}\right) \tag{8.34}$$

As it is rather difficult to specify a wavelength in a wave whose plane of polarization is changing, the rotation per unit length is more useful and is given by

$$\psi = \frac{\zeta}{\lambda} = \tfrac{1}{2}(\beta^+ - \beta^-) \tag{8.35}$$

This rotation of the plane of polarization was first observed by Faraday with the rotation of the plane of polarization of light through paramagnetic liquids. Here the rotation was very small, being a few degrees in many wavelengths, but with magnetic substances such as ferrites at microwave frequencies rotations of 90° are easily obtainable in fractions of a wavelength. Because the rotation was first observed by Faraday, the phenomenon is called *Faraday rotation.*

8.10. Small Field Approximation

If we assume that there is a linear relationship between μ and κ, and the static magnetic field below saturation, then there is a simple approximation which shows that rotation is linearly proportional to the magnetic field.

Assume that the field is much smaller than that required for resonance, hence

$$\gamma H_0 \ll \omega; \qquad \frac{\gamma M_0}{\mu_0} \ll \omega$$

Then from eqns. (8.6) and (8.9)

$$\mu \approx \mu_0; \qquad \kappa \approx -\frac{\gamma M_0}{\omega}$$

The phase constant is given by

$$\beta^{\pm} \approx \omega\sqrt{(\varepsilon\mu_0)}\left(1 \pm \frac{\gamma M_0}{2\omega\mu_0}\right)$$

and the rotation by

$$\psi \approx \tfrac{1}{2}\gamma M_0\sqrt{\left(\frac{\varepsilon}{\mu_0}\right)} \tag{8.36}$$

As for low fields, M_0 is proportional to H_0, the rotation is proportional to the field. This direct relationship between magnetic field and rotation for low fields is seen in Fig. 8.7. An infinite block of ferrite is a practical impossibility but it will be seen that the fields at the centre of circular waveguide for the TE_{11}-mode approximate to a plane wave. Hence a narrow rod at the centre of the waveguide and magnetized in the direction of propagation might be expected to obey this simple relationship.

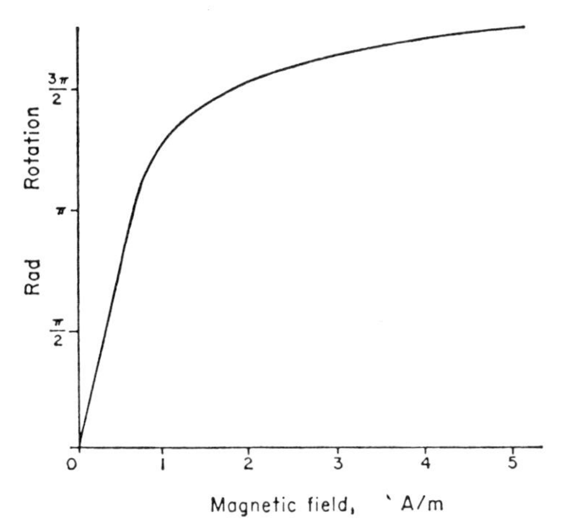

FIG. 8.7. Change of rotation of the linearly polarized TE_{11}-mode in 22.8 mm (0.9 in.) diameter circular waveguide, due to a 6.25 mm (0.25 in.) diameter rod of ferrite 50.8 mm (2 in.) long, against magnetic field at 9.37 GHz ($\lambda_0 = 32$ mm).

8.11. Ferrite in Waveguide

There are a very large number of different ways in which ferrite material may be incorporated into a length of waveguide in order to make practical microwave devices, and a few will be mentioned here in order to outline the principles on which the devices operate. First consider a rotation device which consists of a ferrite rod axially situated at the centre of circular

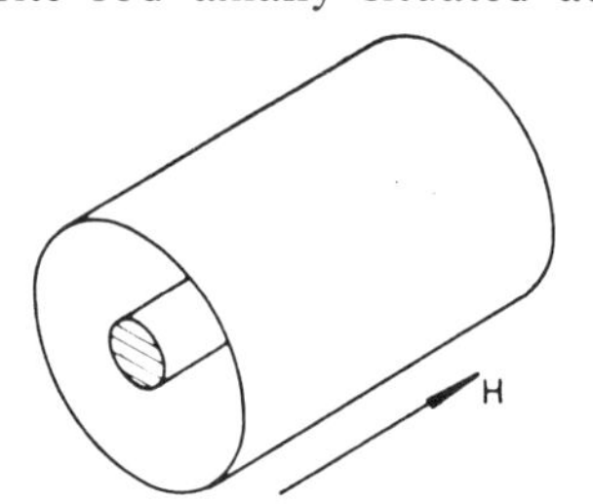

FIG. 8.8. Ferrite and Faraday rotator.

waveguide as shown in Fig. 8.8. It may be seen from Fig. 6.5 that the fields at the centre of circular waveguide for the TE_{11}-mode approximate to those of a plane wave. Hence a ferrite rod magnetized in the direction of propagation can cause rotation of the plane of polarization of a linearly polarized TE_{11}-mode in circular waveguide. Such a device is called a *rotator*. For the circularly polarized TE_{11}-modes in circular waveguide, the effective permeability of the ferrite is approximately μ^+ and μ^- as given in Fig. 8.5. The ferrite can be used to provide variable phase change or resonance absorption depending on the strength of the static magnetic field.

For the dominant TE_{10}-mode in rectangular waveguide, there is a plane at a distance about a quarter of the way across the waveguide where the magnetic field is circularly polarized in the plane of the broad face of the waveguide. This may be deduced from Fig. 5.3, but is clearer in Fig. 8.9, which shows a plot of the magnetic field in the plane of the broad face of the waveguide. The size of the arrow denotes the strength of the field. It is seen that the magnetic field is circularly polarized in the plane of the plot along the line *AA* and is circularly polarized with the opposite hand along the line *BB*.

If a slab of ferrite is placed in the waveguide at the position of circular polarization and is magnetized perpendicular to the broad face of the waveguide as shown in Fig. 8.10, the effective permeability of the ferrite is also that given in Fig. 8.5. The magnetic field is circularly polarized with the opposite hand of rotation on the opposite side of the waveguide, so that two slabs of ferrite magnetized as shown in Fig. 8.11(a) have twice the effect of the single ferrite slab shown in Fig. 8.10. Further study of the field patterns of the waveguide mode shows that, if a forward flowing wave presents a positive circularly polarized magnetic field to the ferrite, a wave flowing in the reverse direction presents a negative circularly polarized magnetic field

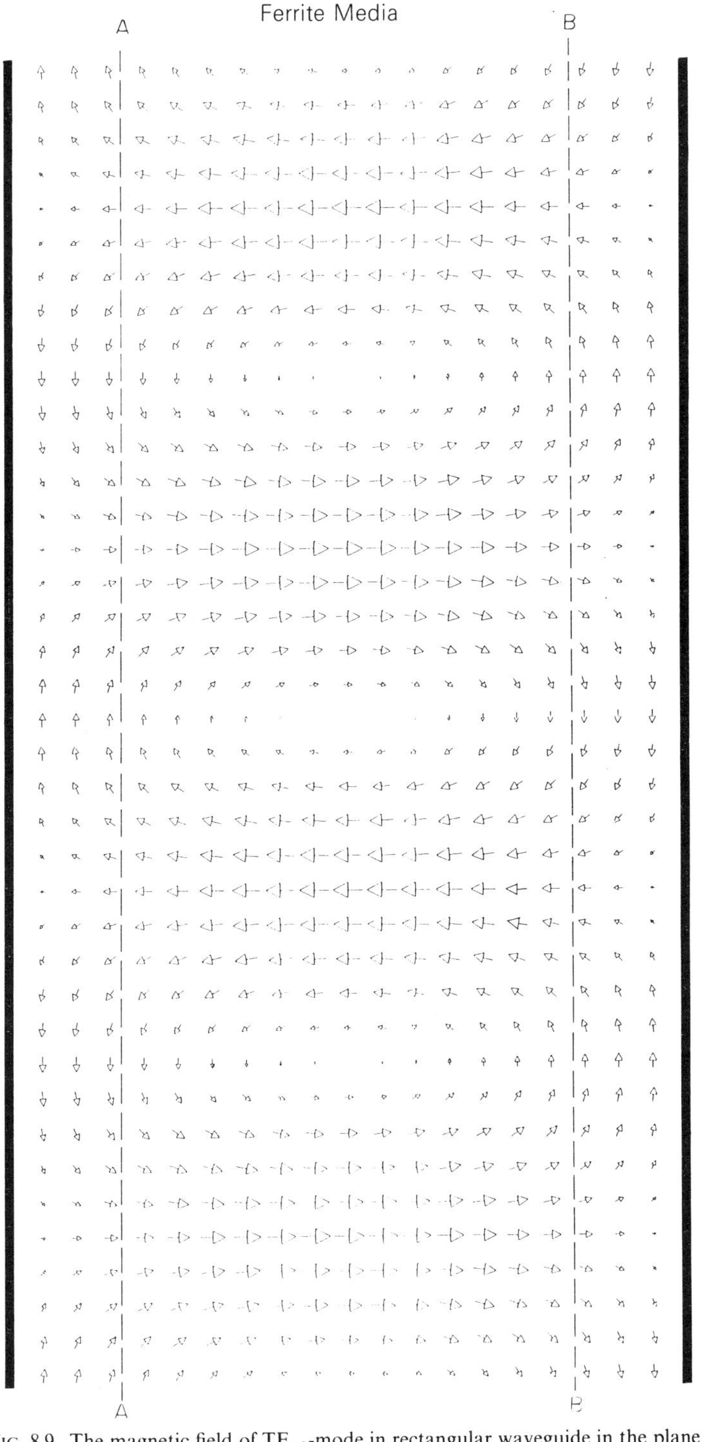

FIG. 8.9. The magnetic field of TE_{10}-mode in rectangular waveguide in the plane of the broadface of the waveguide. *AA* and *BB* are the planes where the magnetic field is circularly polarized.

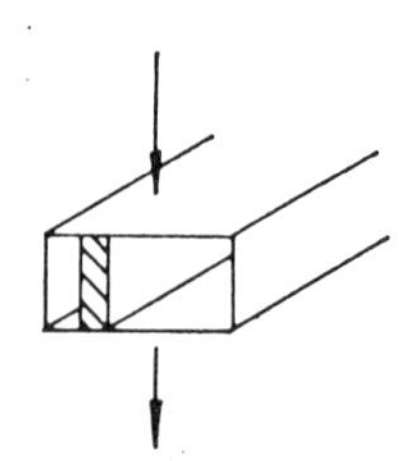

FIG. 8.10. Transversely magnetized ferrite slab in rectangular waveguide.

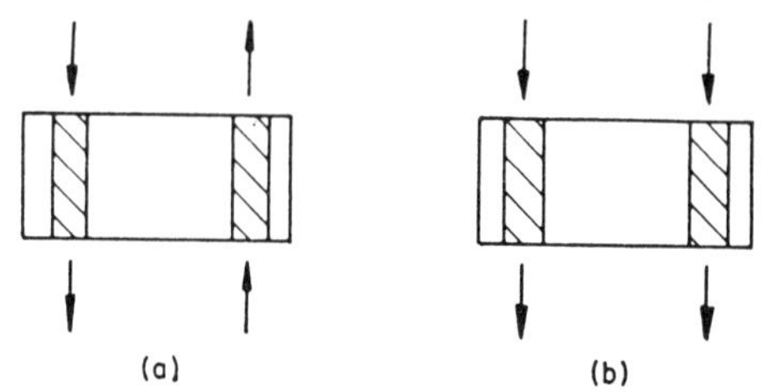

FIG. 8.11. Ferrite phase changer. (a) Nonreciprocal. (b) Reciprocal.

to the ferrite. The effective permeability of the ferrite to the forward and the reverse waves is μ^+ and μ^- respectively and the two waves are affected differently. This provides another nonreciprocal device.

Provided the ferrite is magnetized well below resonance, the rectangular waveguide ferrite device acts as a nonreciprocal phase changer since the electrical length of the ferrite device is different from an equivalent length of rectangular waveguide. If the ferrite is magnetized to resonance, a forward wave is absorbed and attenuated by the ferrite whereas a reverse wave is relatively unattenuated. This makes a device called an Isolator which is described in section 13.6. If a reciprocal ferrite phase changer is required, the two slabs may be magnetized as shown in Fig. 8.11 (b). Here the effective permeability of each slab is different for either a forward or reverse wave but the total effect is the same for each wave.

8.12. Summary

8.1. An electron may be considered as a ***spinning magnetic top***.

8.2. If disturbed, it will ***precess*** about the equilibrium position.

8.3. ***Resonance absorption*** occurs when an electromagnetic field causes a forcing disturbance at the precession frequency.

8.4. The magnetization equation is

$$\frac{d\boldsymbol{M}}{dt} = \gamma \boldsymbol{M} \times \boldsymbol{H} \tag{8.4}$$

8.5. If the ferrite is statically magnetized in the z-direction, its microwave ***permeability tensor*** is given by the relationship

$$B = \begin{vmatrix} \mu & j\kappa & 0 \\ -j\kappa & \mu & 0 \\ 0 & 0 & \mu_0 \end{vmatrix} H \tag{8.11}$$

where

$$\mu = \frac{\gamma^2 H_0 B_0 - \omega^2 \mu_0}{\gamma^2 H_0^2 - \omega^2} = \mu_0 \left(1 + \frac{\omega_0 \omega_m}{\omega_0^2 - \omega^2}\right) \tag{8.9}$$

and

$$\kappa = -\frac{\omega \gamma M_0}{\gamma^2 H_0^2 - \omega^2} = -\mu_0 \frac{\omega \omega_m}{\omega_0^2 - \omega^2} \tag{8.6}$$

8.6. For a plane wave propagating in the z-direction through statically magnetized ferrite

$$\beta^2 = \omega^2 \varepsilon(\mu \pm \kappa) \tag{8.17}$$

leading to two possible solutions

$$\beta^+ = \omega\sqrt{[\varepsilon(\mu + \kappa)]} \tag{8.18}$$

$$\beta^- = \omega\sqrt{[\varepsilon(\mu - \kappa)]} \tag{8.19}$$

which apply to two circularly polarized waves of opposite hand.

8.7. The effective permeability of a ferrite material is given in Fig. 8.5.

8.8. In cylindrical polar coordinates, β^+ is associated with the solution $n = 1$ and β^- is associated with the solution $n = -1$. The circularly polarized waves are the fundamental modes of propagation in magnetized ferrite.

8.9. ***Faraday rotation*** is the rotation of the plane of polarization of a linearly polarized wave caused by the different propagation constants of the two circularly polarized waves.

Rotation per unit length $\psi = \frac{1}{2}(\beta^+ - \beta^-)$ (8.35)

8.10. For small values of the static magnetic field,

$$\psi = \tfrac{1}{2}\gamma M_0 \sqrt{\left(\frac{\varepsilon}{\mu_0}\right)} \tag{8.36}$$

8.11. A ***rotator*** consists of a length of ferrite in circular waveguide magnetized so as to cause rotation of an incident linearly polarized wave.

Problems

8.1. Work out the units in each of the eqns. (8.1) to (8.4) and check that they are dimensionally correct.

8.2. Check the working and substitutions from eqn. (8.4) to eqn. (8.8).

8.3. A typical microwave ferrite might have the properties: $\gamma = 28\,\text{GHz/T}$, saturation magnetization 0.21 T, $\varepsilon_r = 12$. Calculate and plot values of μ and κ against magnetizing field at 10 GHz up to a field of 0.5 T. Hence plot the effective permeability $\mu \pm \kappa$ against magnetizing field.

8.4. Write down the form of the permeability tensor of a ferrite medium magnetized parallel to the x-axis. Hence write the $\boldsymbol{B}$, $\boldsymbol{H}$ relationship in its component form in rectangular coordinates.

8.5. Following a procedure similar to that given in section 8.6, prove that the propagation constants for a plane wave propagating through an infinite ferrite medium magnetized perpendicular to the direction of propagation are

$$\beta = \omega\sqrt{(\mu_0\varepsilon)} \quad \text{and} \quad \beta = \omega\sqrt{\left[\frac{\varepsilon(\mu^2 - \kappa^2)}{\mu}\right]}$$

8.6. A plane wave in free space is normally incident onto the plane face of a semi-infinite ferrite medium magnetized normally to the plane face. Find an expression for the VSWR of the standing wave in the free space.

$$\left[S = \frac{\eta(\eta^+ + \eta^-) - 2\eta^+\eta^-}{2\eta^2 - \eta(\eta^+ + \eta^-)}\right]$$

8.7. For the ferrite material given in problem 8.3, calculate:

(a) the wavelength of the two hands of circular polarization,
(b) the wavelength of the equivalent linearly polarized wave,
(c) the rotation per unit length,

for a 10 GHz wave for a few different magnetic fields up to 0.3 T and compare the rotation with that calculated using the approximate formula given in eqn. (8.36).

8.8. Repeat the calculations of problems 8.3 and 8.7 for a 1 GHz wave.

8.9. Some microwave ferrite materials exhibit a large power absorption when operated below saturation at low microwave frequencies, called the low-field loss. In the light of the results of problem 8.8, comment on the fact that this ferrite material is not suitable for use at 1 GHz. Could any ferrite devices be devised using this ferrite material at 1 GHz? If so, how?

8.10. It is inconvenient to provide a magnetic flux density greater than about 1.3 T between the poles of a permanent magnet suitable for producing a transverse magnetic field in rectangular waveguide. In the light of this fact, comment on the use of resonance isolators (see section 13.6) at the higher microwave frequencies.

9

Plasma and Electron Beam

9.1. Properties of Plasma

A plasma consists of charged particles which in most cases have been produced by the ionization of a gas. In a gas, a plasma can arise in an electric discharge, when it is usually produced intentionally, or it can arise from excessive heating such as the plasma sheath that surrounds a space vehicle during re-entry into the atmosphere. The plasma in a gas consists of positively charged gas ions and negatively charged free electrons. By definition, a plasma is assumed to be electrically neutral and to consist of an equal density of positive ions and negative electrons. A plasma also exists inside a conductor since by definition every part of a conductor is electrically neutral except possibly at the boundaries. In most conductors, the current density is so high that the theory to be here expounded is not really applicable. In semiconductors, however, the concentration of conductors is similar to the concentration of electrons and ions in ionized gas so that the electromagnetic properties of a plasma to be expounded in this chapter may also be applicable to semiconductor materials.

The theory of plasma is based on certain properties of an ionized gas. The plasma is assumed to consist of an equal number of mobile light electrons and heavy static ions. Because the difference in mass and hence in mobility between the electrons and ions is large, it is assumed that the ions provide a static charged medium and that only the electrons are mobile. If the electrons were completely free to move in the medium without any hindrance, there would be no transfer of energy from the electrons to the surrounding heavy ions and gas molecules. The plasma is said to be lossless. However, there are elastic and inelastic collisions between the electrons and the other particles in the plasma and this causes the electrons to lose some energy. The total loss of energy due to collisions is accounted for in an *effective collision frequency*. It is the equivalent number of collisions, completely stopping the electrons, occurring in unit time that would extract the same total energy from the electrons as happens in practice.

The current density in the plasma is given by the product of the charge and the number of charge carriers and the mean velocity of the charge carriers, so that for the electrons

$$\boldsymbol{J} = -ne\boldsymbol{v} \tag{9.1}$$

where n is the number of electrons in unit volume,
e and m are the electronic charge and mass, and
$\boldsymbol{v}$ is the mean velocity.

If υ is the effective collision frequency, the force impeding the motion due to collisions is the loss of momentum,

$$m\upsilon\boldsymbol{v}$$

Therefore the equation of motion of an electron in the presence of an electromagnetic field is

$$m\frac{d\boldsymbol{v}}{dt} = -m\upsilon\boldsymbol{v} - e(\boldsymbol{E} + \boldsymbol{v} \times \boldsymbol{B}) \tag{9.2}$$

where the terms on the right-hand side of this equation are, the force due to the stopping effect of collisions, and the Lorentz relationship giving the force due to the interaction between a negative electron and the electromagnetic field. For the normal fields in an electromagnetic wave, the force due to the magnetic flux may be neglected compared with the force due to the electric field, so that the last term in eqn. (9.2) may be omitted.

$\boldsymbol{v}$ must be considered to be both a function of space as well as time, hence

$$\frac{d\boldsymbol{v}}{dt} = \frac{\partial \boldsymbol{v}}{\partial z}\frac{dz}{dt} + \frac{\partial \boldsymbol{v}}{\partial t} \tag{9.3}$$

However, dz/dt is the low frequency or steady movement of the electrons through the plasma, which may be assumed to be zero or negligible, therefore

$$\frac{dz}{dt} = 0$$

and

$$\frac{d\boldsymbol{v}}{dt} = \frac{\partial \boldsymbol{v}}{\partial t}$$

Hence if the time-dependence $\exp j\omega t$ is assumed, eqn. (9.2) becomes

$$j\omega m\boldsymbol{v} + m\upsilon\boldsymbol{v} = -e\boldsymbol{E} \tag{9.4}$$

Substituting a value for $\boldsymbol{v}$ given by eqn. (9.1) into eqn. (9.4) gives

$$\boldsymbol{J}(j\omega + \upsilon) = \frac{ne^2}{m}\boldsymbol{E} \tag{9.5}$$

For a lossless plasma, $\upsilon = 0$ and eqn. (9.5) becomes

$$j\omega\boldsymbol{J} = \frac{ne^2}{m}\boldsymbol{E} \tag{9.6}$$

If there are no external fields, or if they are small compared with that generated by the electron movement, the left-hand side of eqn. (2.7) is negligible and

$$\boldsymbol{J} = -j\omega\varepsilon_0\boldsymbol{E} \tag{9.7}$$

Eliminating $\boldsymbol{J}$ between eqns. (9.6) and (9.7) gives

$$\omega^2\varepsilon_0\boldsymbol{E} = \frac{ne^2}{m}\boldsymbol{E}$$

Therefore

$$\omega^2 = \frac{ne^2}{m\varepsilon_0} \tag{9.8}$$

In a neutral plasma under the influence of no external fields, it is found that the electrons have a natural frequency of oscillation given by eqn. (9.8) called the *plasma frequency*, which is given by

$$2\pi f_p = \omega_p = \sqrt{\left(\frac{ne^2}{m\varepsilon_0}\right)} \tag{9.9}$$

9.2. Electromagnetic Properties

The plasma behaves as a conducting medium with a conductivity that is a function of frequency and for the lossless plasma it is a conducting medium where there is no loss of power in the medium. In order to investigate the properties of electromagnetic wave propagation through a plasma, it is necessary to obtain an expression for the effective permittivity of the plasma. Substitution of the plasma frequency into eqn. (9.5) gives

$$\boldsymbol{J} = \frac{\varepsilon_0\omega_p^2}{v + j\omega}\boldsymbol{E} = \frac{\varepsilon_0\omega_p^2}{\omega^2 + v^2}(v - j\omega)\,\boldsymbol{E} \tag{9.10}$$

For propagation in a conducting medium, the effective permittivity has already been defined in eqn. (7.2). Therefore

$$j\omega\varepsilon_{\text{eff}}\boldsymbol{E} = \boldsymbol{J} + j\omega\varepsilon_0\,\boldsymbol{E} \tag{9.11}$$

and substitution from eqn. (9.10) for $\boldsymbol{J}$ gives

$$\varepsilon_{\text{eff}} = \varepsilon_0\left[1 - \frac{\omega_p^2}{\omega^2 + v^2}\right] - j\frac{\varepsilon_0 v\omega_p^2}{\omega(\omega^2 + v^2)} \tag{9.12}$$

or by comparison with eqn. (7.3)

$$\varepsilon'_{\text{eff}} = \varepsilon_0\left[1 - \frac{\omega_p^2}{\omega^2 + v^2}\right] \tag{9.13}$$

$$\omega\varepsilon''_{\text{eff}} = \sigma = \frac{\varepsilon_0 v\omega_p^2}{\omega^2 + v^2} \tag{9.14}$$

Equation (9.12) gives the effective permittivity for a neutral plasma consisting of mobile electrons in a sea of fixed charged ions. It is assumed that the parent gas of the plasma has no magnetic properties to have any effect on an electromagnetic wave so that the other property of the plasma is the permeability constant μ_0.

9.3. Plane Wave in Unmagnetized Plasma

Having defined the properties of the plasma in terms of an effective permittivity, it is now only necessary to proceed with mathematical analysis by a method similar to that used in Chapter 7 for plane wave propagation through a conducting media. Substitution of values for ε' and σ given by eqn. (9.12) into eqn. (7.4) gives

$$\gamma = \alpha + j\beta = j\omega\sqrt{\left\{\mu_0\varepsilon_0\left[\left(1 - \frac{\omega_p^2}{\omega^2 + v^2}\right) - j\frac{v\omega_p^2}{\omega(\omega^2 + v^2)}\right]\right\}} \tag{9.15}$$

A plasma which has a large effective conductivity is going to behave like a conducting medium and the electromagnetic wave will be strongly attenuated in traversing the plasma. The effective conductivity is a function of the effective collision frequency, showing, in another way, that the collision frequency is a measure of the lossiness of the plasma. For a lossless or nearly lossless plasma, $v \ll \omega$ and eqn. (9.15) can be simplified to

$$\gamma = \alpha + j\beta = j\omega\sqrt{\left[\mu_0\varepsilon_0\left(1 - \frac{\omega_p^2}{\omega^2}\right)\right]} \tag{9.16}$$

If $\omega > \omega_p$ there is normal propagation and

$$\beta = \omega\sqrt{\left[\mu_0\varepsilon_0\left(1 - \frac{\omega_p^2}{\omega^2}\right)\right]}; \qquad \alpha = 0 \tag{9.17}$$

If $\omega < \omega_p$ the wave is cut off and

$$\alpha = \omega\sqrt{\left[\mu_0\varepsilon_0\left(\frac{\omega_p^2}{\omega^2} - 1\right)\right]}; \qquad \beta = 0 \tag{9.18}$$

Equation (9.18) means that although a plane wave is propagating through an infinite medium, the wave behaves as if it were inside cut-off waveguide. There is no sinusoidal variation of the field quantities in the propagation direction, but the fields decay exponentially with distance. Hence, this second condition cannot be considered as a propagating medium. The plasma appears to be *cut off*. Further manipulation of eqn. (9.17) shows that this is a cut-off phenomenon and that it affects the propagating condition. If λ is the plane wave wavelength for propagation through the plasma, if λ_0 is the characteristic wavelength of the wave and if λ_p is the characteristic wavelength appropriate to the plasma frequency, eqn. (9.17) becomes

$$\lambda = \frac{\lambda_0}{\sqrt{\left[1 - \left(\frac{\lambda_0}{\lambda_p}\right)^2\right]}}$$

This equation is similar to eqn. (3.4), the equation for the waveguide wavelength, except that the free space wavelength in the plasma has replaced the waveguide wavelength and the plasma wavelength has replaced the cut-off wavelength. Hence, electromagnetic wave propagation through an infinite plasma behaves similarly to electromagnetic propagation along waveguide where the plasma frequency replaces the cut-off frequency of the waveguide.

If $\omega = \omega_p$, a TEM-mode cannot exist at all in the medium; however, there is a longitudinal electric wave, sometimes called a plasma wave, that can exist but this wave will not be discussed here. Apart from the waveguide type propagating conditions, the properties of the plane wave that propagates through a lossless plasma are the same as those of a plane wave propagating through a non-conducting media. The fields are given by eqn. (2.28) except that ε is replaced by its modified value given in eqn. (9.12).

9.4. Magnetically Biased Plasma

If a static magnetic field is applied to a plasma, the plasma becomes electrically anisotropic to electromagnetic waves. This is a similar phenomenon to the electromagnetic properties of magnetically biased ferrites. The equation of motion of an electron in a lossless plasma is given by eqn. (9.2) with $v = 0$, hence

$$m\frac{d\boldsymbol{v}}{dt} = -e(\boldsymbol{E} + \boldsymbol{v} \times \boldsymbol{B}_0) \tag{9.19}$$

where B_0 is the static magnetic flux density. Here the static magnetic field is sufficiently large to contribute to the forces whereas the magnetic field of the electromagnetic wave does not make an appreciable contribution. If it is assumed that the static magnetic field acts in the z-direction in a rectangular system of coordinates and that the time dependence of any varying quantities is $\exp j\omega t$, then eqn. (9.19) becomes

$$\left.\begin{aligned} j\omega v_x &= -\frac{e}{m}E_x - \frac{e}{m}B_0 v_y \\ j\omega v_y &= -\frac{e}{m}E_y + \frac{e}{m}B_0 v_x \\ j\omega v_z &= -\frac{e}{m}E_z \end{aligned}\right\} \quad (9.20)$$

Define a quantity similar to ω_m in eqns. (8.9a) and (8.6a), called the gyrofrequency of the electrons, given by

$$\omega_g = \frac{e}{m}B_0 \quad (9.21)$$

Equations (9.20) may be solved to obtain expressions for the components of the velocity of the electrons in terms of the components of the electric field. The relationships are

$$\left.\begin{aligned} (\omega_g^2 - \omega^2)v_x &= -j\omega\frac{e}{m}E_x + \omega_g\frac{e}{m}E_y \\ (\omega_g^2 - \omega^2)v_y &= -\omega_g\frac{e}{m}E_x - j\omega\frac{e}{m}E_y \\ j\omega v_z &= -\frac{e}{m}E_z \end{aligned}\right\} \quad (9.22)$$

Substituting for ω_p into eqn. (9.1) gives

$$\boldsymbol{J} = -\omega_p^2\varepsilon_0\frac{m}{e}\boldsymbol{v}$$

which when substituted into eqn. (9.22) gives

$$\left.\begin{aligned} J_x &= j\omega\frac{\varepsilon_0\omega_p^2}{\omega_g^2 - \omega^2}E_x - \omega_g\frac{\varepsilon_0\omega_p^2}{\omega_g^2 - \omega^2}E_y \\ J_y &= \omega_g\frac{\varepsilon_0\omega_p^2}{\omega_g^2 - \omega^2}E_x + j\omega\frac{\varepsilon_0\omega_p^2}{\omega_g^2 - \omega^2}E_y \\ J_z &= -j\varepsilon_0\frac{\omega_p^2}{\omega}E_z \end{aligned}\right\} \quad (9.23)$$

9.5. Permittivity Tensor

It is seen from eqn. (9.23) that there is a two-dimensional relationship between $\boldsymbol{J}$ and $\boldsymbol{E}$ which leads to a tensor form for the microwave permittivity which is similar to the permeability tensor of the magnetically biased ferrite material. If $\boldsymbol{\varepsilon}$ is the permittivity tensor, then

$$j\omega\boldsymbol{\varepsilon}\boldsymbol{E} = \boldsymbol{J} + j\omega\varepsilon_0\boldsymbol{E} \tag{9.24}$$

where

$$\boldsymbol{D} = \boldsymbol{\varepsilon}\boldsymbol{E} \tag{9.25}$$

If the elements of $\boldsymbol{\varepsilon}$ are ε_t, ε_z and η_t, eqn. (9.25) may be expanded to give

$$\left.\begin{aligned} D_x &= \varepsilon_t E_x + j\eta_t E_y \\ D_y &= -j\eta_t E_x + \varepsilon_t E_y \\ D_z &= \varepsilon_z E_z \end{aligned}\right\} \tag{9.26}$$

Substitution from eqn. (9.23) into eqn. (9.24) gives the elements of the permittivity tensor to be

$$\varepsilon_t = \varepsilon_0\left(1 - \frac{\omega_p^2}{\omega^2 - \omega_g^2}\right) \tag{9.27}$$

$$\eta_t = -\frac{\varepsilon_0\omega_p^2\omega_g}{\omega(\omega^2 - \omega_g^2)} \tag{9.28}$$

$$\varepsilon_z = \varepsilon_0\left(1 - \frac{\omega_p^2}{\omega^2}\right) \tag{9.29}$$

where in this chapter η_t is the cross-diagonal component of the permittivity tensor defined by eqn. (9.28), and it is not the intrinsic impedance, which quantity is associated with η in the rest of this book.

The permittivity equation, eqn. (9.26), may be written

$$\boldsymbol{D} = \begin{vmatrix} \varepsilon_t & j\eta_t & 0 \\ -j\eta_t & \varepsilon_t & 0 \\ 0 & 0 & \varepsilon_z \end{vmatrix} \boldsymbol{E} \tag{9.30}$$

or the permittivity alone may be given by the tensor

$$\boldsymbol{\varepsilon} = \begin{vmatrix} \varepsilon_t & j\eta_t & 0 \\ -j\eta_t & \varepsilon_t & 0 \\ 0 & 0 & \varepsilon_z \end{vmatrix}$$

The plasma is a gyroelectric material which exhibits the properties of Faraday rotation and resonance. These properties have already been extensively discussed in connection with the properties of ferromagnetic materials in Chapter 8. The gyroelectric properties are due to the circular orbits of the electrons around the axis of the static magnetic field.

9.6. Plane Wave in Magnetized Plasma

We now consider the effect of electromagnetic wave propagation through a magnetically biased plasma. For simplicity, plane wave propagation through an infinite plasma medium is considered, because although this is not a situation that is likely to happen in practice, it gives a useful insight into the properties of plasma. The direction of propagation is the same as the direction of the magnetic field.

The permittivity relationship is given by eqn. (9.26) which is rewritten here

$$\left.\begin{aligned} D_x &= \varepsilon_t E_x + j\eta_t E_y \\ D_y &= -j\eta_t E_x + \varepsilon_t E_y \\ D_z &= \varepsilon_z E_z \end{aligned}\right\} \tag{9.26}$$

Substitution of eqn. (9.26) into eqn. (2.7) gives

$$\left.\begin{aligned} \frac{\partial H_z}{\partial y} - \frac{\partial H_y}{\partial z} &= j\omega\varepsilon_t E_x - \omega\eta_t E_y \\ \frac{\partial H_x}{\partial z} - \frac{\partial H_z}{\partial x} &= \omega\eta_t E_x + j\omega\varepsilon_t E_y \\ \frac{\partial H_y}{\partial x} - \frac{\partial H_x}{\partial y} &= j\omega\varepsilon_z E_z \end{aligned}\right\} \tag{9.31}$$

which is one of Maxwell's curl equations and the other is given by eqn. (2.24). If the conditions for a plane wave propagating in the z-direction

$$\frac{\partial}{\partial x} = \frac{\partial}{\partial y} = 0; \quad \frac{\partial}{\partial z} = -j\beta$$

are substituted into eqns. (9.31) and (2.24), we get

$$\left.\begin{aligned} \beta E_y &= -\omega\mu_0 H_x \\ \beta E_x &= \omega\mu_0 H_y \\ \beta \mathrm{H}_y &= \omega\varepsilon_t E_x + j\omega\eta_t E_y \\ \beta H_x &= j\omega\eta_t E_x - \omega\varepsilon_t E_y \\ H_z &= E_z = 0 \end{aligned}\right\} \tag{9.32}$$

The solution to eqn. (9.32) is obtained by a method similar to that used to find the solutions given by eqn. (8.17). Hence the solutions are similar;

$$\beta^2 = \omega^2\mu_0(\varepsilon_t \pm \eta_t) \tag{9.33}$$

and similar notation is used for the two modes

$$\beta^+ = \omega\sqrt{[\mu_0(\varepsilon_t + \eta_t)]} \tag{9.34}$$

$$\beta^- = \omega\sqrt{[\mu_0(\varepsilon_t - \eta_t)]} \tag{9.35}$$

9.7. Rotation

The two propagating modes are circularly polarized modes of opposite hand similar to the modes which propagate in magnetized ferrite. The rotation per unit length is given by eqn. (8.35) which is rewritten here

$$\psi = \tfrac{1}{2}(\beta^{+} - \beta^{-})$$

If the frequency is high compared with the characteristic frequencies of the plasma, some useful approximations may be made to the previous expressions. The conditions are

$$\omega_g \ll \omega \quad \text{and} \quad \omega_p \ll \omega$$

If these conditions are substituted into eqns. (9.27) and (9.28) then

$$\varepsilon_t \approx \varepsilon_0$$

$$\eta_t \approx -\varepsilon_0 \frac{\omega_p^2 \omega_g}{\omega^3}$$

Substituting these values into eqn. (9.34) and (9.35) gives

$$\beta^{+} = \omega\left(1 - \tfrac{1}{2}\frac{\omega_p^2\omega_g}{\omega^3}\right)\sqrt{(\varepsilon_0\mu_0)}$$

$$\beta^{-} = \omega\left(1 + \tfrac{1}{2}\frac{\omega_p^2\omega_g}{\omega^3}\right)\sqrt{(\varepsilon_0\mu_0)}$$

Therefore

$$\psi = -\tfrac{1}{2}\,\omega_g\left(\frac{\omega_p}{\omega}\right)^2\sqrt{(\varepsilon_0\mu_0)} \tag{9.36}$$

As ω_g is proportional to the biasing magnetic field, eqn. (9.36) shows that the rotation is proportional to the applied field.

A free gas plasma is not used for its gyromagnetic properties in waveguide devices because its properties cannot be controlled sufficiently. In most cases, a plasma from a gas discharge is unstable and cannot be used as a propagating medium. Interest is centred on the properties of plasmas because the ionosphere is a plasma and because satellites become surrounded by a plasma during re-entry into the atmosphere. However, the conductors in a semiconductor form a plasma. In ferrite materials it is impossible to provide a reasonable gyromagnetic effect at the higher microwave frequencies, whereas in a semiconductor the gyroelectric effect can be useful. At high frequencies, waveguide structures are small so that a small slab of semiconductor material is still a reasonable size in the waveguide. Therefore, plasma in a slab of semiconductor material can provide nonreciprocal and control functions in waveguide.

9.8. Electron Beam Dynamics

The electron beam is used in all electron tubes, and at microwave frequencies tubes are useful for generation and amplification. A consideration of the dynamics of an electron beam helps to explain the operation of some of the microwave tubes.

Consider an electron beam which consists of a relatively dense electron region with the electrons travelling under the influence of a static electric field. The properties of the beam are described in terms of the velocity, $\boldsymbol{v}$, charge density, ρ, and current density, $\boldsymbol{J}$. It is assumed that all these quantities do not vary throughout the cross-section of the beam that is being considered but that there are superimposed oscillations of a travelling wave nature along the beam. In so far as it is necessary to specify a coordinate system, let the beam axis be coincident with the z-axis of either a rectangular or a cylindrical coordinate system.

Let the expressions for the beam properties be

$$v_z = v_0 + v_1 \exp j(\omega t - \beta z) \tag{9.37}$$

$$\rho = \rho_0 + \rho_1 \exp j(\omega t - \beta z) \tag{9.38}$$

$$J_z = J_0 + J_1 \exp j(\omega t - \beta z) \tag{9.39}$$

with the transverse components so small that they may be neglected. In all these equations it is assumed that the constant term is much larger than the travelling wave portion. The beam current density is given by

$$J_z = \rho v_z$$

hence

$$J_z = \rho_0 v_0 + (\rho_1 v_0 + \rho_0 v_1) \exp j(\omega t - \beta z) \tag{9.40}$$

where the term which is the product of two small quantities has been neglected. Hence eqns. (9.39) and (9.40) give

$$J_1 = \rho_1 v_0 + \rho_0 v_1 \tag{9.41}$$

The equation of motion of the electrons is similar to eqn. (9.2) neglecting the loss term and any magnetic field

$$\frac{dv_z}{dt} = -\frac{e}{m} E_z \tag{9.42}$$

v_z is given by eqn. (9.3)

$$\frac{dv_z}{dt} = \frac{\partial v_z}{\partial z}\frac{dz}{dt} + \frac{\partial v_z}{\partial t} \tag{9.43}$$

and differentiating eqn. (9.37) gives

$$\frac{\partial v_z}{\partial z} = -j\beta v_1 \exp j(\omega t - \beta z)$$

$$\frac{\partial v_z}{\partial t} = j\omega v_1 \exp j(\omega t - \beta z)$$

$$\frac{dz}{dt} = v_0$$

whence substituting into eqns. (9.42) and (9.43) and simplifying

$$\frac{e}{m} E_z = -jv_1 v_0 \left(\frac{\omega}{v_0} - \beta\right) \exp j(\omega t - \beta z) \tag{9.44}$$

the laws of conservation of charge give

$$\nabla . \boldsymbol{J} = -\frac{\partial \rho}{\partial t} \tag{9.45}$$

or considering only the one-dimensional divergence of the beam current density,

$$\frac{\partial J_z}{\partial z} = -\frac{\partial \rho}{\partial t} \tag{9.46}$$

Differentiating eqns. (9.38) and (9.39) gives

$$j\beta J_1 = j\omega \rho_1 \tag{9.47}$$

From eqn. (9.41)

$$v_1 = \frac{J_1 - \rho_1 v_0}{\rho_0}$$

which with substitution for ρ_1 from eqn. (9.47) gives

$$v_1 = \frac{J_1}{\rho_0}\left(1 - \frac{\beta v_0}{\omega}\right) \tag{9.48}$$

Substitution for v_1 from eqn. (9.48) into eqn. (9.44) gives

$$E_z = -j\frac{mv_0^2 J_1}{e\rho_0 \omega}\left(\beta - \frac{\omega}{v_0}\right)^2 \exp j(\omega t - \beta z) \tag{9.49}$$

9.9. Beam Current Wave

It is assumed that, at the centre of the beam, the electromagnetic fields approximate to a plane wave. Then writing Maxwell's curl equations for a plane wave (similar to eqn. (9.32) except that now a current term is required):

$$\left.\begin{aligned}\beta E_y &= -\omega\mu_0 H_x\\ \beta E_x &= \omega\mu_0 H_y\end{aligned}\right\} \tag{9.50}$$

$$0 = H_z \tag{9.51}$$

$$\left.\begin{aligned}j\beta H_y &= j\omega\varepsilon E_x + J_x\\ j\beta H_x &= -j\omega\varepsilon E_y - J_y\end{aligned}\right\} \tag{9.52}$$

$$0 = j\omega\varepsilon E_z + J_z \tag{9.53}$$

The assumption has already been made that the electron beam is uniform in cross-section. Hence there is no flow of current perpendicular to the beam and $J_x = J_y = 0$. Equations (9.50) to (9.52) now comprise a complete set of field equations for the normal plane wave propagating in free space. However, eqn. (9.53) gives the characteristic equation for the beam current wave. Substitution from eqns. (9.39) and (9.49) into eqn. (9.53) gives

$$\left[\frac{\omega\varepsilon_0 m v_0^2 J_1}{e\rho_0\omega}\left(\beta - \frac{\omega}{v_0}\right)^2 + J_1\right]\exp j(\omega t - \beta z) = 0$$

which simplifies to

$$\left[\left(\frac{v_0}{\omega_p}\right)^2\left(\beta - \frac{\omega}{v_0}\right)^2 - 1\right]J_1 \exp j(\omega t - \beta z) = 0 \tag{9.54}$$

The characteristic relationship for the beam current wave is given by a solution to eqn. (9.54).

$$J_1 \exp j(\omega t - \beta z) = 0$$

is a trivial solution and the characteristic equation is

$$\left(\beta - \frac{\omega}{v_0}\right)^2 - \left(\frac{\omega_p}{v_0}\right)^2 = 0 \tag{9.55}$$

Therefore

$$\beta - \frac{\omega}{v_0} = \pm\frac{\omega_p}{v_0}$$

or

$$\beta = \frac{\omega \pm \omega_p}{v_0} \tag{9.56}$$

Equation (9.56) represents two waves of current and space charge density on the electron beam that exist irrespective of any external excitation of the beam. They are termed the slow wave and the fast wave. In travelling wave type electron tubes, these electron-beam waves interact with the slow travelling electromagnetic waves in a slow wave structure.

9.10. Summary

9.1. ***Plasma*** consists of charged particles, mobile light electrons and relatively stationary heavy ions.

Effective collision frequency v

$$\textbf{\textit{Plasma frequency}} \quad \omega_p = \sqrt{\left(\frac{ne^2}{m\varepsilon_0}\right)} \tag{9.9}$$

9.2. Components of the effective permittivity

$$\varepsilon'_{\text{eff}} = \varepsilon_0\left[1 - \frac{\omega_p^2}{\omega^2 + v^2}\right] \tag{9.13}$$

$$\omega\varepsilon''_{\text{eff}} = \sigma = \frac{\varepsilon_0 v \omega_p^2}{\omega^2 + v^2} \tag{9.14}$$

9.3. For plane wave propagation in a lossless ($v \ll \omega$) plasma

$$\gamma = j\omega\sqrt{\left[\mu_0\varepsilon_0\left(1 - \frac{\omega_p^2}{\omega^2}\right)\right]} \tag{9.16}$$

If $\omega > \omega_p$ there is normal propagation

$$\beta = \omega\sqrt{\left[\mu_0\varepsilon_0\left(1 - \frac{\omega_p^2}{\omega^2}\right)\right]}; \qquad \alpha = 0 \tag{9.17}$$

but if $\omega < \omega_p$ the wave is cut off

$$\alpha = \omega\sqrt{\left[\mu_0\varepsilon_0\left(\frac{\omega_p^2}{\omega^2} - 1\right)\right]}; \qquad \beta = 0 \tag{9.18}$$

9.4. The magnetically biased plasma has gyroelectric properties.

$$\text{Gyrofrequency} \qquad \omega_g = \frac{e}{m}B_0 \tag{9.21}$$

9.5. The ***permittivity tensor*** is given by the relationship

$$\boldsymbol{D} = \begin{vmatrix} \varepsilon_t & j\eta_t & 0 \\ -j\eta_t & \varepsilon_t & 0 \\ 0 & 0 & \varepsilon_z \end{vmatrix} \boldsymbol{E} \tag{9.30}$$

where

$$\varepsilon_t = \varepsilon_0\left(1 - \frac{\omega_p^2}{\omega^2 - \omega_g^2}\right) \tag{9.27}$$

$$\eta_t = -\frac{\varepsilon_0\omega_p^2\omega_g}{\omega(\omega^2 - \omega_g^2)} \tag{9.28}$$

$$\varepsilon_z = \varepsilon_0\left(1 - \frac{\omega_p^2}{\omega^2}\right) \tag{9.29}$$

9.6. The permittivity tensor gives rise to nonreciprocal properties and two possible modes of propagation associated with the two hands of circular polarization. The propagation constants are

$$\beta^+ = \omega\sqrt{[\mu_0(\varepsilon_t + \eta_t)]} \tag{9.34}$$

$$\beta^- = \omega\sqrt{[\mu_0(\varepsilon_t - \eta_t)]} \tag{9.35}$$

9.7. Rotation of the plane of polarization of a linearly polarized wave occurs, and at high frequencies the rotation per unit length is given by

$$\psi = -\tfrac{1}{2}\,\omega_g\left(\frac{\omega_p}{\omega}\right)^2\sqrt{(\varepsilon_0\mu_0)} \tag{9.36}$$

9.9. In an electron beam with a mean speed v_0 there are two waves of current and space charge density whose propagation constants are given by

$$\beta = \frac{\omega \pm \omega_p}{v_0} \tag{9.56}$$

Problems

9.1. Calculate the plasma frequency of an electron beam having an electron density of $1.2 \times 10^{14}\,\mathrm{m}^{-3}$.

Calculate the plasma frequency of the ionosphere with an electron density of $1.12 \times 10^{11}\,\mathrm{m}^{-3}$.

Calculate the gyrofrequency of the ionosphere with the earth's magnetic field strength of $50\,\mu\mathrm{T}$. [109 MHz, 3.0 MHz, 1.4 MHz]

9.2. Substitute units for each of the quantities and prove that eqns. (9.9) and (9.21) defining the plasma frequency and gyrofrequency of the electrons are correct dimensionally.

9.3. Substitute eqn. (9.12) into eqns. (2.1) to (2.7) and prove eqn. (9.15) from first principles.

9.4. Obtain expressions for α and β of the lossy plasma similar to those given in eqns. (7.5) and (7.6) and show that these expressions simplify to eqns. (9.17) and (9.18) for the lossless condition

9.5. In the light of microwave theory, discuss the fact that the ionosphere appears to be a perfect reflector of radio waves at low frequencies whereas at high frequencies it is perfectly transparent.

9.6. Calculate the electron density necessary to give a TEM-mode wavelength 1 per cent longer than the characteristic wavelength in a uniform lossless plasma at 100 MHz and 10 GHz. [$2.5 \times 10^{12}\,\mathrm{m}^{-3}$, $2.5 \times 10^{16}\,\mathrm{m}^{-3}$]

9.7. Discuss whether a uniform plasma can be considered to be exactly similar to a uniform dielectric material of relative permittivity less than one. In particular consider whether the group velocity is the same as the phase velocity. Plot the relationship between plane-wave wavelength and frequency and compare it with the relationship between waveguide wavelength and frequency for air-filled waveguide.

9.8. Solve eqns. (2.24) and (9.31) and show that eqn. (9.33) gives the correct solution for the propagation constant.

9.9. Using the properties of the ionosphere as specified in problem 9.1, calculate the rotation of the plane of polarization of a microwave signal at 1 GHz propagating through the ionosphere. Assume that the ionosphere consists of a uniform region 200 km thick, and that the earth's magnetic field acts parallel to the direction of propagation.

Repeat the calculation for a 10 GHz signal. [3°, 0.03°]

9.10. Plot the variation of ε_t and η_t against magnetic field and hence the effective permittivity to the two hands of circular polarization for a magnetically biased plasma where $\omega_p = 0.9\,\omega$ and the direction of propagation is the same as the direction of the static magnetic field.

10

Oscillators and Amplifiers

10.1. Introduction

At the lower microwave frequencies, conventional lumped circuit oscillators and amplifiers can be made without the need to use special microwave techniques in their design. However, in the rest of the microwave frequency range, conventionally designed oscillators and amplifiers do not work. In *p–n* junction diodes and transistors, the transit time of the charge carriers in the active region is greater than the period of the microwave signal, so that no rectification or amplification can occur, unless the active region is made very thin. Various rectifying diodes especially developed for use at microwave frequencies are described in section 10.2. The very small clearances possible between the electrodes of the interdigital planar FET, as shown in Fig. 10.1, do enable transit-time effects to be reduced, so that high frequencies of operation are possible. Any associated circuit is designed using transmission

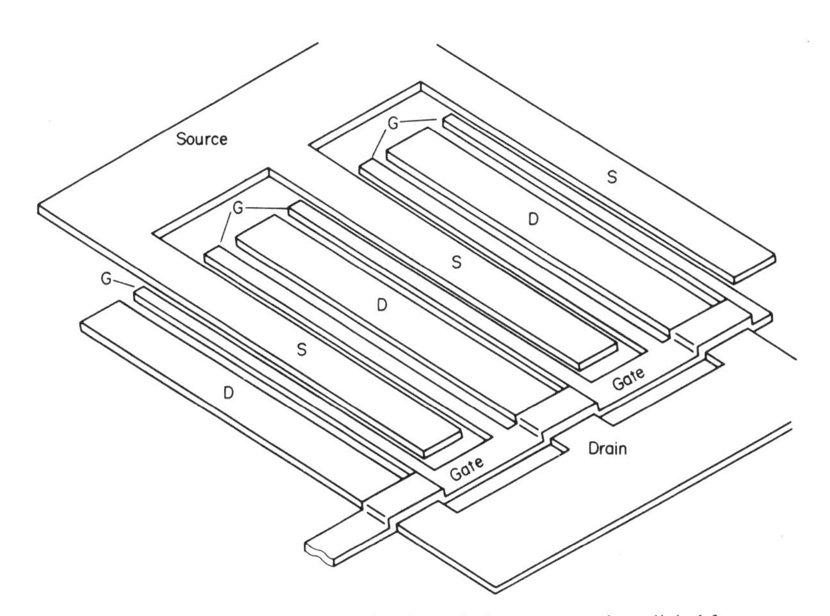

FIG. 10.1. Diagram showing the electrode layout on an interdigital f.e.t.

line theory and the microwave circuit techniques described in Chapter 4. The other devices designed for operation at microwave frequencies make use of the transit time of the conductors in the active region to provide oscillation or amplification. Most of the devices described in this chapter are transit time devices.

Because the active region of microwave devices is very small, the devices have limited power handling capability. As the frequency of operation is raised, so the power capability is reduced. For high power, either a number of semiconductor devices are operated in parallel using combining circuits or multielement antennas, or electron tubes must be used. The three commonest types of electron tubes for use at microwave frequencies are described at the end of this chapter. Again at microwave frequencies, conventional triodes and other electronic tubes becomes useless. The transit time of the electrons between the electrodes is longer than the period of the oscillations that are to be amplified. Specially constructed triodes, having a very close spacing between the electrodes, are used at the lower microwave frequencies but the other microwave electronic tubes utilize the transit time of the electrons between the electrodes.

10.2. Diodes

In this section we describe briefly some of the characteristics of the simple semiconductor diode that has been developed for microwave use, starting with the rectifier diode that can be used either as a detector or as a mixer.

The point-contact diode has seen continuous use since about 1940. It consists of a metal semi-conductor Schottky-barrier junction formed by the sharpened tip of a tungsten wire making a pressure contact onto the surface of a piece of semiconductor material. It is similar to the cat's-whisker detector that was used in the radio sets of the 1920s. The effective thickness of a metal-semiconductor junction can be extremely small so that the transit time is smaller than the period of any microwave signal. A thin barrier gives rise to a relatively large shunt capacitance unless the junction area is also made small. This capacitance provides a path for the microwave signal shunting the rectifying barrier so that point-contact structures are used to give a very small area of junction. The d.c. rectified current output is proportional to the square of the r.f. voltage across the diode so that the current output is proportional to the r.f. power. The point-contact diode uses relatively low resistance semiconductor material so that even at small power levels the device provides low impedance rectification with a good square law characteristic. This square law characteristic also leads to good performance as a mixer and the point-contact diode can be designed to give low noise performance. However, it is not very robust because the point contact is maintained by pressure, and is also particularly susceptible to destruction through excessive r.f. power.

The Schottky-barrier diode. Planar techniques developed for the manufacture of transistors can be used to provide metal semiconductor rectifying junctions of very small area so that the effect of the junction capacitance does not swamp the rectifying action at microwave frequencies. Mechanically the junction is more robust than the point contact and the performance of devices is more consistent. The Schottky-barrier metal silicon diode makes a very good mixer diode and a rectifier diode for relatively high power levels. Although in a point-contact diode the metal-semiconductor junction is a Schottky-barrier junction, the term *Schottky-barrier diode* has only been used for devices where the metal forming the junction is deposited onto the semiconductor material. Epitaxial techniques provide a film of high resistivity material for the junction on a low resistivity substrate. This gives the Schottky-barrier diode a higher junction impedance but a lower series resistance than the point-contact diode. For detection at low power levels, the impedance is unacceptably high and it becomes necessary to apply a d.c. bias current in the forward direction so that the diode is operating on a lower impedance part of its characteristic. Careful choice of junction materials can reduce the potential barrier to provide diodes suitable for use without bias, hence the *zero-bias Schottky-barrier diode.*

The backward diode. High doping levels in the semiconductor at the junction of a p–n junction diode give a situation where near zero the reverse current is larger than the forward current and a good square law characteristic is obtained for rectification in the reverse direction. The backward diode is suitable for use at low power levels and provides a very low impedance junction. Used as detectors, they have high sensitivity for zero bias.

The varactor diode. In a reverse biased p–n diode, the junction capacitance is a function of the bias voltage. The device can be used as a variable capacitance in a tuned circuit. It is used to provide electronically tunable microwave oscillators. It behaves as a low-loss voltage-dependent capacitor.

The capacitance of a reverse biased semiconductor p–n junction diode is given by the expression

$$C = \frac{C_0}{\left(1 + \dfrac{V}{\phi}\right)^m} \tag{10.1}$$

where C_0 is the capacitance at zero bias,

ϕ is the diffusion voltage which is the voltage developed across the junction in the absence of any applied voltage,

m depends on the grading of the doping in the semiconductor, being $\frac{1}{2}$ for an abrupt junction and less than this for other dopings down to $\frac{1}{3}$ for a linearly graded junction, and

V is the voltage across the junction.

The PIN diode. The impedance of a *p–i–n* semiconductor junction diode depends on the bias current for frequencies above 0.1 GHz. The PIN diode, as it is called, is a silicon junction diode whose *p* and *n* regions are separated by a layer of intrinsic *i* semiconductor. At frequencies below 0.1 GHz the PIN diode rectifies in the same way as a simple junction diode. However, at higher frequencies rectification ceases to occur due to the stored charge in the intrinsic layer and the diode conducts in both directions. Its effective resistance is inversely proportional to the amount of charge in the layer. An increase in forward bias current increases the stored charge and decreases the effective resistance of the diode. There is an approximately inverse relationship with current,

$$R \approx \frac{K}{I} \tag{10.2}$$

where K is a constant. With reverse bias, the stored charge is depleted and the effective resistance becomes a maximum. In a microwave circuit, the PIN diode acts as a variable resistance but it is most often used as a switch. With full forward bias the diode appears to be a short circuit and with negative bias the diode appears to be an open circuit to microwave signals.

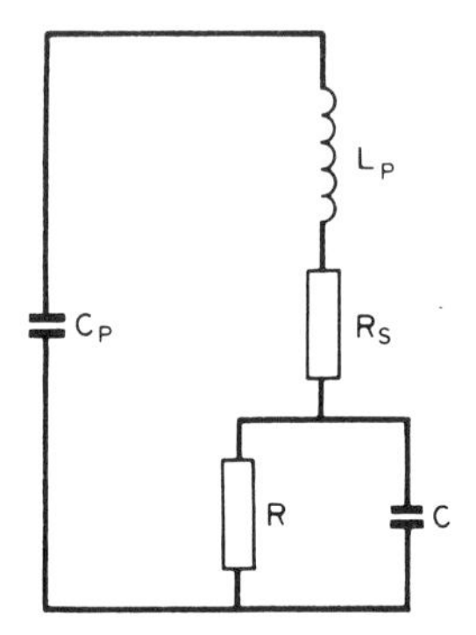

FIG. 10.2. Semiconductor diode equivalent circuit.

The diode equivalent circuit. An equivalent circuit for a *p–n* or *p–i–n* junction diode is shown in Fig. 10.2. The junction is shown as a resistance in parallel with the junction capacitance. For a PIN-diode the junction acts as a variable resistance with approximately a fixed junction capacitance and for a varactor diode the junction is a variable capacitance in parallel with a large resistance. R_s is the series resistance of the bulk semiconductor adjacent to the junction. L_p is the lead inductance which has to be considered because at microwave frequencies any short length of wire has an appreciable impedance. C_p is the case capacitance. The subscript p is used to denote the effects of the package. In some miniature circuits the semiconductor chips are used unencapsulated so that there will be no case capacitance but there is probably a bonding lead whose inductance needs to be considered.

10.3. Transistor oscillator or amplifier

Using epitaxial planar techniques, the width of the base region in transistors has been reduced and transistors can be made with useful gain at microwave frequencies. In order to provide useful power in devices with very narrow base regions, the electrodes have an interdigital layout similar to that shown in Fig. 10.1. The bipolar transistor operates up to about 10 GHz. The FET has been found preferable for higher frequencies. Because the electron mobility in gallium arsenide material (GaAs) is about five times that in silicon, the FET constructed in GaAs is found to give the best performance. The transistor, usually in the form of an unencapsulated chip, is mounted in a miniature microstrip circuit to make an oscillator or amplifier. Alternatively, the circuit components or transmission lines may be constructed with the transistors on a silicon or GaAs substrate to make an integrated circuit. An artist's impression of various components in an integrated circuit on a GaAs substrate is shown in Fig. 10.3.

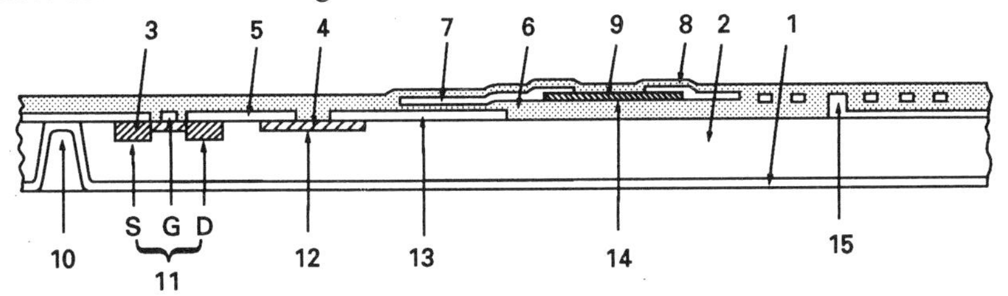

FIG. 10.3. Diagram of some integrated circuit components on a gallium arsenide substrate. (1) Ground level metallization, (2) semi-insulating GaAs substrate, (3) $n+$ implant, (4) n active region, (5) first level metallization, (6) interlayer dielectric, (7) second level metallization, (8) protection and passivation dielectric, (9) thin film resistance material, (10) through substrate via hole connection to earth, (11) GaAs FET with central gate finger, (12) GaAs resistor, (13) overlay capacitor, (14) thin-film resistor, (15) spiral inductor using first level metallization to connect to the centre of the spiral.

A transistor amplifier using transmission lines and matching stubs such as that shown in Fig. 4.11 would be constructed in the form of a miniature microstrip circuit using unencapsulated chips bonded to the circuit for the transistors and diodes. A transistor amplifier using miniature circuit components such as that shown in Fig. 4.13 would be constructed in one unit on a semiconductor substrate as an integrated circuit.

A transistor oscillator needs some tuning mechanism in the circuit. All the usual low frequency oscillator circuits may be realised at microwave frequencies but they are not usually the most convenient. The simple LC tuned circuit is simple to realise at microwave frequencies. A lumped component parallel tuned circuit in microstrip is shown in Fig. 4.4. However, any lead or strap connecting the semiconductor chip to the microstrip line often has sufficient inductance which may be tuned by the variable capacitance provided by a varactor diode. Alternatively, the oscillator may be tuned at a fixed frequency by a microwave tuned cavity as described in section 6.13. A small disk of high-permittivity low-loss material will oscillate in the TM_{010}-mode if it is inserted into the substrate near to a microstrip line. The microwave field in the cavity couples to that of the line and can be used to control the frequency of oscillation of an oscillator. A similar magnetically tunable resonant cavity is the ferrite resonant cavity described in section 13.9 which can also be used to control the frequency of a transistor oscillator. Only the varactor diode is suitable to be part of an integrated circuit.

The economics of integrated circuit production mean that they are particularly advantageous when vast quantities of the same circuit are required, such as the microwave stage of a satellite TV receiver or the components of a phased array radar. For a receiver, all the microwave functions may be integrated into a single chip. A receiver chip might consist of a low-noise microwave amplifier, a microwave local oscillator, a mixer and intermediate frequency amplifier. One element in a multielement phased array radar might consist of a variable frequency oscillator, complicated phase control circuits and a power amplifier.

The rest of this chapter is devoted to devices having properties peculiar to the microwave frequency range.

10.4. Avalanche Oscillator

Different names have been given to the variations of the avalanche oscillator but the most general name is the IMPATT diode. The name is derived from the initial letters of IMPact Avalanche Transit Time. The IMPATT diode is a specially doped *p–n* junction diode which is biased in the reverse direction so that a very high electric field intensity exists across a narrow region at the junction. Avalanche breakdown occurs in this narrow region and causes oscillation in the microwave range of frequencies. Avalanche diodes need a supply of about 70 V from a constant current d.c. source. They are mostly used when maximum power is required.

The electric field intensity across a fully depleted *p–n* junction in an IMPATT diode is shown in Fig. 10.4. The *p*-region is very heavily doped so that the *p* depletion region is very narrow and is often ignored. The electrical field intensity rises to a peak at the junction, and as the voltage across the diode is increased, avalanche breakdown occurs and limits the maximum electric field intensity. In its active state, the diode consists of a narrow avalanche region at the junction adjacent to a relatively large depletion region through which any charge generated in the avalanche region drifts. If the diode is biased to the limit of avalanche breakdown, the addition of a microwave signal causes packets of charge to be generated in the diode. During the positive half cycle the breakdown voltage is exceeded and charge carriers are generated in the avalanche region. During the negative half cycle, these charge carriers drift through the depletion region and arrive at the

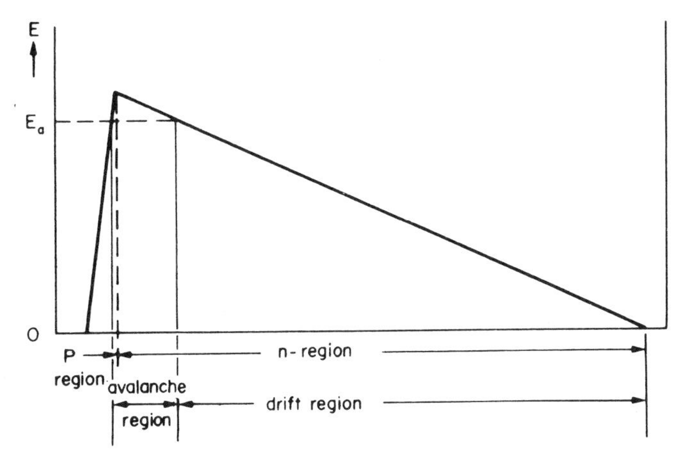

FIG. 10.4. Electric field intensity in an IMPATT diode.

cathode. Due to the avalanche effect, the maximum charge generation occurs at the end of the positive half cycle so it is a quarter cycle out of phase with the voltage. Time in passing through the drift region causes a further quarter cycle delay so that the output current is out of phase with the voltage across the diode, and the device is an oscillator or an amplifier. For small a.c. potentials, the generated current is sinusoidal anti-phase with the voltage. For larger a.c. potentials, there is an exponential increase in the current in the avalanche region; this current pulse is flattened during passage through the drift region and gives a flat topped output current pulse which is anti-phase with the voltage. The IMPATT oscillator operates into the mm-wave region.

The semiconductor diode is used in a resonant circuit or resonant cavity. The operating frequency and the frequency stability of the device are determined by the characteristics of the resonant circuit or cavity. If they are tunable, the diode can be made to oscillate over a wide range of frequencies. A varactor diode or a ferrite resonant cavity can be used to provide electrical control of the oscillator frequency.

10.5. Transferred Electron Oscillator

The transferred electron oscillator is often called the Gunn diode oscillator after the man who first monitored the effect. Most of the semiconductor devices with which the electronics engineer is familiar are junction devices. That is, the useful properties of the device are the properties of a *p–n* junction in the device. The Gunn diode oscillator, however, does not depend on the properties of any junction between two differently doped semiconductor materials but on the properties of the semiconductor material itself. The microwave oscillations occur in the bulk of the semiconductor material rather than in a narrow junction. The material used to make Gunn oscillators is gallium arsenide. It is not the only material to exhibit the required properties but it is the only one to show promise for general use. Gunn oscillations depend on the fact that in some semiconductors conduction electrons can exist in more than one stable state with different mobilities or different effective masses, called the two valley effect. The electron velocity–electric field relationship for the two states is shown in Fig. 10.5.

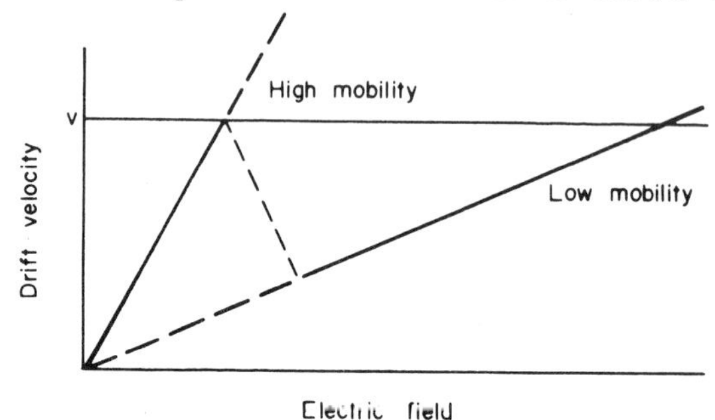

FIG. 10.5. Drift velocity versus electric field in gallium arsenide.

At low applied electric fields, the drift velocity of the electrons increases linearly with electric field. As the electric field is increased above a certain threshold value, the electron drift velocity ceases to increase and some of the electrons transfer to the low mobility state. The low mobility electrons require a larger electric field for the same drift speed than the high mobility electrons so that there is a situation where a short length of the sample has a large electric field gradient while the rest of the sample has a lower electric field gradient. The high field domain travels through the sample at the drift velocity, and collapses at the end of the sample giving rise to the simultaneous generation of another domain at the negative electrode. The sudden collapse of the high electric field domain gives rise to a current pulse at the anode. If the transit time through the specimen is such that these current pulses occur with a frequency in the microwave region, the generation of power at microwave frequencies occurs. The material converts a direct current into microwave radiation. The frequency of the microwave radiation is determined by the dimensions of the specimen. To give oscillations in the microwave region, the specimen must be very thin, $\sim 10^{-5}$ m, so that the potential difference across the oscillator is of the order of a few volts. One method of producing a very thin active layer of gallium arsenide is to deposit a thin layer of the material epitaxially on a thicker slice of the same material. The active element is a high resistivity layer on a low resistivity substrate.

The microwave oscillation occurs due to the pulses of current when the high field domain collapses at the anode. The supply required is about 7 V d.c. so that the transferred electron Gunn oscillator is compatible with microwave transistor circuits and makes battery operated microwave systems feasible.

The semiconductor element may be mounted in a cavity or resonant circuit. Then the oscillation frequency and frequency stability of the device are determined by the characteristics of the resonant circuit. If the resonant circuit is such that there is a large voltage swing across the active region, the p.d. will fall below the threshold voltage during the cycle and a new domain will then fail to be generated until the potential rises again. If the potential falls to such an extent that an existing domain is quenched before it reaches the anode, it immediately gives rise to an output current pulse. Therefore we see that operation is also possible at frequencies well above that governed by the transit time of the high field domain through the active region of the device.

10.6. Parametric Amplification

The parametric amplifier is a device in which energy is transferred to a signal at one frequency f_s from power at another frequency f_p by means of the variation of one of the "parameters" of the system at the frequency f_p. The microwave parametric amplifier is dependent on the use of the varactor diode as a variable reactance.

A simple description of the parametric amplifier may be obtained by referring to Fig. 10.6. The capacitor in this circuit may be varied at will. If an oscillating signal of maximum voltage V is applied to the terminals of the circuit, and the capacitance has a value C, then the maximum charge on the capacitor is given by

$$Q = CV \tag{10.3}$$

and the energy stored in the capacitor is given by

$$W = \tfrac{1}{2}CV^2 = \tfrac{1}{2}QV$$

If the separation between the plates is increased, so that the capacitance is decreased to $C - \delta C$ when the voltage is a maximum irrespective of its polarity, the charge on the plates will remain constant. The voltage across

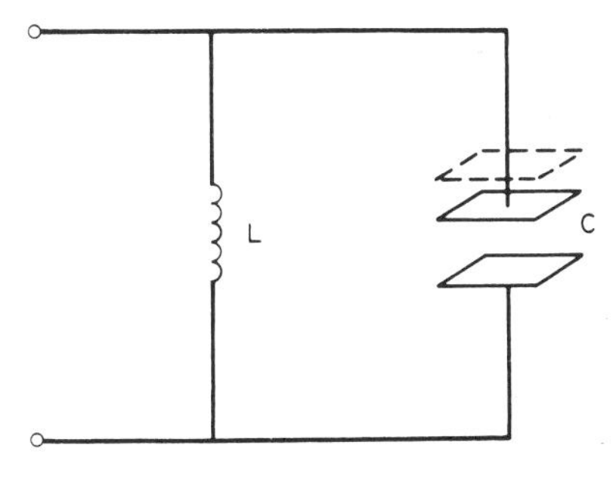

FIG. 10.6. Resonant circuit incorporating a variable capacitor.

the capacitor increases to $V + \delta V$, and

$$Q = (C - \delta C)(V + \delta V) \tag{10.4}$$

Then the stored energy at the frequency of the oscillating signal is increased to

$$(W + \delta W) = \tfrac{1}{2}(C - \delta C)(V + \delta V)^2 = \tfrac{1}{2}Q(V + \delta V) \tag{10.5}$$

The increase in energy is equivalent to the work done in separating the plates against the force of attraction between them. If the plates are returned to their original spacing when the instantaneous stored charge is zero, no work is done and there is no effect on the oscillating electrical signal. If the separation is again made at the next maximum of the signal, further energy can be transferred to the electrical signal. This pumping action of the signal is illustrated in Fig. 10.7.

In practice, the system described in this section is too simple. The capacitance variation is sinusoidal from an electrical signal operating on the voltage-dependent properties of a varactor diode. It is impossible to maintain the exact 2:1 ratio of pump frequency to signal frequency because the amplifier is often required to amplify signals occupying a band of frequencies.

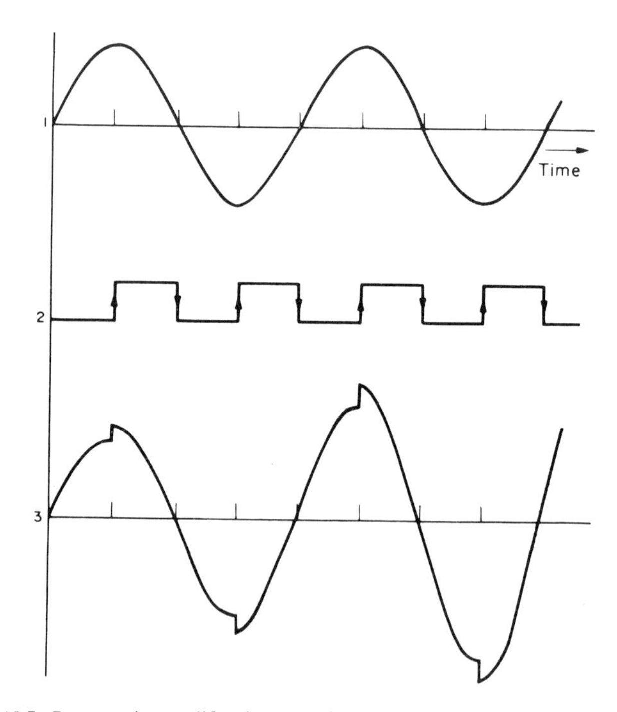

FIG. 10.7. Parametric amplification waveforms with instantaneous variation of capacitance. (1) Applied voltage. (2) Movement of capacitor plate. (3) Amplified voltage.

The varactor diode, however, can be considered to have a non-linear relationship between voltage and capacitance and, if two signals at different frequencies are applied to such a diode, signals are produced at all the sum and difference frequencies involving all the harmonics of the original signal frequencies. The relationships between the powers at all these frequencies are given by the Manley–Rowe equations

$$\sum_{m=0}^{\infty} \sum_{n=-\infty}^{\infty} \frac{mP_{mn}}{m\omega_s + n\omega_p} = 0 \tag{10.6}$$

$$\sum_{m=-\infty}^{\infty} \sum_{n=0}^{\infty} \frac{nP_{mn}}{m\omega_s + n\omega_p} = 0 \tag{10.7}$$

where

ω_s is the angular frequency of the signal,
ω_p is the angular frequency of the pump,
P_{mn} is the flow of power into the varactor at a frequency of $\pm(m\omega_s + n\omega_p)$, and
m and n are harmonic coefficients.

To satisfy the Manley–Rowe equations, if power is entering the varactor at certain frequencies, there must be a flow of power out of the varactor at other frequencies. In most practical amplifiers, the pump is at a higher frequency than the signal and gain is obtainable provided power is allowed to flow at one of the side frequencies. Power flow at all the other harmonic frequencies may be suppressed. The side frequencies usually chosen are the difference frequency, defined by

$$\omega_{p-s} = \omega_p - \omega_s$$

or the sum frequency, defined by

$$\omega_{p+s} = \omega_p + \omega_s$$

Substituting the permissible values of m and n into eqns. (10.6) and (10.7) gives the power relationships for the varactor system in which either the sum or the difference frequency has been an allowed frequency. The relationships become

$$\frac{P_s}{\omega_s} \pm \frac{P_{p\pm s}}{\omega_{p\pm s}} = 0 \tag{10.8}$$

$$\frac{P_p}{\omega_p} + \frac{P_{p\pm s}}{\omega_{p\pm s}} = 0 \tag{10.9}$$

where the plus sign is taken for the sum frequency relationships and the minus sign is taken for the difference frequency relationships. Equation (10.9) shows that if power is entering at the pump frequency, then power must be generated at the sum/difference frequency. Hence

$$-\frac{P_{p\pm s}}{P_p}=\frac{\omega_{p\pm s}}{\omega_p} \tag{10.10}$$

If no power is removed from the parametric amplifier at this sum/difference frequency, it is often called the idler. It is necessary if parametric amplification is to occur for the idler frequency to be allowed to be generated in the device. Combination of eqns. (10.8) and (10.10) shows that if the difference frequency is allowed to exist, there is a net flow of power out of the device at the signal frequency and the device is a straight amplifier. If the sum frequency is allowed to exist, there can be no power flow out of the device at the signal frequency, but there can be amplification of a signal entering at the signal frequency and leaving at the sum frequency. The simplest parametric amplifier is one in which the pump frequency is twice the signal frequency and the difference frequency is the signal frequency.

The practical parametric amplifier will house the varactor diode in a cavity that is resonant to the pump and signal frequencies and to such idler frequencies are as needed. The diode is positioned so as to be at a position of maximum electric field strength. Suitable filters are required to keep the pump and idler signals out of the output line of the main signal. Because of the restricted bandwidth of resonant structures, some of the resonant cavities are very low Q and some filters may also be used. Very wide bandwidths can be obtained by using a travelling wave structure incorporating a number of diodes. Parametric amplifiers provide low noise amplification without cooling because they consist of an entirely reactive circuit and there is little resistance to provide noise. They can be very broadband and can be driven by Gunn or IMPATT oscillators.

10.7. Harmonic Generator

The varactor diode having a non-linear relationship between current and voltage can also be used for frequency doubling and frequency mixing. In a way it could be said that the parametric amplifier described in the previous section is a mixer. This section is concerned with the generation of microwave power by the generation of suitable harmonics of a lower frequency signal.

For harmonic generation without any idling circuits, the varactor is mounted in a circuit similar to that shown in Fig. 10.8 where the filters F

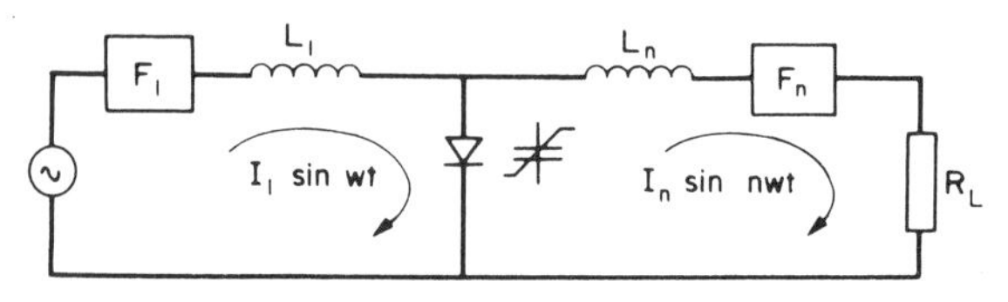

FIG. 10.8. Harmonic generator circuit without an idler circuit for generation of power at frequency $n\omega$ from an input at frequency ω.

can be considered to be a short circuit at the frequency of the signal in the circuit and an open circuit at other frequencies. The efficiency of generation of the higher harmonics is increased if idler circuits allow some of the intermediate harmonics to flow. A typical circuit of a varactor multiplier with one idler is shown in Fig. 10.9, when n and p are integers and $p < n$. The introduction of an idler increases the circuit complexity and it may be more economic to operate at a lower efficiency without idlers. It may also be more economic to operate a number of doubler stages in cascade rather than to operate one stage of high harmonic multiplication. Harmonic generation can be used to provide phase synchronized and controlled oscillations from lower frequency synchronized transistor oscillators.

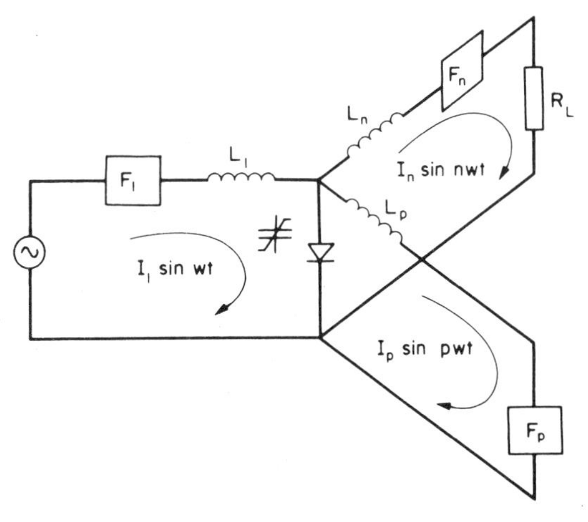

FIG. 10.9. Harmonic generator circuit with one idler. Input frequency ω, output frequency $n\omega$, and idler frequency $p\omega$.

10.8. Laser

The word laser is derived from the initial letters of the full name of the device, *light amplification by stimulated emission of radiation.* The device particularly for use at microwave frequencies is called the maser, where *microwave* replaces the word *light.* It is found that any material has characteristic frequencies for the absorption and emission of electromagnetic radiation. Most people are familiar with the characteristic light radiation associated with sodium or mercury. The electrons of any element may be in any of a number of discrete energy states. At the absolute zero of temperature the electrons occupy the lowest possible energy states but at any higher temperature the electrons become excited and move temporarily to higher energy states. They then return to the lower energy states and emit the characteristic radiation in the process. The frequency of radiation of the transition from energy state W_2 to energy state W_1 is

$$f_{12} = \frac{W_2 - W_1}{h} \tag{10.11}$$

where h is Planck's constant.

The probability of a transition from a higher to a lower energy state occurring is proportional to the population of electrons at the higher state and inversely proportional to the relaxation time. Absorption of radiation of the correct frequency causes a transition from a low energy state to a higher one.

In thermal equilibrium, the numbers of electrons in two energy states is given by the Boltzmann relationship

$$\frac{N_2}{N_1} = \exp - \frac{W_2 - W_1}{kT} \tag{10.12}$$

where N_2 and N_1 are the numbers in the upper and lower states respectively, k is Boltzmann's constant and T is the absolute temperature. Under normal conditions, the number of transitions to the higher energy state is equal to the number from the higher state to the lower. If equilibrium is disturbed by absorption of external radiation, it will be quickly restored since the rate of transition is proportional to the population at any state.

For a system with only two energy states, it might be possible to increase the population in the higher energy state with a pulse of microwave power and then immediately afterwards to observe the generation of microwave radiation of the same frequency. This two-level system, however, does not lead to a practical source of electromagnetic waves. The three-level laser system overcomes the disadvantages of the two-level system. A pictorial

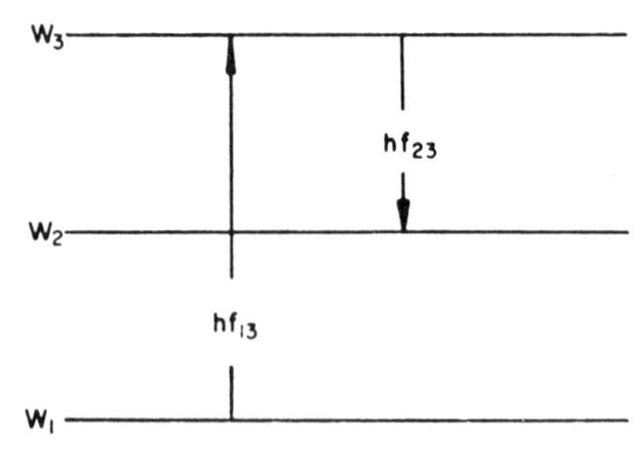

FIG. 10.10. Energy levels in the three-level laser.

description of the energy states of a three-level laser system is shown in Fig. 10.10. The incident radiation is called the pump power. In the system described in Fig. 10.10, the pump operates at the frequency f_{13} and increases the population at the energy state W_3. There will then be spontaneous transitions of $W_3 - W_2$, $W_3 - W_1$ and $W_2 - W_1$. If a small microwave signal is now introduced at a frequency of f_{23}, it is found that the transition

$W_3 - W_2$ is stimulated and there is more radiation at the frequency f_{23} than was used for stimulation. It is also found that the level of the stimulated radiation is proportional to the level of the stimulating radiation and the device is an amplifier.

The pump power must be kept out of the signal system by making the two circuits frequency selective. If the laser material is situated in a resonant cavity, the cavity must be resonant at both the signal and pump frequencies, but it does enable the system to be made selective against other possible frequencies of laser operation. Maser amplifiers have been made with quite broad bandwidths by building them in non-resonant travelling wave structures. For maser materials at room temperatures the relaxation time is extremely small and equilibrium is quickly reached. It would require an excessive amount of external radiation to cause appreciable variation in populations compared with the equilibrium condition. However, at temperatures near to the absolute zero, the relaxation time becomes large and the equilibrium population at the higher energy states becomes sufficiently small, for appreciable variation in populations to be caused by microwave radiation. Because the maser has to be kept near to the absolute zero of temperature, it is essentially a low noise device and it has made possible the reception of very low-level microwave signals in astronomy and space research.

The semiconductor junction diode laser used in many optical communications systems operates in a slightly different way. The lasing transition is between the conduction band which provides the excited state and the valence band which provides the ground state. Pumping occurs by direct electrical means. By forward biasing the diode, the external circuit increases the population of electrons in the conduction band so that laser action occurs.

10.9. Klystron

The klystron is one of the commonest transit time electron tube microwave oscillators and amplifiers. A schematic diagram of a two-cavity klystron is shown in Fig. 10.11. An electron beam leaves the cathode K through an accelerating anode A_1 and passes through two cavities to the final anode A_2. The cavities are microwave resonators which are so constructed that there is a strong electric field in the narrow central region. The walls of this region are composed of a wire mesh that is transparent to the electron beam. As the electrons pass through the resonator, they are accelerated or decelerated depending on the phase of the microwave electric field across the cavity. When the electrons enter the drift space, they gather into bunches because the faster electrons overtake the slower electrons. Provided the second resonator is tuned to the same frequency as the first, the electron bunches excite a microwave field in the resonant cavity. If the second

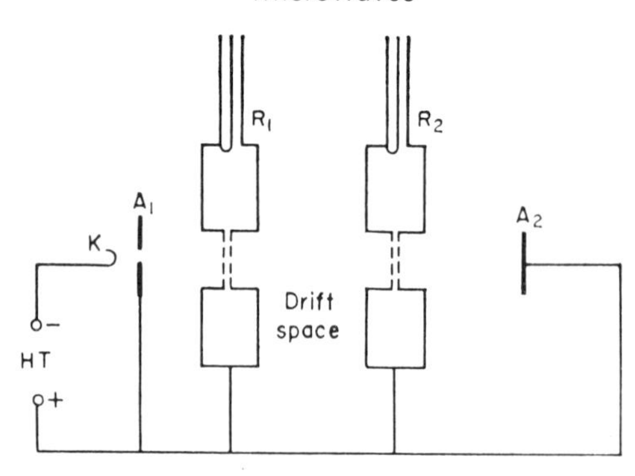

FIG. 10.11. Two-cavity klystron.

resonator is placed so as to intercept the electrons when they are closely packed into bunches and before they have separated again, microwave power can be extracted by means of the coaxial line leading from the resonator.

The electric field in R_2 is such that it retards each bunch of electrons as they cross the gap in the cavity. The energy of the electron beam is given up to the microwave field and the device acts as an amplifier. The effectiveness of the klystron as an amplifier at any particular frequency is dependent on both the resonant cavities being tuned to the correct frequency, and on the anode–cathode potential difference being such that optimum bunching occurs in the drift space. Hence the klystron is a tuned device which only operates at one particular frequency.

If the klystron is to be used as an oscillator, the input and output cavities may be combined. Instead of having a drift space between two cavities, a reflector is placed in the drift space so that the electrons return to the first resonator and the second is not required. This reflex klystron is shown diagrammatically in Fig. 10.12. The electrons leaving the cathode are accelerated by the anode and pass through the resonator where the beam bunching process is initiated. The beam then emerges into the retarding field due to the negative potential of the reflector electrode. The reflecting potential is adjusted so that the beam is reflected and returns to the resonator with a drift time giving the optimum beam bunching conditions. The frequency of operation of the klystron is again governed by the tuned frequency of the resonant cavity and by the reflector potential which governs the length of the drift space.

If the reflector potential is such that the electrons do not arrive in the optimally bunched condition, there is some difference of phase between the induced current in the cavity and that of the optimum retarding field. This phase difference increases or decreases the frequency of oscillation according to its sign, and is accompanied by a corresponding loss of efficiency and hence by a loss of power output. This means that the reflex klystron is voltage tunable by varying the reflector potential. The tuning is limited by

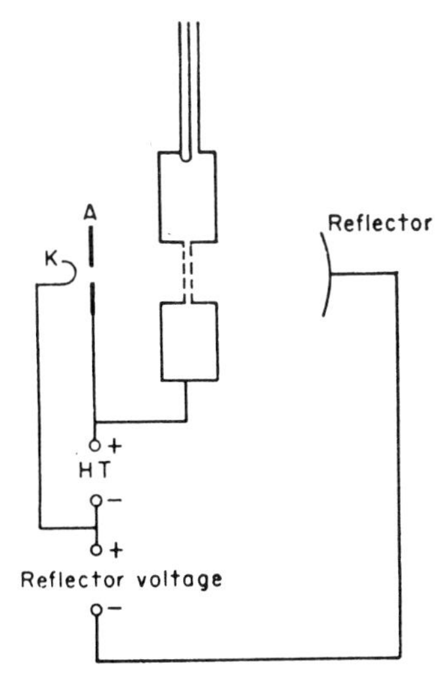

FIG. 10.12. Reflex klystron.

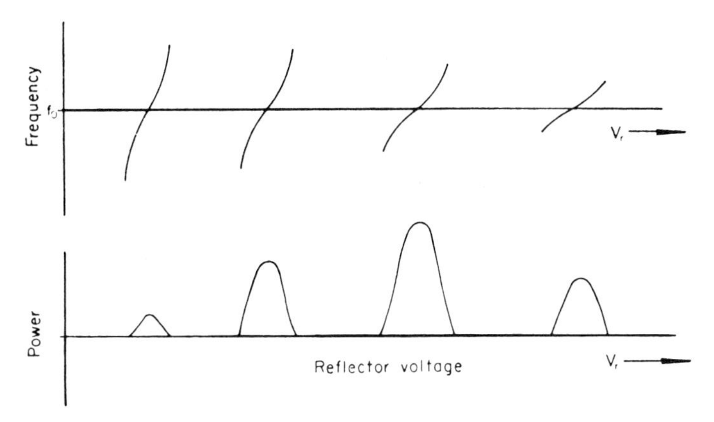

FIG. 10.13. Voltage tuning characteristics of a reflex klystron.

the bandwidth and hence the Q of the resonator cavity. Typical voltage tuning characteristics of a reflex klystron are shown in Fig. 10.13 where V_r is the reflector potential specified in Fig. 10.12. Operation of the klystron in several modes is possible as shown by Fig. 10.13. Each mode corresponds to a different time spent by the electrons in the drift space.

Because the resonant cavity is connected to a coaxial line or waveguide output the anode of the klystron must be connected to earth. The cathode is at a negative potential compared with earth and the reflector is at a further negative potential compared with the cathode. The low-power reflex klystron is cheap and can be built with a small mechanical tuning range. High-power continuous wave transmitters use large high-power amplifier klystrons.

10.10. Magnetron

The magnetron was the first high-power microwave oscillator to be developed. It was the invention of the magnetron that made the microwave radar systems of World War II possible. A schematic diagram of a magnetron oscillator is shown in Fig. 10.14.

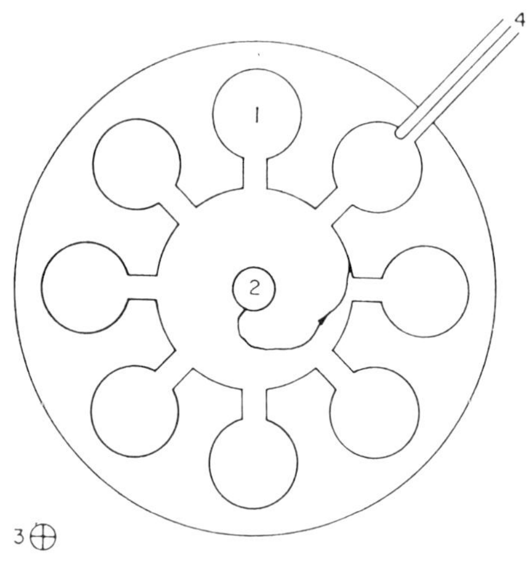

FIG. 10.14. Cavity magnetron. (1) Anode cavity. (2) Cathode. (3) Direction of static magnetic field. (4) Output.

In the klystron oscillator, the electron beam interacts with the high-frequency fields only during the short time that it is passing through the resonator cavity. The magnetron is one of a large class of microwave tubes in which the electrons interact with microwave fields over an extended region. Consider the magnetron as shown in Fig. 10.14. There is a large magnetic field acting perpendicular to the paper. The central electrode is the cathode which emits electrons and the outer electrode is the anode which includes a number of coupled resonator cavities. The electron stream flows radially outward from the cylindrical cathode to the anode. In the absence of any disturbing fields, the electrons leave the cathode and proceed in circular paths under the influence of the magnetic field either to reach the anode or with a stronger field, to return to the cathode.

The electromagnetic fields in the anode-cathode region may be considered as a travelling wave which moves round the inside surface of the anode. Each anode cavity behaves as an individual cavity resonator with an aperture onto the anode–cathode space. Because the aperture lowers its Q, each cavity only oscillates in its lowest or fundamental mode. The fields in the aperture excite a travelling wave in the anode–cathode space and the travelling wave couples the fields in all the cavities together.

Consideration of the motion of an electron under the static magnetic field and a static electric radial field shows that if the magnetic field is sufficiently strong, the electron never reaches the anode and returns to the cathode. The magnetron oscillator is usually operated under these conditions so that when there is no microwave oscillation, there is very little anode current. If there are electromagnetic fields in the anode-cathode space, some of the electrons are retarded by the microwave electric field and follow a path similar to that shown in Fig. 10.14. Before reaching the anode, the electron will have given up most of its energy to the electromagnetic fields. If the phase of the electromagnetic fields is such that the electron is accelerated soon after leaving the cathode, it quickly returns to the cathode and there is little interaction with the fields. This electron bombards the cathode, and in a typical magnetron about a twentieth of the anode power is used in this way to heat the cathode.

The frequency of the magnetron oscillations is sensitive to changes in the load impedance. This effect is called pulling and can be a source of trouble in magnetron applications. The power capability of a magnetron is limited by the ability to remove heat from the cathode which is situated in the middle of an evacuated space. This requirement means that magnetrons are usually operated in pulse conditions when the peak power is high but the mean power can be kept quite low. Most simple radar systems operate with a pulsed microwave source and the magnetron is the ideal radar oscillator.

10.11. Travelling Wave Tube

It has already been shown that an electron beam can support travelling wave type oscillations (section 9.9). If the electron beam is enclosed by a structure which propagates an electromagnetic wave at approximately the same velocity as the beam current wave, there is an energy interchange between the beam and the wave and the wave is amplified.

Since the phase velocity in waveguide is faster than that of light, normal waveguide transmission is not suitable for a travelling wave tube. A *slow wave* structure is required. The first slow wave structure to be used in

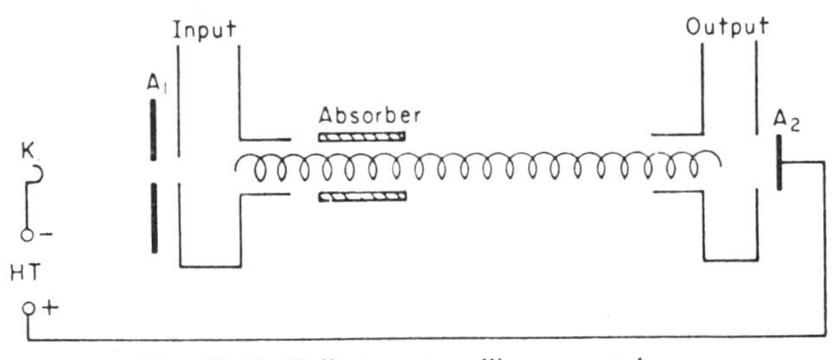

FIG. 10.15. Helix-type travelling wave tube.

travelling wave tubes was the helix. Figure 10.15 shows a schematic diagram of a helix-type travelling wave tube. To a first approximation, it is assumed that the electromagnetic wave propagates along the wire of the helix with the speed of light. Hence its speed in the axial direction is governed by the pitch of the helix. There are many other slow-wave structures that have been used in different applications but only the helical travelling wave tube is described here. Often a magnetic field is applied parallel to the axis of the structure to prevent spreading of the electron beam as it travels.

Mathematical analysis shows that of the possible waves propagating in the electron-beam slow-wave structure coupled system, there are two propagating at a speed which is slightly less than that for propagation in the slow-wave structure alone. One of these waves is attenuated exponentially and the other grows exponentially as it travels. It is this last wave which is used to provide amplification in travelling wave tubes.

A mode converter is used at the input of the tube to transform the electromagnetic signal from its waveguide or coaxial line input into the helix without causing reflection of the input signal. The signal is then amplified as it proceeds along the slow-wave structure and is transformed into the waveguide or coaxial line for the output. It is impossible to design the input and output structures to be reflectionless and consideration must be given to preventing spontaneous oscillations arising from reflections at the output. The energy reflected at the output travels back to the input and can provide a spurious feedback signal which is amplified as well as the desired signal. This feedback can be suppressed by introducing an absorber into the helical slow wave system as shown in Fig. 10.15. The reflected wave is absorbed whereas the forward wave is only partially attenuated because it is carried on the electron beam. For this reason the absorber is situated near to the input of the tube at a position where the electron beam has already been modulated by the signal but before there has been appreciable power transfer from the beam to the signal. Obviously the absorber does reduce the maximum gain available from any tube, and some travelling wave tubes have ferrite isolators (see section 13.6) built into the slow-wave structure instead to absorb the reflected signal.

The travelling wave tube is inherently a broadband device because it does not have any resonant characteristics on which it relies for operation. A typical helical travelling wave tube has an operating bandwidth of 2:1 with a gain between 20 and 40 dB which is constant over the band within 6 dB. The helix structure itself is particularly suited to broadband operation so that the helix travelling wave tube is used where broadband operation is required. Other slow-wave structures may have preferable characteristics for other applications.

10.12. Summary

10.2. The diodes used as rectifiers or mixers are ***point-contact***, ***Schottky-barrier*** or ***backward*** diodes.

The ***varactor diode*** is used as a variable capacitor.

The ***PIN diode*** is used as a variable resistor or switch.

10.3. ***Transistors*** can be made with very small spacing between electrodes so that they operate in the microwave region. They are mounted in miniature microstrip circuits or as part of integrated circuits to make oscillators or amplifiers.

10.4. ***Avalanche oscillator*** or IMPATT oscillator. Avalanche breakdown in a reverse biased p–n junction diode can give microwave oscillations.

10.5. ***Transferred electron oscillator*** or Gunn diode oscillator. The direct generation of microwave oscillations in the bulk of a semiconductor material having a suitable direct current.

10.6. ***Parametric amplification.*** The relationships between power levels at different frequencies incident upon any non-linear device are given by the Manley–Rowe equations

$$\sum_{m=0}^{\infty} \sum_{n=-\infty}^{\infty} \frac{mP_{mn}}{m\omega_s + n\omega_p} = 0 \tag{10.6}$$

$$\sum_{m=-\infty}^{\infty} \sum_{n=0}^{\infty} \frac{nP_{mn}}{m\omega_s + n\omega_p} = 0 \tag{10.7}$$

The simplest parametric amplifier has a pump operating at twice the signal frequency.

10.7. The non-linear characteristic of a varactor diode is used to generate harmonics of the input signal. The ***harmonic generator*** is used to generate microwave power from a lower-frequency oscillator.

10.8. ***Laser*** (light amplification by stimulated emission of radiation). In a three-level laser, the electrons are excited to the energy state W_3 from the ground W_1 by absorption of radiation at the pump frequency and emit radiation at the signal frequency corresponding to a relaxation from the excited state W_3 to an intermediate energy state W_2. The emission of radiation at the signal frequency can be stimulated by incident radiation at the same frequency and the device acts as an amplifier.

10.9. The ***klystron amplifier***. The electromagnetic field in the input resonant cavity modulates an electron beam, which bunches in the drift space and excites an amplified electromagnetic signal in the output resonant cavity.

The ***reflex klystron*** combines the input and output resonant cavities into one cavity. A reflector electrode in the drift space, maintained at a negative potential difference compared to the cathode, reflects the electron beam back to the resonant cavity.

10.10. The ***magnetron***. Electrons flowing radially from a cylindrical cathode excite microwave oscillations in the cavities of the surrounding anode when a static magnetic field is applied perpendicular to the direction of electron flow. The magnetron is a high-power pulsed microwave source.

10.11. In the ***travelling wave tube*** there is continuous interaction between the electromagnetic fields and the electron beam. A slow-wave structure is used so that the phase velocity of the electromagnetic wave is the same as that of the electron beam waves.

11

Waveguide Components

11.1. Waveguide Components and Devices

In the earlier chapters of this book we have discussed the theoretical considerations describing electromagnetic wave propagation in transmission lines and waveguide. In the next three chapters we shall discuss some of the components and devices that might be used in microwave systems. In particular this chapter contains a description of some waveguide components. These are lengths of waveguide which are modified in some way so as to control the electromagnetic wave. The devices described in Chapters 12 and 13 consist of lengths of transmission line or waveguide within which non-metallic materials control the electromagnetic wave. This separation between components and devices is confined to this book. Normally either word is used indiscriminately to describe any waveguide component or device.

The theory of the earlier chapters help in an understanding of the mode of operation of the components and devices to be described. In particular, it must be remembered that they all operate in the dominant mode, that is the TE_{10}-mode in rectangular waveguide and the TE_{11}-mode in circular waveguide, and that these two modes have similar field distributions in the waveguide. The theory of more complicated modes has been given to provide a basis for the understanding of more complicated components and devices that may be part of any particular microwave system. Most microstrip circuits are enclosed in a sealed metal box. Care must be taken that waveguide modes cannot exist within any such box or enclosure. The components and devices here described are the simplest and commonest and form the basis of the system of waveguide test equipment that is described in Chapter 14. It is hoped that the reader will obtain some understanding of microwave practice to provide a practical base for the electromagnetic theory. The treatment of the material in these last four chapters is of necessity brief and is largely descriptive. A start is made with the waveguide itself and its interconnections.

11.2. Waveguide

The dimensions of rigid rectangular waveguide have been standardized. A list of the standard sizes together with their British, American and

International standard nomenclature and recommended operating frequency range is given in Table 11.1. They can be made of silver, high-conductivity copper, brass or aluminium. Brass has the lowest conductivity but it is still a popular material for waveguide manufacture because, in many applications, the losses due to the waveguide transmission are negligible compared to the losses due to the devices in the system. Aluminium has the advantage of lightness but it is difficult to joint by soldering and welding may cause distortion of the waveguide, making nonsense of close dimensional tolerances. There are also some standard sizes of circular waveguide, but the frequency range of operation of the dominant mode is much smaller than in rectangular waveguide so that for many applications each manufacturer decides on the most convenient size irrespective of any standards.

For certain applications it is not possible to have all the parts of a waveguide system rigidly connected and then use can be made of *flexible waveguide*. It consists of approximately rectangular pipe corrugated perpendicular to its length with a thin metal wall. It is protected with rubber covering the outside. The construction is such as to give the pipe great flexibility. Most commercially available flexible waveguide is twistable as well as flexible. The dimensions are chosen so that the flexible waveguide has the same impedance as the standard waveguide with which it is to be compatible so that it causes the minimum disturbance to the electromagnetic fields.

11.3. Couplings

For interconnection, the waveguide is fitted with flanges or couplings which are bolted together. The cross-section through a typical flange assembly connecting two lengths of waveguide is shown in Fig. 11.1. Two plane flanges with perfectly flat faces connecting two lengths of waveguide together only provide a good joint if there is perfect continuity of electrical conductivity at the inside surface of the waveguide. Often there is an intermittent open-circuit at the join which causes some mismatch in the waveguide. The choke flange compensates for the bad fitting. It is designed to provide a discontinuity at the point of the join on the inside surface of the waveguide. The bottom of the choke ditch in the flange is arranged to be a half-wavelength away from the inside surface of the waveguide so that the short circuit in the choke

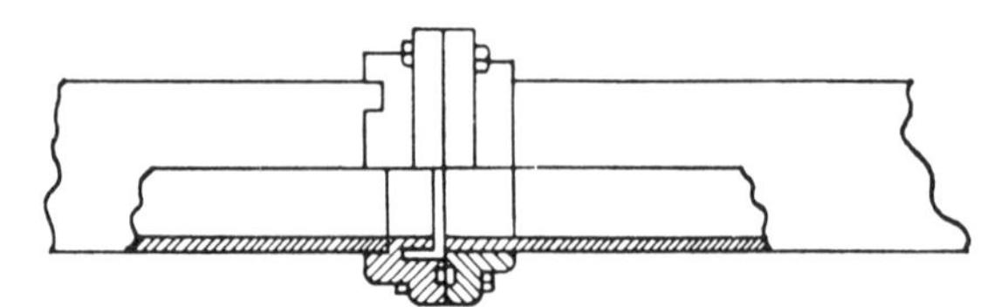

FIG. 11.1. A part section through a waveguide joint using a choke-plain flange combination.

TABLE 11.1. *Standard Sizes of Rectangular Waveguide*

Nomenclature			Dimensions (in.)						Performance (GHz)	
			Inside dimensions				Outside dimensions			
International Standard 153 IEC–R	American WR–	British WG	Width a	Height b	Ratio a/b	Wall thickness	Width	Height	Cut-off frequency	Recommended operating frequency range
3	2300	0.0	23.000	11.500	2.000	0.125	23.250	11.750	0.2565838	0.32–0.45
4	2100	0	21.000	10.500	2.000	0.125	21.250	10.750	0.2810203	0.35–0.50
5	1800	1	18.000	9.000	2.000	0.125	18.250	9.250	0.3278571	0.45–0.63
6	1500	2	15.000	7.500	2.000	0.125	15.250	7.750	0.3934285	0.50–0.75
8	1150	3	11.500	5.750	2.000	0.125	11.750	6.000	0.5131676	0.63–0.97
9	975	4	9.750	4.875	2.000	0.125	10.000	5.125	0.6052746	0.75–1.15
12	770	5	7.700	3.850	2.000	0.125	7.950	4.100	0.7664191	0.97–1.45
14	650	6	6.500	3.250	2.000	0.080	6.660	3.410	0.9079119	1.15–1.72
18	510	7	5.100	2.550	2.000	0.080	5.260	2.710	1.157143	1.45–2.20
22	430	8	4.300	2.150	2.000	0.080	4.460	2.310	1.372425	1.72–2.60
26	340	9A	3.400	1.700	2.000	0.080	3.560	1.860	1.735714	2.20–3.30
32	284	10	2.8400	1.3400	2.1194	0.080	3.000	1.500	2.077967	2.60–3.95
40	229	11A	2.2900	1.1450	2.000	0.064	2.418	1.273	2.577042	3.30–4.90
48	187	12	1.8720	0.8720	2.1468	0.064	2.000	1.000	3.152472	3.95–5.85
58	159	13	1.5900	0.7950	2.000	0.064	1.718	0.923	3.711589	4.90–7.05
70	137	14	1.3720	0.6220	2.2058	0.064	1.500	0.750	4.301332	5.85–8.20
84	112	15	1.1200	0.4970	2.2575	0.064	1.250	0.625	5.259739	7.05–10.0
100	90	16	0.9000	0.4000	2.2500	0.050	1.000	0.500	6.557141	8.20–12.4
120	75	17	0.7500	0.3750	2.000	0.050	0.850	0.475	7.868569	10.0–15.0
140	62	18	0.6220	0.3110	2.000	0.040	0.702	0.391	9.487825	12.4–18.0
180	51	19	0.5100	0.2550	2.000	0.040	0.590	0.335	11.57143	15.0–22.0
220	42	20	0.4200	0.1700	2.4706	0.040	0.500	0.250	14.05102	18.0–26.5
260	34	21	0.3400	0.1700	2.000	0.040	0.420	0.250	17.35714	22.0–33.0
320	28	22	0.2800	0.1400	2.000	0.040	0.360	0.220	21.07653	26.5–40.0
400	22	23	0.2240	0.1120	2.000	0.040	0.304	0.192	26.34566	33.0–50.0
500	19	24	0.1880	0.0940	2.000	0.040	0.268	0.174	31.39057	40.0–60.0
620	15	25	0.1480	0.0740	2.000	0.040	0.228	0.154	39.87451	50.0–75.0
740	12	26	0.1220	0.0610	2.000	0.040	0.202	0.141	48.37235	60.0–90.0
900	10	27	0.1000	0.0500	2.000	0.040	0.180	0.130	59.01427	75.0–112
1200	8	28	0.0800	0.0400	2.000	0.030	0.140	0.100	73.76784	90.0–140
1400	7	29	0.0650	0.0325	2.000	0.030	0.125	0.0925	90.79119	112–172
1800	5	30	0.0510	0.0255	2.000	0.030	0.111	0.0855	115.7143	140–220
2200	4	31	0.0430	0.0215	2.000	0.030	0.103	0.0815	137.2425	172–260
2600	3	32	0.0340	0.0170	2.000	0.030	0.094	0.0770	173.5714	220–330

flange is reflected to act as a short circuit at the inside surface of the waveguide. The usual choke–plain flange combination only introduces a mismatch equivalent to a VSWR of 1.01. For high-precision measurements, it may be possible to achieve performances better than this for the combination of two plain flanges provided they are not damaged. If there is any possibility of scratches or tarnish on the face of the flanges, the choke–plain combination gives a better performance. Under high-power conditions, there is always the possibility that the joint between two plain flanges is not perfect and sparking may occur. Sparking causes reflection of most of the microwave power back down the waveguide to the generator. The choke–plain combination does not cause sparking. There are standard designs of couplings to fit most standard sizes of waveguide.

11.4. Bends and Twists

In any waveguide system it will be necessary at some time to change the direction of the waveguide run. The theory of rectangular waveguide in Chapter 5 always assumes that the waveguide is perfectly straight. Small deviations from straightness have negligible effect on the propagating conditions in the waveguide, so that bends and twists occupying a great many wavelengths of waveguide would be satisfactory electrically, although they would usually be impossible practically. As a separate waveguide component, a bend or twist is usually half a wavelength or a wavelength long. Provided the cross-section of the waveguide is undistorted, it is found that the simple radius bend or regular twist makes a satisfactory component. The propagation constant in the curved or twisted section of waveguide is slightly different from that in undistorted waveguide of the same cross-section so that there is a slight mismatch at the change from straight to curved or twisted waveguide. In order to minimize the effect of this mismatch, the curved or twisted section of waveguide is made an integral number of half wavelengths long so that the reflections from the two ends of the device cancel. A typical bend and twist are shown in Figs. 11.2 and 11.3.

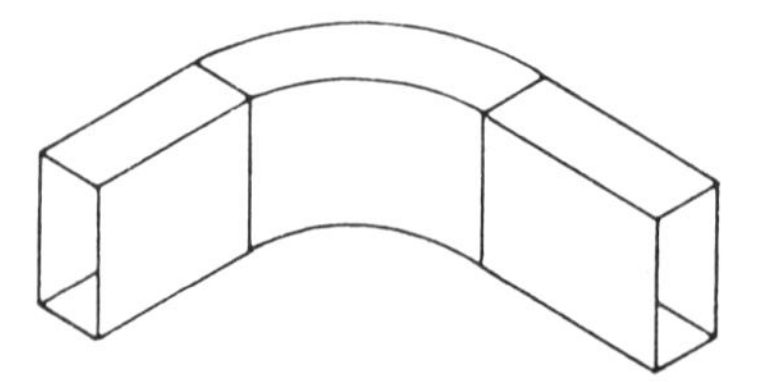

FIG. 11.2. Waveguide radius bend, showing a curved section of waveguide, abutting onto two straight sections of waveguide.

There is another kind of bend which has been quite popular for use in larger sizes of waveguide. That is the mitre bend shown in Fig. 11.4. The plane section set into the face of the corner acts as a mirror to reflect the

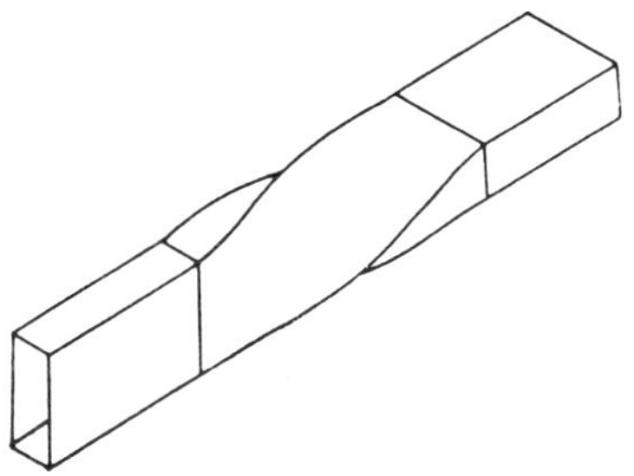

FIG. 11.3. Waveguide twist, showing a twisted section of waveguide, abutting onto two straight sections of waveguide.

wave round the corner. There is an optimum position for the mirror section on the corner and this position is dependent on the frequency.

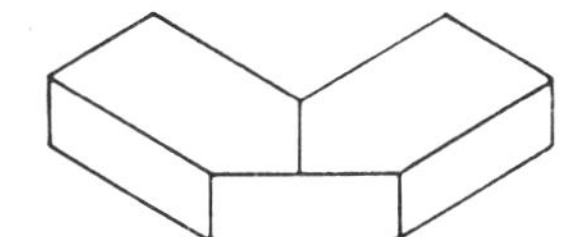

FIG. 11.4. Waveguide mitre bend.

11.5. Directional Coupler

The branch line directional coupler shown in Fig. 4.16 may be realized very easily in waveguide. Consider two waveguides running adjacent to each other, then holes provide the coupling between them. An arrangement of only two coupling holes similar to that shown in Fig. 11.5(a) will only work

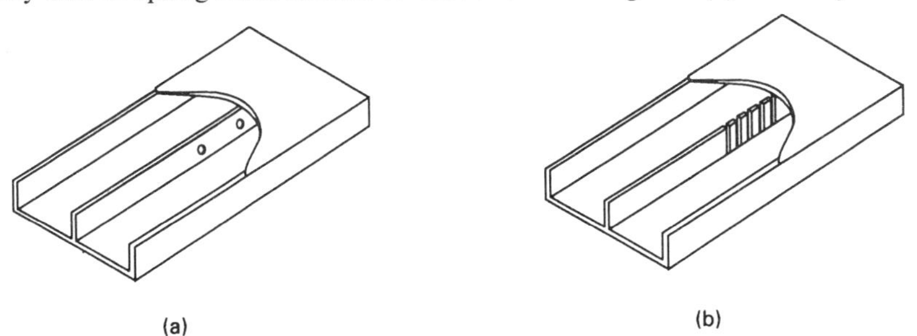

FIG. 11.5. Side-wall coupled waveguide directional couplers, (a) two-hole coupler, (b) multislot 3 dB coupler.

at one particular frequency when the holes are a quarter of a wavelength apart. To increase the frequency band over which the coupler works, the number of holes is increased, while each hole couples a smaller portion of the total power. The coupling holes are not necessarily equal. Their coupling may be arranged to be in a binomial series (i.e. five holes arranged for coupling in ratio 1:4:6:4:1) or according to some Tchebyshev ratio. To provide larger coupling ratios, the holes may be opened up into full-height slots as shown in Fig. 11.5(b). There are also many different methods of providing the coupling holes between the waveguides. They can be in either the broad or the narrow wall of the waveguide, and there are even half-power couplers where a section of the waveguide wall between two waveguides is removed altogether. There are a very large number of different ways in which the coupling between two waveguides may be arranged so as to make a directional coupler but they are all four-port devices and the power ratios between the ports is defined by the coupling factor and directivity.

11.6. T-junctions

The E-plane and the H-plane T-junctions are shown in Fig. 11.6. In both these junctions, if power enters the junction by the arm labelled 1, then it is equally split between the other two arms. For the H-plane junction, the signals are in phase at an equal distance from the centre of the junction and for the E-plane junction, the signals are in anti-phase, in the two output arms. As explained in section 4.6 when these two T-junctions are constructed of empty waveguide they are not matched devices. If the two output arms are terminated in matched impedances, the third arm does not present a matched impedance to the source.

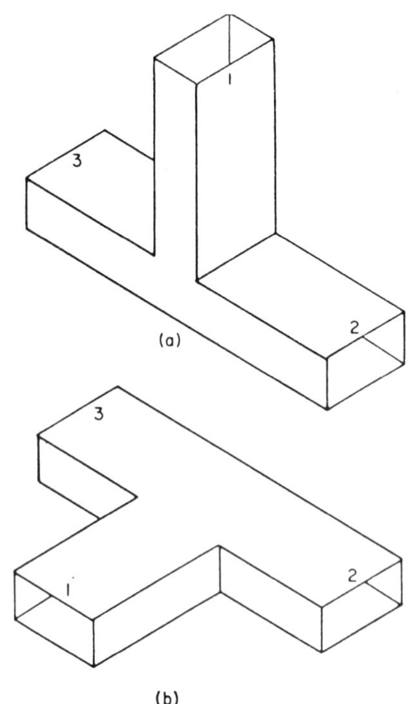

FIG. 11.6. T-junctions. (a) E-plane. (b) H-plane.

A combination of the E- and H-plane T-junction in the form shown in Fig. 11.7 is called the *hybrid T-junction*. This device has a number of useful applications. A wave entering arm 1 excites equal waves of equal phase in arms 2 and 3 and a wave entering arm 4 excites equal waves of opposite phase in arms 2 and 3. It is seen from the geometry of the device, that a wave in arm 1 will not excite any dominant mode wave in arm 4 and vice versa. That is, there is no direct transmission between arms 1 and 4. Further, it can be shown that, if the E- and H-plane arms of the junction are matched, the other two arms are matched and there is also no transmission between arms 2 and 3. Under these conditions, a wave entering arm 2 is equally split between arms 1 and 4, and a wave entering arm 3 is equally split between arms 4 and 1. Conversely, the sum of two equal waves entering arms 1 and 4 appears in arms 2 and 3 depending on the phase, and the sum of two waves in arms 2 and 3 appears in arms 1 or 4. Such a matched hybrid T is often called a *magic T*. The magic T has the properties of a hybrid ring, as described in section 4.9, or of a 3 dB or equal power split directional coupler.

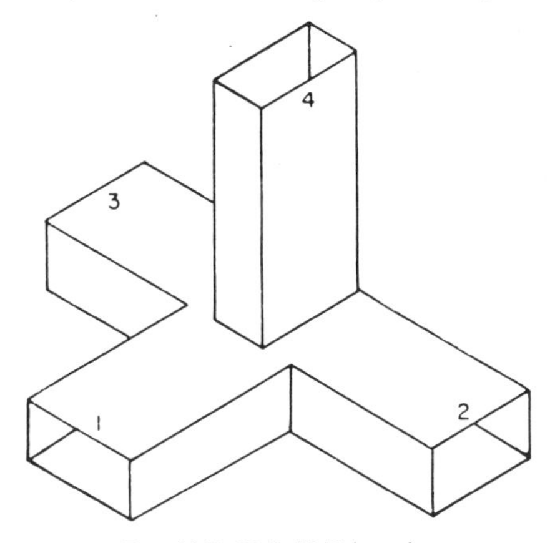

FIG. 11.7. Hybrid T-junction

11.7. Matched Termination

The matched termination in any transmission line system is used to terminate the line in its characteristic impedance. It absorbs the electromagnetic energy incident on the line without causing any reflection of electromagnetic power from the termination. The matched termination is made of a material which absorbs electromagnetic power. The theory of reflection of a plane wave from a plane conducting surface, being similar to the theory of blooming of lenses, shows that a precise thickness of bulk-absorbing material used as a surface covering for the plane conducting surface acts as a matched termination for a plane wave. Such sheet absorbers are used to hide obstacles on the ground near to radar equipments which might otherwise render the radar sets partially useless. Similar slabs of absorber have been used in waveguide to provide a termination which is matched over a narrow band of frequencies. More usually, the absorber is tapered with the point towards the generator so that all the power is absorbed without reflection. The absorbers are made of different materials for different applications. A very easy and convenient absorber for experimental use consists of a length of wood about four wavelengths long. The wood often used is prime quality beech but no wood will act as a very precise matched termination because its absorptive qualities depend on its moisture content which varies with the humidity. Although wooden matched terminations are unlikely to be available commercially, their performance can be as good as any high-precision termination.

A vane absorber, consisting of a thin sheet of conducting material, can be mounted centrally in the waveguide parallel to the electric field vector to make a matched termination. The same vane absorber material is used in the vane attenuator (see section 12.4). Any microwave absorbing material made into a wedge shape and mounted in the waveguide can be used to make matched terminations. For low-power applications the absorbing material might be an epoxy resin loaded with iron powder or a ferrite material with a high microwave loss. For higher powers, the absorbing material must be a ceramic such as carborundum, or water which is a good absorber of microwave power may be circulated through the waveguide in ceramic or glass tubes.

Whatever material is used to absorb the microwave power, its shape and position in the waveguide must be so designed that there is a minimum of reflected power. A good high-precision grade 1 termination giving a reflection coefficient of 0.003 in its worst condition has an absorber with a gradual taper and consequently the device is long. The short termination has more reflected power, possibly a reflection coefficient of 0.05.

11.8. Short Circuit

In low-frequency circuit theory, there are two circuit conditions that are very easy to produce. They are the short circuit and the open circuit. The open circuit cannot be produced in transmission lines because an open-ended transmission line radiates some of the power in the forward direction and consequently behaves as if it is terminated in some load resistance. The short circuit, however, can be produced by placing a perfect conductor between the wires of the transmission line or across the end of a waveguide. Many devices are terminated by a short circuit which is positioned so that the power reflected from the short circuit is of such a phase that it cancels the power reflected from the device itself. For this purpose, the short circuit consists of a plug of metal which is soldered into the waveguide or a metal plate that is fixed over the end of the waveguide.

For measurement purposes, however, it is necessary to have a short circuit that is variable in position. The variable short circuit usually consists of a plunger that is made to be a slide fit inside the waveguide. There are difficulties due to the possibility of intermittent contact between the plunger and the waveguide. The contact difficulties are overcome by making the plunger non-contacting and by arranging a choking system so that there is a further short circuit reflected into the gap between the plunger and the

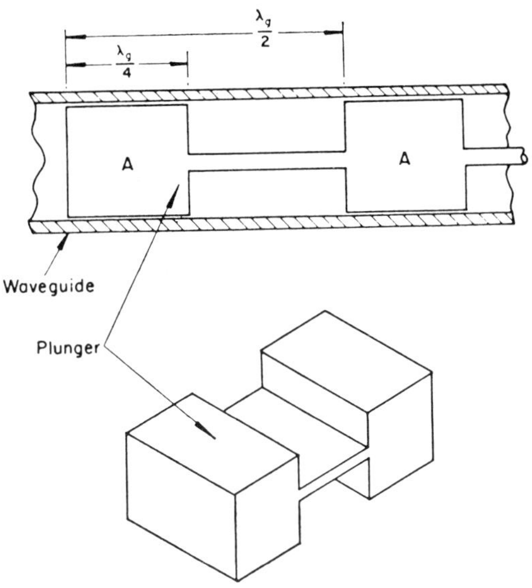

FIG. 11.8. Schematic diagrams of a non-contacting waveguide short circuit.

waveguide. The system is shown in Fig. 11.8. It is found that if the narrow section of waveguide (section A in the figure) is made a quarter of a waveguide wavelength long, then the short-circuit effect of the plunger is enhanced. The high impedance step at the back of the plunger is transformed by the quarter wavelength section into a low impedance at the front face of the plunger. If the first narrow section is followed by another narrow section situated half a wavelength behind the front face of the first section A, any microwave power passing the first part of the plunger is reflected by the second part with such a phase that it appears as a short circuit at the face of the plunger, further contributing to the efficacy of the short circuit.

A further refinement adopted by some manufacturers is to have vertical slots in the front face of the short-circuit plunger. These are parallel to the narrow wall of the waveguide. A study of Figs. 5.2 and 5.3 shows that all possible waveguide modes except the dominant TE_{10}-mode have a component of the electric field parallel to the broad wall of the waveguide. The vertical slots provide a high impedance path to the currents generated by these higher order modes and discourage any tendency for the short circuit to generate extra waveguide modes.

To provide an insulating surface between the plunger and the waveguide, the plunger is made of anodized aluminium. Anodizing provides a good wear-resistant insulating surface. The plunger is attached to some positioning device so that it can be accurately set and locked in the waveguide.

11.9. Standing-wave Meter

A probe moving along the waveguide is used to detect the standing wave pattern inside the waveguide. It is used to measure the voltage standing wave ratio. A slot in the centre of the broad face of the waveguide, parallel to the axis of the waveguide, does not cut any current streamlines of the dominant mode so that the slot should not radiate any power or disturb the field pattern inside the waveguide. A small probe inserted through the slot will couple to the electric field in the waveguide. The probe is connected to a detector crystal so that the output from the crystal is a direct current proportional to the mean power at that position in the waveguide. As the position of the probe is moved along the waveguide, it gives an output proportional to the standing wave pattern inside the waveguide.

The standing-wave meter consists of an accurately machined waveguide with a narrow slot in the centre of the broad face. The waveguide dimensions are critical because the standing-wave meter can be used for waveguide wavelength measurements. The probe is in a carriage which moves on the outside surface of the waveguide. It is necessary that the motion of the probe exactly follows the inside surface of the waveguide since any variation of probe insertion would lead to variation of output for the same microwave power in the waveguide. Hence in most standing-wave meters, the top outside surface of the waveguide is machined to be exactly parallel with the top inside surface. A section through a waveguide standing-wave meter is shown in Fig. 11.9. The precision waveguide section with the slot in the centre of the broad face is often called a *slotted measuring section.*

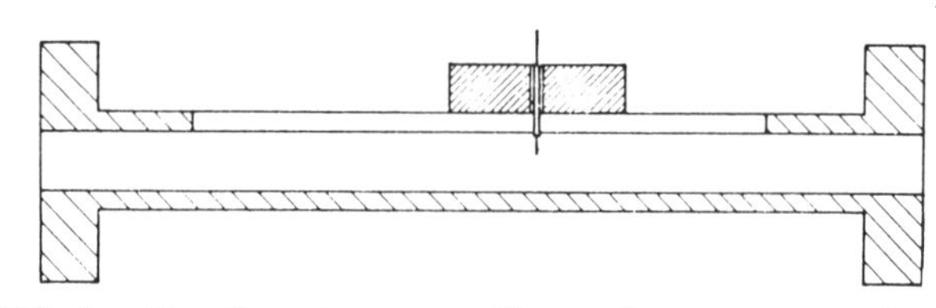

FIG. 11.9. A section through a waveguide standing-wave meter, showing the slotted line and carriage with probe but omitting other details.

The detector crystal used in the standing-wave meter has an approximately square law relationship between the output current and the input microwave electric field so that the output is assumed to be the square of the input voltage.

11.10. Probes

Probes can be used to couple microwave signals between a waveguide or cavity and a coaxial line. An electric probe was described in the last section as part of the standing-wave meter. It consists of a straight wire in the waveguide which is parallel to the electric field of the waveguide mode. Usually a coaxial line terminates at a hole in the waveguide wall and the inner of the coaxial line projects into the waveguide. In the standing-wave meter the probe is used to couple a small amount of power into the detector. Probes can also be used to couple power into the waveguide, and are often used to make a connection between an oscillator and waveguide. An electric probe designed to couple to the maximum electric field of the dominant mode in rectangular waveguide is shown in Fig. 11.10.

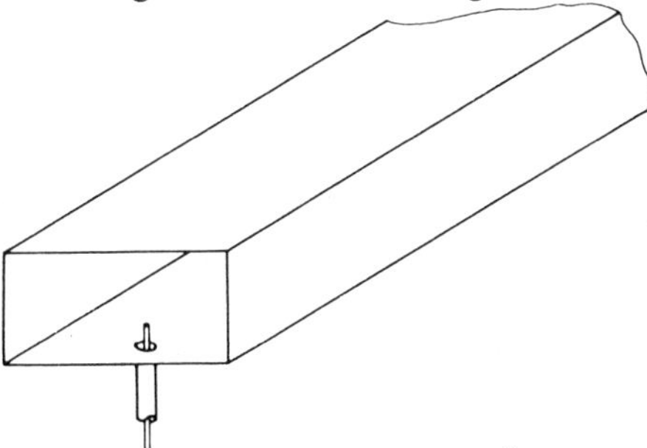

FIG. 11.10. An electric probe coupling to the maximum electric field of the dominant mode in rectangular waveguide.

For a magnetic probe, the inner of the coaxial line is bent into a loop and connected to the waveguide wall. The coupling loop is orientated so that the plane of the loop is perpendicular to the magnetic field in the waveguide. A magnetic probe coupling to the magnetic field of the dominant mode in rectangular waveguide is shown in Fig. 11.11.

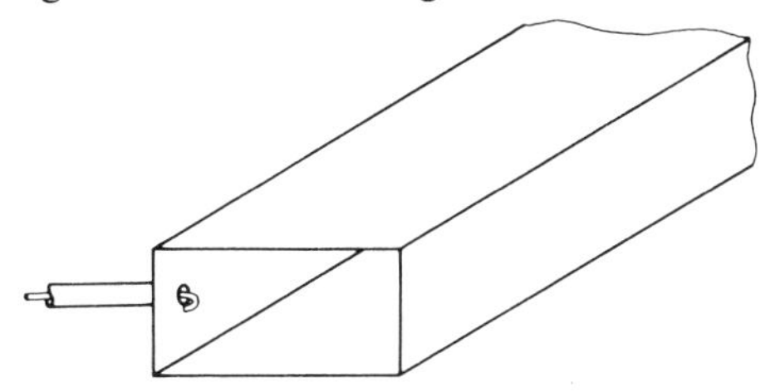

FIG. 11.11. A magnetic probe coupling to the magnetic field of the dominant mode in rectangular waveguide.

Other methods of coupling between a waveguide and a coaxial line are the crossbar transformer and the door-knob transformer. In the crossbar transformer, the electric probe in Fig. 11.10 is extended into the waveguide and connects with a metal rod connected to and perpendicular to each of the narrow sides. In the door-knob transformer, the electric probe extends fully across the waveguide and increases in diameter and connects with the opposite wall. The crystal receiver shown in Fig. 12.11 uses a door-knob transformer to the coaxial line in which the crystal is mounted.

11.11. Mode Filters

A mode filter is designed to discriminate between different modes of propagation in the waveguide, often discriminating in favour of the dominant mode in oversize waveguide. The vertical slots in the front face of the short-circuit plunger described in section 11.8 are a mode filter in favour of the dominant mode in rectangular waveguide. An exception is the mode filter described in section 6.11 which is used to discriminate in favour of the TE_{01}-mode in circular waveguide because that mode can be used to provide very low loss waveguide transmission. The simplest mode filter in favour of the dominant mode in rectangular waveguide is a metal plate or an absorbing vane perpendicular to the electric field of the dominant mode which will reflect or absorb all other modes of propagation in the waveguide. Such a vane is shown in Fig. 11.12. Because the dominant mode in circular waveguide is similar to the dominant mode in rectangular waveguide, a similar vane is used to absorb the perpendicular component of the dominant mode in the rotary attenuator described in section 12.5.

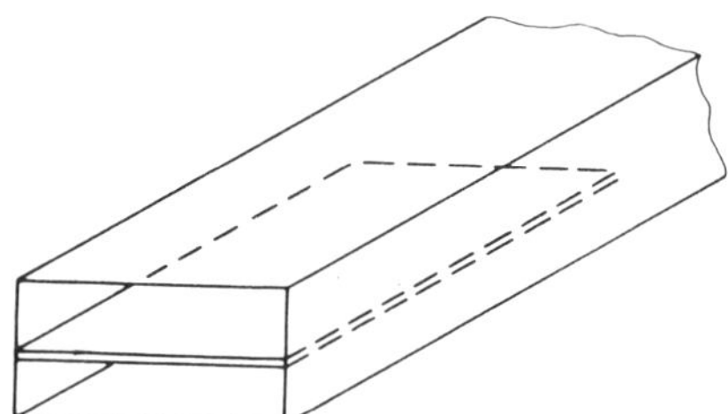

FIG. 11.12. A mode filter. The metal vane discriminates in favour of the dominant mode in rectangular waveguide.

In square waveguide, there are two dominant modes whose electric field vectors are perpendicular to one another. A mode filter is shown in Fig. 11.13 which discriminates between the two perpendicular modes in square waveguide. The metal vanes allow one mode free passage into the taper to the rectangular waveguide and the perpendicular mode, which would be reflected by the taper, is deflected into the side arm rectangular waveguide. Such a mode filter, or mode duplexer, in circular waveguide is used in the Faraday rotation circulator described in section 13.3.

11.12. Stub Tuner

The effects of any mismatch in a waveguide system can be cancelled at any particular frequency by introducing another mismatch elsewhere in the

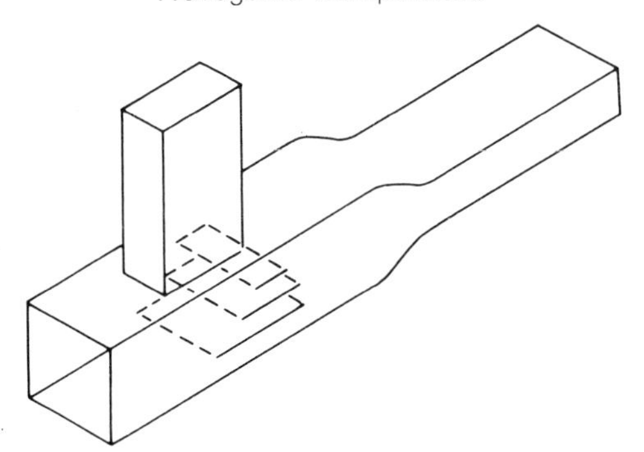

FIG. 11.13. A mode filter which separates the two perpendicular dominant modes in square waveguide.

system whose reflection coefficient is in anti-phase with that of the original mismatch, hence cancelling the reflected wave. The device used to introduce the additional mismatch into the system is called a *stub tuner* or a *matching section*. The stub tuner needs to produce a variable mismatch with a variable phase. In transmission line systems, the stub tuner consists of a short-circuited length of transmission line of variable length. However, the theory given in sections 1.10 and 4.3 shows that the stub tuner is providing a variable susceptance across the transmission line. In waveguide, a variable capacitance can be provided by a variable length of post inserted into the centre of the broad face of the waveguide. This provides the variable mismatch. The variable phase is produced by varying the position of the post along the axis of the waveguide. The *variable mismatch unit* provides this facility. A carriage moves along a waveguide having a longitudinal slot in the centre of its broad face, and carries a post protruding into the waveguide through the slot. It is similar in construction to the standing wave detector shown in Fig. 11.9. A simpler system is to use a number of posts at fixed positions in the waveguide. Then a suitable combination of post insertions will give the required mismatch in the correct phase. Theory shows that any mismatch could be cancelled using three stubs, but some manufacturers use four or five stubs to give greater versatility. In its simplest form, the stub tuner consists of the required number of screws equally spaced along the centre line of the broad face of the waveguide. As the screw is inserted, it provides the mismatch in the waveguide. It then only requires some provision for locking the screws in position when they are set as required. This device is sometimes called a *screw matching section*, and is shown diagrammatically in Fig. 11.14.

11.13. Wavemeter

The wavemeter is a waveguide frequency meter. Frequency can be measured precisely by electronic methods, but these are beyond the scope

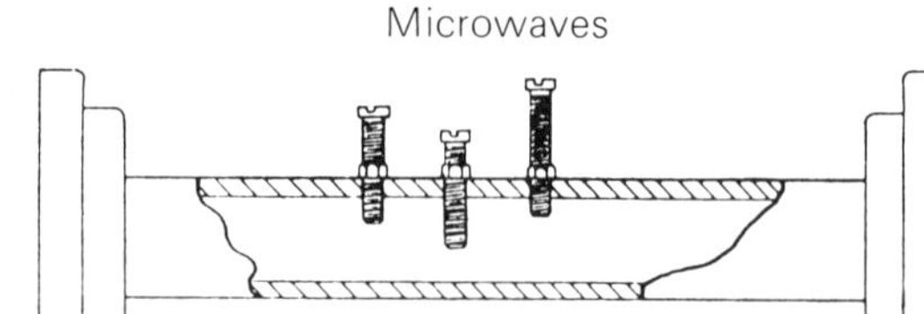

FIG. 11.14. Three-stub tuner or screw matching section.

of this book. The wavemeter uses a microwave resonant cavity which is made variable in length and consequently variable in resonant frequency. If the cavity is coupled to a waveguide system, on tune, the cavity abstracts a small amount of power from the microwave system, whereas at other frequencies it has no effect. A cylindrical cavity wavemeter is shown diagrammatically in Fig. 11.15. The *absorption wavemeter* acts by absorbing a small amount of the microwave power at its resonant frequency causing a small dip in the indicated power in the waveguide system. The *transmission* or *indicating wavemeter* also absorbs a small amount of power at its resonant frequency, but it couples this power to an output waveguide or detector. When the wavemeter is on tune, there is an indicated output from it but at all other frequencies there is no output.

The absorption wavemeter causes a dip in the output power of a waveguide system. It is used to set a microwave oscillator to the required frequency or it can be used to provide an indicating dip on a swept frequency display. It

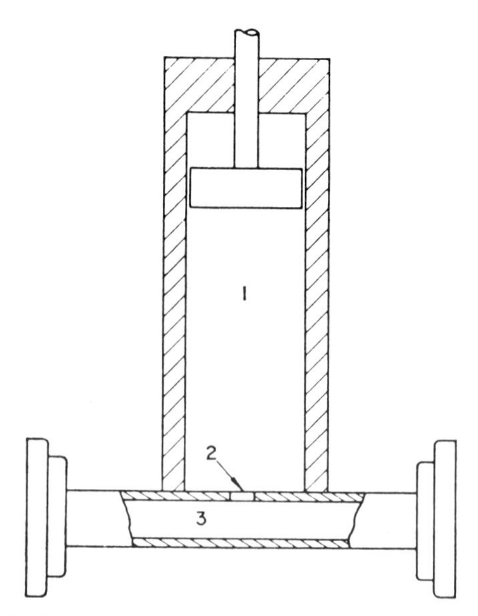

FIG. 11.15. Cylindrical cavity wavemeter. (1) Cavity. (2) Coupling hole. (3) Waveguide.

cannot be used to provide a continuous monitor of the frequency of a fixed frequency system. The absorption wavemeter is simple and cheap and is usually of relatively low Q-factor. The transmission wavemeter can be so loosely coupled to the waveguide system that it does not cause any appreciable loss of power in the main output when it is on tune. It provides an output which can be continuously monitored if required. The transmission wavemeter is complicated and more expensive, especially if it incorporates its own detector, but it is usually of high Q-factor giving better frequency discrimination than many absorption wavemeters.

The simplest wavemeter cavity is a length of rectangular waveguide which is then direct reading in waveguide wavelength. This is a low Q cavity. Higher Q cavities are usually circular in cross-section and may operate on modes other than the fundamental. The position of the variable short circuit is controlled by a micrometer screw. Although the resonant frequencies of all cavities can be calculated from the dimensions, it is found that wavemeters need calibration against some fundamental frequency standard. This is because the finite conductivity of the metal (usually copper) of the cavity modifies the waveguide wavelength and because the design of the variable short circuit sometimes means that the electrical position of the short circuit is not coincident with the front end of the plunger. These variations only become noticeable in the design of high Q resonant cavities. A grade 1 precision wavemeter has an accuracy of 1 part in 10^4.

11.14. Summary

11.2. ***Rigid rectangular waveguide*** comes in a number of standard sizes covering the whole of the frequency range, 0.3–300 GHz. A list of 34 standard sizes is given in Table 11.1.

11.3. ***Couplings*** are used to connect lengths of waveguide. A choke coupling ensures a reflectionless joint when there is an electrical discontinuity.

11.5. The ***directional coupler*** is used to couple a portion of the power in the main waveguide run into a side arm. Its performance is given by eqns. (4.24) and (4.25).

11.6. A ***magic-T*** is a matched ***hybrid T-junction*** which behaves like an equal power split directional coupler. Input at any one arm is split equally between the two perpendicular arms and there is no output at the fourth arm.

11.7. The ***matched termination*** is used to provide a reflectionless termination for a length of waveguide.

11.8. The ***short circuit*** provides a termination which reflects all the incident microwave power.

11.9. The ***standing-wave meter*** has a variable position probe, coupled to the electric field in the waveguide. It is used to detect the standing wave pattern in the waveguide and to measure the voltage standing wave ratio.

11.10. A ***probe*** is used to couple microwave signals between a waveguide or cavity and a coaxial line.

In an ***electric probe***, the inner of the coaxial line projects into the waveguide electric field of the waveguide mode.

In a ***magnetic probe***, the inner of the coaxial line terminates in a loop linking with the magnetic field of the waveguide mode.

A ***crossbar transformer*** or a ***door-knob transformer*** can also be used to couple a microwave signal between waveguide and a coaxial line.

11.11. A ***mode filter*** is designed to discriminate between different modes of propagation in the waveguide by absorbing or reflecting the unwanted modes.

11.12. The ***stub tuner*** is used to provide a mismatch in a waveguide system of such an amplitude and phase, that it can be used to cancel an undesired standing wave already existing in the system. It may consist of a single stub which is variable in position along the waveguide and is called a ***variable mismatch unit*** or it may consist of three (or more) stubs at fixed positions in the waveguide called a ***three*** (or more) ***stub tuner***. If the variable insertion stubs are simple screws, the unit is sometimes called a ***screw matching section***.

11.13. The ***wavemeter*** is a variable tuning resonant cavity used as a frequency meter.

The ***absorption wavemeter*** on tune absorbs some of the power in a waveguide system and causes a corresponding dip in the indicated output of the waveguide system.

The ***transmission wavemeter*** on tune provides an output in a side arm which may or may not have an integral detector.

12

Devices

12.1. Phase Changer

In any transmission line system, the simplest method of providing additional phase change over and above that naturally occurring due to any lengths of line already in the system, is to provide additional lengths of transmission line. They are switched into the system as required. The switches may be mechanical or they may be electrical using transistors or PIN diodes. The size of the additional lengths of transmission line may be reduced by forming them of a high permittivity dielectric. Switched phase changers are compatible with a control system having a digital output.

A continuously variable phase changer may be provided by having a trombone section of transmission line or waveguide. Such a trombone section is not usually very precise due to the difficulty of providing sliding contacts that do not produce distortion of the fields in the transmission line or waveguide. An alternative is to use a 3 dB directional coupler and two variable position short circuits as shown in Fig. 12.1. The input signal is divided in half by the directional coupler and gives an output of two equal signals having equal phase. Provided the two short circuits are coupled

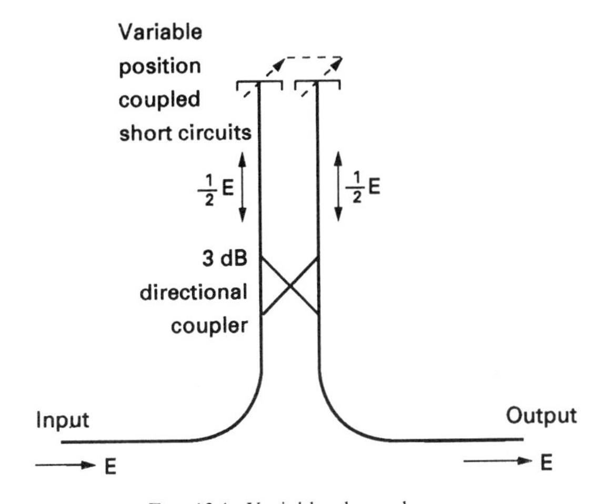

FIG. 12.1. Variable phase changer.

together, the two signals are reflected back with equal phase to the directional coupler and combine to give an output whose phase is variable compared to the input. If the device is realized in low-loss transmission line or waveguide, there is no change in the amplitude of the output signal as the phase is changed and the device is also suitable for high-power use. In waveguide, everything is constructed in waveguide metal and the high-power capability of the device is limited by the high-power capability of the directional coupler or of the variable short circuits.

In rectangular waveguide, variable phase change may be provided by moving a slab of dielectric material across the waveguide, giving a compact variable phase changer in a fixed length of waveguide. The slab is made of some low-loss material with a permittivity greater than that of air. The dielectric slab affects the electric field distribution in the broad dimension of the waveguide so that it is distorted from sinusoidal to that indicated in Fig. 12.2. This shows that the dielectric slab has the same effect as increasing the broad dimension of the waveguide and it reduces the wavelength in the waveguide. Hence it alters the electrical length of the phase changer compared with that of an equal length of empty waveguide. The amount of phase change depends on the position of the dielectric slab in the waveguide. The slab is moved across the waveguide from the position of minimum electric field strength at the wall to the position of maximum electric field strength at the centre. The slab has the least effect when it is adjacent to the narrow wall of the waveguide and it has the maximum effect when it is in the centre of the waveguide. The slab may be mounted on two rods in order to move it similar to the mechanism shown in Fig. 12.6 for the vane attenuator.

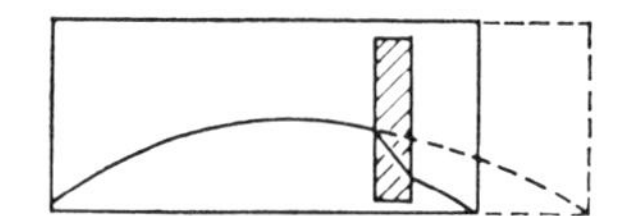

FIG. 12.2. A graph of the electric field strength inside rectangular waveguide containing a dielectric slab and, dotted, that of the equivalent empty waveguide.

There are different degrees of accuracy claimed for waveguide phase changers but these are only concerned with the accuracy of the mechanism that is used to position the slab of dielectric in the waveguide. The phase changer needs to be calibrated against a standard. It does not have a linear relationship of phase change with position nor is its calibration constant with change of frequency.

12.2. Rotary Phase Changer

The rotary phase changer provides continuously variable phase change for the TE_{11}-dominant mode in circular waveguide. It is necessary to use a rectangular to circular waveguide transformer at each end when connecting the rotary phase changer into a rectangular waveguide system. It is a self calibrating instrument whose phase change is independent of frequency. A diagram of the rotary phase changer is shown in Fig. 12.3. The waveguide transformer changes the TE_{10}-mode in rectangular waveguide into the TE_{11}-mode in circular waveguide. A study of Fig. 5.3 and Fig. 6.5 show that the field patterns of the two dominant modes in the two different shapes of waveguide are similar so that a simple taper provides a suitable transformer between the modes in the two different waveguides. The input linearly polarized wave is changed into a circularly polarized wave in circular waveguide. The circular polarizer consists of a quarter-wave plate which is described in section 6.12. The action of the half-wave plate on a linearly polarized wave is also described in section 6.12. On a circularly polarized wave, the half-wave plate reverses the hand of polarization and makes a variable phase change equal to twice the angle through which it has been rotated. Therefore the rotary phase changer is self-calibrating with the phase change given in terms of angle and with no change of attenuation.

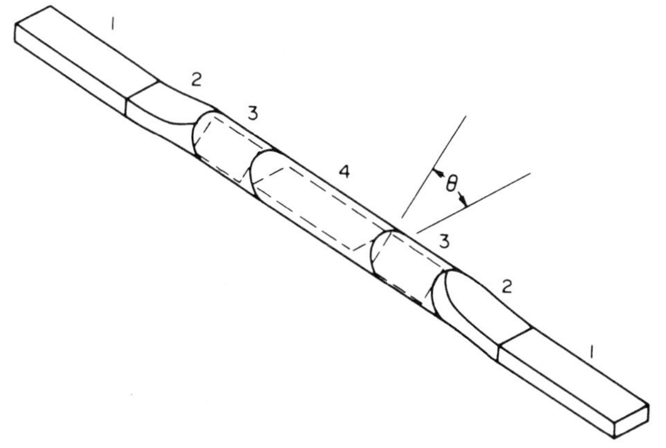

FIG. 12.3. Rotary phase changer. (1) Input rectangular waveguide. (2) Rectangular to circular waveguide transformer. (3) Circular polarizer. (4) Rotating section of circular waveguide containing a half-wave plate.

12.3. Pad Attenuator

Signal attenuation is caused by suitable resistive networks. At low frequencies, lumped resistor networks can be used as attenuators. At higher frequencies, T or π resistive networks operate satisfactorily if the shunt elements are thin film discs and the series elements are thin tubular films. The resistance material is a thin film of conducting material such as nichrome deposited onto a dielectric support. A resistive T-network in coaxial line is shown in Fig. 12.4 where the elements of the T-network are identified from Fig. 1.3. Considering the matched T-network of Fig. 1.3, if V_1 and V_2 are the

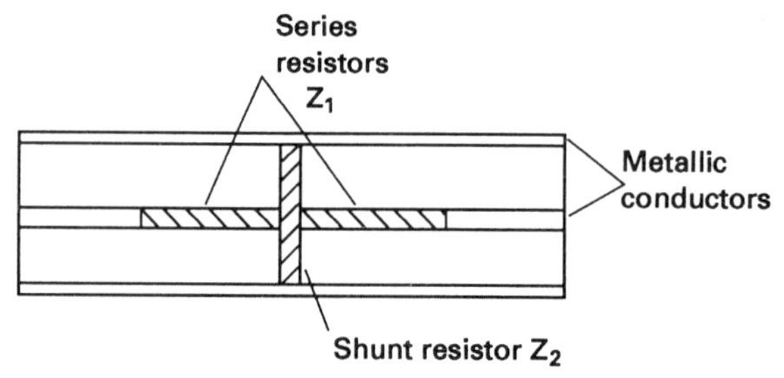

FIG. 12.4. Resistive T-network attenuator in coaxial line.

input and output voltages respectively, the voltage drop through the T-network is given by

$$V_1 - V_2 = \frac{Z_1}{Z_0}(V_1 + V_2) \tag{12.1}$$

Define the voltage attenuation,

$$A = \frac{V_1}{V_2}$$

then from eqn. (12.1), the normalized value of the series element is

$$\frac{Z_1}{Z_0} = \frac{A - 1}{A + 1} \tag{12.2}$$

and from eqn. (1.3), the normalized value of the shunt element is

$$\frac{Z_2}{Z_0} = \frac{Z_0}{2Z_1} - \frac{Z_1}{2Z_0} = \frac{2A}{A^2 - 1} \tag{12.3}$$

For even higher frequencies of operation, these thin film resistors are no longer lumped elements. The resistive network has to be replaced by its distributed equivalent, consisting of a series and shunt resistance distributed along the line. These elements could be provided by a high resistance conductor for the inner of the coaxial line and a low resistance material for the insulator between the conductors of the line. However, a much simpler approximation to these lossy components of the coaxial line is a plane resistive film completely spanning the inside of the coaxial line. Then the series resistance is provided by the centre section of the resistive film and the shunt resistance by the outer sections of the film. Figure 12.5 shows such a film mounted in a coaxial line having outer and inner radii a and b respectively. Remembering that the resistive element is a plane sheet, the series resistance is given by

$$R = \frac{\rho}{2b}\,\Omega/\text{m}$$

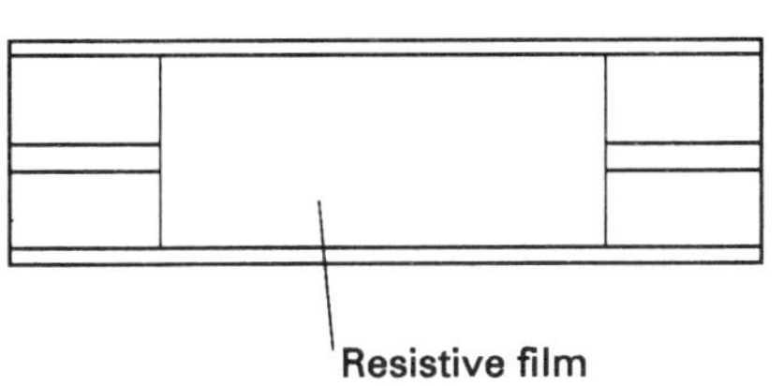

FIG. 12.5. Resistive film attenuator in coaxial line.

and the shunt conductance by

$$G = \frac{2}{\rho(a - b)} \text{ S/m}$$

where ρ is the surface resistivity of the resistive sheet. For a transmission line, where the attenuation terms dominate, we see from eqn. (1.22) that the attenuation constant is given by

$$\alpha = \sqrt{(RG)} \tag{12.4}$$

and from eqn. (1.28) the characteristic impedance is given by

$$Z_0 = \sqrt{\left(\frac{R}{G}\right)} \tag{12.5}$$

Substitution of values for R and G for the planar film into eqn. (12.4) show that the attenuation is independent of the resistivity of the film. Therefore such a distributed attenuator has an attenuation which is independent of temperature.

In microstrip circuits or integrated circuits, such resistive networks may be switched electrically using transistors or PIN diodes. Continuously variable attenuation is difficult to provide using resistive networks. One very good mechanical switch for use with resistive attenuation networks is provided by thin strip conductors of spring metal which may be switched between two stable positions and which provide a minimum disturbance to the fields on the line. The line itself is the strip of spring metal which has two alternative positions switching resistive elements into or out of the circuit as required.

A high-precision attenuator for use at the lower microwave frequencies is the *piston attenuator*. If a section of waveguide is operated below its cut-off frequency, the microwave field decays exponentially in the waveguide and provides calculable attenuation of the wave. For the cut-off condition, when $\beta = 0$ and $\gamma = \alpha$, eqn. (5.9) becomes

$$\alpha = k_z = \pm\sqrt{(k_c^2 - k^2)} \tag{12.6}$$

When the operating frequency is much less than the cut-off frequency, $k_c > k$

and eqn. (12.6) becomes $\alpha \approx k_c$. The power ratio in a length of cut-off waveguide is

$$\frac{P_2}{P_1} = \exp(-2\alpha z) = \exp(-2k_c z) \tag{12.7}$$

Then the attenuation (in dB) is

$$10 \log_{10} \frac{P_1}{P_2} = 2k_c z \log_{10} e = 8.68\, k_c z \text{ dB} \tag{12.8}$$

The attenuation given by eqn. (12.8) is a very simple relationship, and is used to provide high precision attenuation at low microwave frequencies. The piston attenuator uses the TE_{11}-mode in circular waveguide, so that accurate dimensions may be achieved easily. The waveguide has a short circuit at each end having coaxial line probes to the magnetic field, as shown in Fig. 11.11, in the middle of each short circuit. One short circuit is a piston which moves along the waveguide controlling its length and therefore the attenuation. In order to eliminate nonlinearities due to end effects in the cut off section of waveguide change in attenuation is measured. Then the change in attenuation in dB is proportional to the movement of the piston.

12.4. Vane Attenuator

The level of microwave power in a waveguide may be attenuated by absorbing a portion of the power in a conducting or absorbing material. Most attenuators use a thin sheet of conducting material, such as nichrome, deposited onto some inert support, such as fibreglass sheet or glass strip. The conducting material is formed into a vane which is inserted into the waveguide parallel to the position of maximum electric field strength, as shown diagrammatically in Fig. 12.6. The absorbing vane is positioned

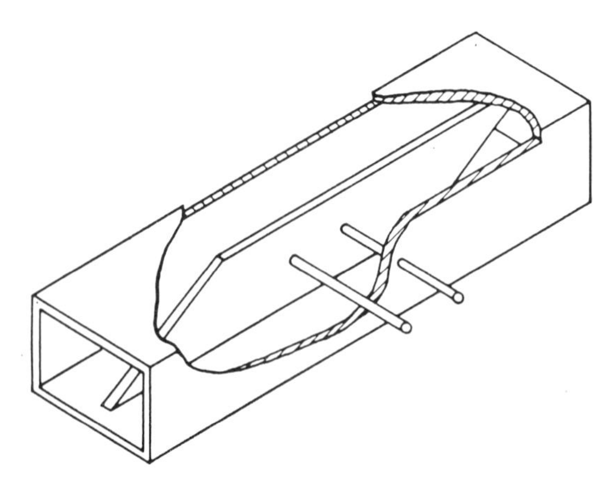

FIG. 12.6. Vane attenuator showing the rods for moving the vane.

parallel to the narrow wall of the waveguide and in order to provide a variation of attenuation it is moved from a position of minimum electric field strength, where it absorbs a minimum of power, to that of maximum electric field strength, where it absorbs a maximum amount of microwave power. Alternatively it can be inserted into the waveguide at the position of maximum electric field strength through a slot in the centre of the broad wall of the waveguide. In this form it is often called a *flap attenuator*.

Because the conducting material has to be supported on a thin sheet of dielectric material to make the absorbing vane, the vane attenuator acts as a variable phase changer as well as a variable attenuator. The theory of its operation as a phase changer is the same as that of the phase changer described in section 12.1. This means that the electrical length of the vane attenuator varies as its attenuation is varied.

There are different degrees of accuracy claimed for vane attenuators but these are only concerned with the accuracy of the mechanism that is used to position the vane in the waveguide. The vane attenuator has to be calibrated against some standard of attenuation. It does not have a linear relationship of attenuation with position nor is its calibration constant with change of frequency. The reason for the variation of calibration with frequency can be seen from the fact that the waveguide wavelength varies with frequency, causing the electrical length of the absorbing vane to vary with frequency.

12.5. Rotary Attenuator

The rotary attenuator provides continuously variable attenuation of the TE_{11}-dominant mode in circular waveguide. Its construction is similar to that of the rotary phase changer described in section 12.2. A diagram of the rotary attenuator is shown in Fig. 12.7. This is a more complicated attenuator than the vane attenuator but it is a self-calibrating instrument whose calibration is independent of frequency. The waveguide transformer changes the TE_{10}-mode in rectangular waveguide into the TE_{11}-mode in circular waveguide. The mode absorber operates similarly to the mode filter described in section 11.11 and shown in Fig. 11.12. It consists of an absorbing vane mounted diametrically across the circular waveguide. The mode absorber attenuates any microwave signal having its electric field parallel to the absorbing vane. The TE_{11}-mode in circular waveguide may be resolved into two mutually perpendicular components. The mode absorber absorbs that component of the TE_{11}-mode having its electric field parallel to the plane of the vane while it does not affect the perpendicular component.

The mode absorber is mounted with the vane parallel to the broad face of the rectangular waveguide. It is used so that if there are any signals present in the circular waveguide which are linearly polarized in a plane perpendicular to that of the rectangular waveguide, they are absorbed rather than reflected

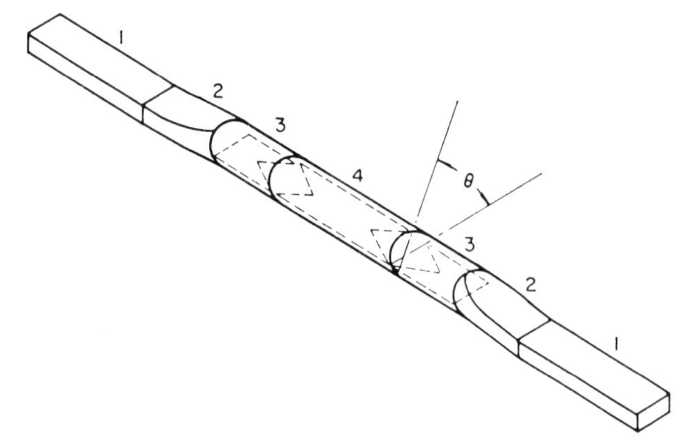

FIG. 12.7. Rotary attenuator. (1) Input rectangular waveguide. (2) Rectangular to circular waveguide transformer. (3) Mode absorber, consisting of an absorbing vane mounted parallel to the broad wall of the rectangular waveguide. (4) Rotating section of circular waveguide also containing a mode absorber.

from the transformer. The centre section of the attenuator also contains a mode absorber and it is free to rotate compared with the rest of the attenuator. Let it make an angle θ with the position of minimum attenuation. The linearly polarized wave in the circular waveguide is resolved into two components parallel to and perpendicular to the mode absorber in the centre section. The parallel mode is absorbed, and the perpendicular mode is transmitted without loss. If the input signal has an amplitude E_0, as shown in Fig. 12.8, then the transmitted signal leaving the rotating section has an amplitude $E_0 \cos\theta$. At the re-entry to the second fixed section, the signal is resolved in a like manner the second time and the final signal out is

$$E_{out} = E_0 \cos^2\theta \tag{12.9}$$

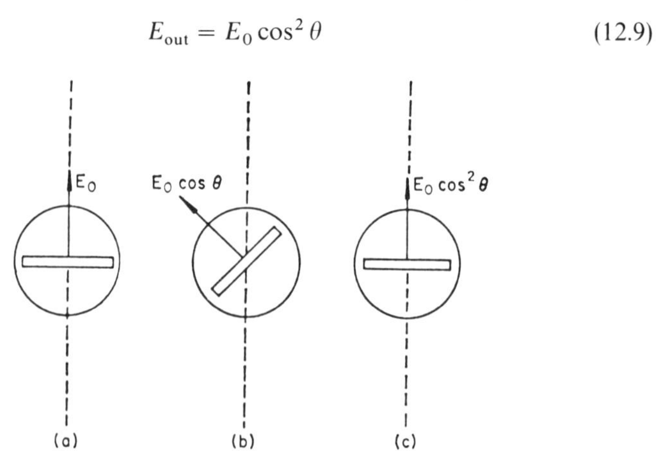

FIG. 12.8. Some relative field strengths in the rotary attenuator. (a) A section through the input mode absorber. (b) A section through the centre section. (c) A section through the output mode absorber.

If the attenuator is calibrated in dB, the attenuation is

$$\text{attenuation} = 40 \log_{10}(\sec\theta) \quad \text{dB} \tag{12.10}$$

where the logarithm is to the base 10 in accordance with the definition of dB.

The rotary attenuator is a device where attenuation is given in terms of an angle. Hence it is self-calibrating provided that care is taken to eliminate sources of error. The sources of error are in the alignment of the absorbing vanes in the mode absorbers, the possibility of reflections from the mode absorbers and the transformers, and in the accuracy of the measurement of the angle θ. The rotary attenuator makes a good precision waveguide attenuator. The rotary attenuator also has the advantage that its electrical length does not vary as its attenuation is varied. It is a constant phase device.

12.6. PIN Diode Attenuator

In microwave circuits, the PIN diode is used most frequently as a switch. However, it may also be used as a variable attenuator. At microwave frequencies, its impedance depends on the bias current. When the diode is forward biased, the diode resistance is low and most of the microwave energy is absorbed in the diode or reflected back down the line. However, when the diode is reverse biased, the diode resistance is high and the electromagnetic wave is transmitted without loss. Diodes may be mounted in a section of ridge waveguide so as to shunt the electromagnetic wave as shown in Fig. 12.9. Three diodes are used so that the mismatch due to one diode may be

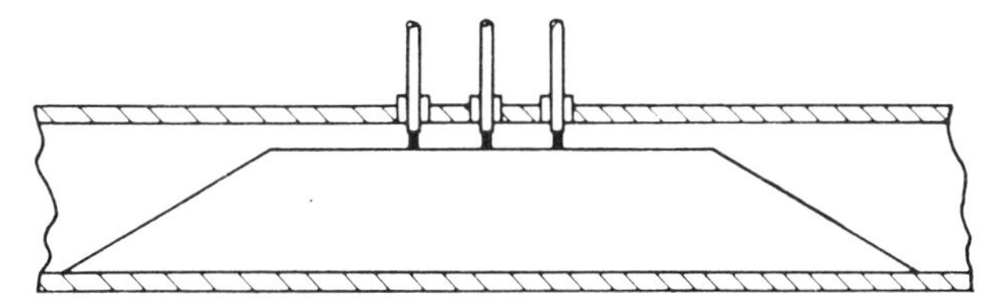

FIG. 12.9. A PIN diode mount showing three diodes mounted in a short section of ridge waveguide with tapers to the rectangular waveguide.

cancelled by the mismatch due to the others. Then when the diodes are reverse biased, the attenuation through the device is very small. Variable attenuation is provided by varying the bias on the diodes. Depending on the bias, some microwave power is absorbed in the diodes. The device may be used as a variable attenuator controlled by the biasing current in the diodes.

12.7. High-Power Attenuator

In all the attenuators described so far, the unwanted microwave power is absorbed in some attenuating element inside the device. Consequently, the power capability of the attenuator is limited by the permissible temperature rise of the attenuating element. In this section, an attenuator is described in which the unwanted microwave power is diverted into a matched termination so that the power capability of the device is limited by the power handling capability of the matched termination. In all previously described attenuators, the power absorbing element is mounted at the centre of a waveguide pipe so that even if it were made of some ceramic capable of withstanding a high temperature, there is no easy path for the dissipation of the absorbed energy. However, inevitably, the high-power attenuator consists of a system of waveguide components connected together so as to control the microwave power at the output. The system is shown in Fig. 12.10. At the input is a 3 dB directional coupler, which provides a half-power split into two parallel lines. In one line is a high-power phase changer as described in section 12.1 and shown in Fig. 12.1. At the output is a second 3 dB directional coupler. In this coupler, the output at one port is the sum of the two input signals and at the other port is the difference between the two signals. As the phase of the signal in the side arm is varied, the proportion of the power directed into the high-power matched termination varies from zero to the whole of the input signal. For medium power applications, the high-power phase changer in Fig. 12.10 could be replaced by a vane type dielectric slab phase changer in rectangular waveguide if the dielectric slab is made of a high-temperature ceramic. This makes a more compact device but, for the highest power capability, the device shown in Fig. 12.10 having no inclusions inside the metallic waveguide is essential.

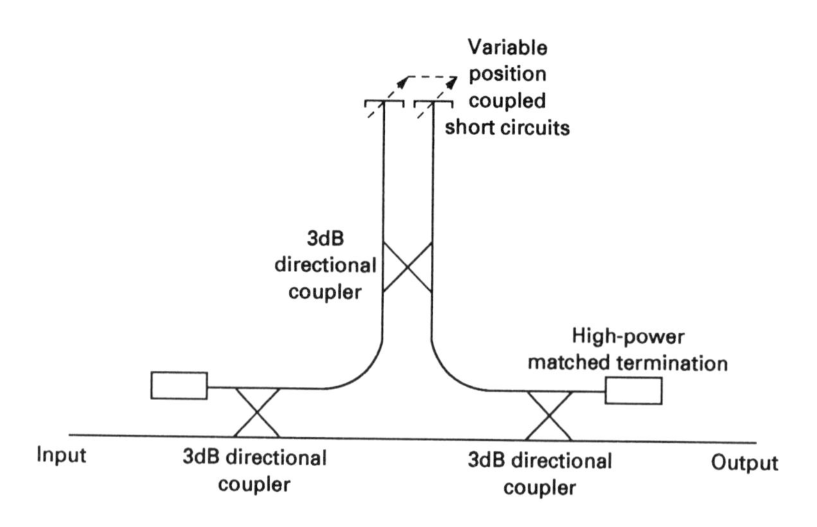

FIG. 12.10. High-power variable attenuator.

12.8. Crystal Receiver Mount

Microwave signals are detected in a crystal receiver by means of a rectifying detector crystal. Many modern microwave receivers are constructed in microstrip or with microwave integrated circuits and the detector or mixer diodes are part of any such circuit. However, there are situations where a detector or mixer diode is needed in a waveguide system and this section describes a waveguide mount for a detector or mixer diode. A microwave diode as described in section 10.2 is enclosed in a mount suitable for use in waveguide or coaxial line. In waveguide the diode can be placed across the centre of the waveguide so that the wires making contact with the crystal are parallel to the electric field, or a coaxially mounted crystal can be placed in a short coaxial line adjacent to the waveguide. A crystal receiver using a door-knob transition to coaxial line is shown diagrammatically in Fig. 12.11.

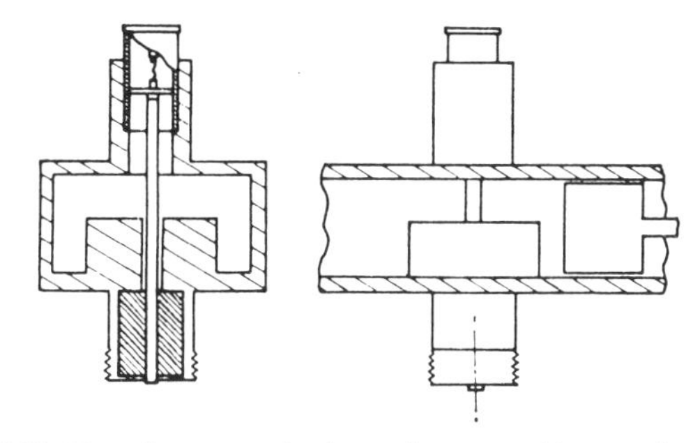

FIG. 12.11. Two diagrammatic views of a waveguide crystal receiver.

The transition is followed by a short circuit in the waveguide so that any power passing the transition is reflected back to increase the power flowing to the crystal and to cancel any power reflected from the transition. The output from the crystal is direct current which can be displayed on a galvanometer. If the microwave signal is amplitude modulated, the output is a direct current with a superimposed alternating component proportional to the amplitude modulation.

The detector crystal has a non-linear characteristic. Therefore it can also be used as a mixer. The operating conditions of a receiver and a mixer are different so that a waveguide device designed as a receiver is not necessarily the same as a waveguide device designed as a mixer. Provided the optimum performance is not required from them, they are interchangeable. The crystal receiver is a non-linear device so that it is usually used to indicate microwave power rather than to measure it. However, the crystal does have an approximately square law characteristic; the output current is approximately proportional to the square of the electric field in the waveguide which makes it proportional to the power in the waveguide. For small changes in power level this approximation is used to make measurements with a crystal receiver. In the standing wave detector, the square law relationship is assumed so that the output is the square of the input voltage.

12.9. Bolometer

It has already been explained that the crystal receiver is not a suitable instrument for absolute measurements of power level. Even if the crystal receiver is calibrated against some other standard, it is not suitable as its characteristics change with time and temperature. At powers greater than 1 W, the power in the waveguide may be measured by means of a water calorimeter. This is a matched termination where water is used as the absorbing medium. The rate of water flow through the termination and its temperature rise give the power being absorbed by the termination and hence power in the waveguide. For lower powers, the measurement is made using bolometric methods. The essence of bolometric power measurement is that when power is dissipated in a resistive element, the element heats up and a change occurs in the element's resistance. The resistive elements are termed *bolometers* and are of two kinds, *thermistors* and *barretters*. The thermistor is a bead of a semiconductor material, mounted between two fine wires, which has a negative temperature coefficient of resistance. The important thing is not that the resistance decreases as the temperature rises but that the change of resistance with temperature is large. The barretter uses an element which has a positive temperature coefficient of resistance which is not as large as that of the thermistor. The barretter consists of a fine wire or a thin film of material deposited onto a glass sheet or as part of a microstrip circuit.

A thin film resistor can be used as a lumped resistor to terminate a microwave transmission line. If the thin film resistor is combined with a temperature sensing element, the device can be used for power measurement. Such a device is the thin film thermoelectric powermeter. The microwave power is absorbed in a terminating resistor made of thermoelectric material so that a small d.c. potential proportional to the mean absorbed power is produced. A typical microstrip circuit is shown in Fig. 12.12.

For waveguide use the bolometer element is housed in a ceramic cartridge, which is mounted across the waveguide so that the wires to the element are parallel to the microwave electric field. If the bolometer mount is followed by a short circuit in the waveguide and the whole is matched, then all the incident power is absorbed in the bolometer element. The resistance of the bolometer element is measured by a wheatstone bridge. The unbalance current in the wheatstone bridge can be used to indicate the incident microwave power. In its most accurate form, the bridge allows considerable direct current to flow through the element. Initially it is balanced with no microwave power incident on the device. Then, when measuring power, the element will warm up, unbalancing the bridge, which can be brought back

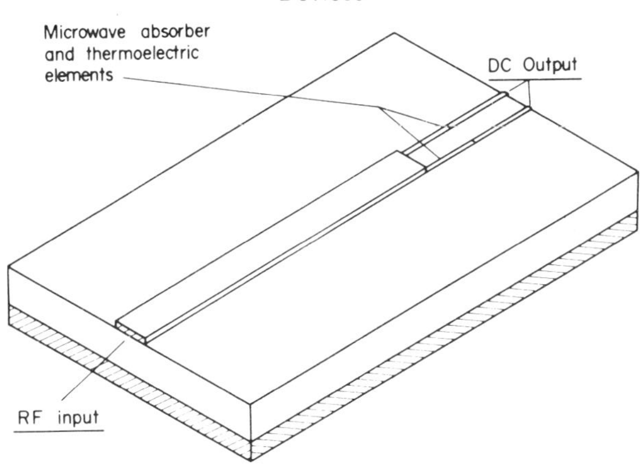

FIG. 12.12. A thin film thermoelectric power measuring device.

to balance by reducing the direct current flowing through the element. The power represented by the reduction in direct current is the same as the microwave power being absorbed by the device.

The detector crystal and the bolometer are both low-power devices and will burn out under high power. If it is required to detect or measure higher powers, it is necessary to absorb some of the power in an attenuator first. The crystal receiver is the device to use whenever detection under working conditions is required as it has the greatest sensitivity and is of more robust construction. The bolometer mount is the device to use when absolute measurement of microwave power is required.

12.10. Summary

12.1. ***Phase change*** is provided by altering the electrical length of a transmission line. Additional lengths of line may be switched into the microwave circuit as required. A combination of a 3 dB directional coupler and two ganged variable position short circuits provide continuously variable phase change.

The ***phase changer*** is a device which alters the electrical length in the waveguide. Variable phase change is obtained by moving a dielectric slab across the waveguide from the position of minimum electric field to the position of maximum field.

12.2. In the ***rotary phase changer***, the phase of a circularly polarized wave is varied by the rotation of a section containing a half-wave plate. The phase change is twice the angle of rotation.

12.3. In the ***pad attenuator*** a suitable combination of resistors is used to attenuate the microwave signal on a transmission line.

Continuously variable attenuation can be provided in waveguide. The ***piston attenuator*** uses a section of a cut-off waveguide as a calculable attenuator,

$$\text{attenuation} = 8.68\, k_c z \text{ dB} \tag{12.8}$$

12.4. The ***vane attenuator*** absorbs some of the power in a conducting vane placed parallel to the electric field vector in the waveguide. The vane is moved from a position of minimum electric field strength to that of maximum field strength in order to vary the electromagnetic power absorbed by the vane.

12.5. The ***rotary attenuator*** resolves the wave into two components by means of an absorbing vane at a variable angle to the plane of the electric field of the incident wave. The angle determines the portion of the wave that is absorbed. The law of the attenuator is

$$\text{attenuation} = 40 \log_{10}(\sec\theta) \quad \text{dB} \tag{12.10}$$

12.6. A ***PIN diode*** presents a variable impedance to a microwave signal. The impedance is controlled by the direct current flowing through the diode. When the diode is mounted in waveguide, it acts as a variable attenuator.

12.7. The ***high-power attenuator*** uses 3 dB directional couplers and a high-power phase changer to divert some of the microwave power into a high-power matched termination.

12.8. A ***crystal receiver*** is a rectifying crystal which is used to produce a mean rectified output from the microwave electric field. The crystal may be mounted in a waveguide and gives an output current approximately proportional to the power in the waveguide.

12.9. The ***bolometer*** is a temperature sensitive resistive element which indicates the microwave power incident on it by its change of temperature. A thin film resistor made of thermoelectric material is used to terminate a microwave transmission line. A small d.c. potential proportional to the microwave power is produced. The bolometer in a waveguide bolometer mount absorbs all the incident microwave power. There are two bolometer elements: thermistors with a negative temperature coefficient of resistance, and barretters with a positive temperature coefficient of resistance. The resistance change is measured in some form of wheatstone bridge.

13

Ferrite Devices

13.1. Introduction

The gyromagnetic properties of ferrites have created a whole family of nonreciprocal and electrically controllable microwave devices, whose performance is dependent on the strength of an external biasing d.c. magnetic field. For the many devices which operate below saturation, the magnetic field is small and may easily be supplied by a small permanent magnet or magnets. The few devices depending on ferromagnetic resonance require a much larger magnetic field and consequently larger permanent magnets. Below saturation, the effective permeability of a ferrite material is directly related to the strength of the biasing magnetic field, enabling ferrite control devices to provide variable attenuation or phase change. Then a variable microwave property is controlled by an electric current flowing in a coil generating a variable magnetic field. Digital control systems may drive latching devices, such as the latching phase changer described in section 13.2, which use the remanent magnetization in the ferrite and have no external biasing magnetic field. In many low-power systems using microstrip or integrated circuits, ferrite nonreciprocal or control components are unsuitable because they are bulky compared with the other circuit components, so that the control functions are provided by transistor or diode switches in the microwave circuit. In medium or high-power systems ferrite control components are used. In very-high-power systems, great care is needed when using any devices having ferrite or anything other than metal in the waveguide.

The two nonreciprocal devices in common use are the isolator and the circulator. These ferrite devices provide microwave functions that cannot be provided any other way. The isolator is a device which allows free passage of the microwave power through it in one direction but absorbs any microwave power passing in the reverse direction. It is often used to protect a microwave oscillator from power reflected from the load. Ferrite isolators are described in sections 13.5 to 13.7. The other nonreciprocal device is the circulator. The circuit symbol for a four-port circulator is shown in Fig. 13.1. A signal input to port 1 is connected to port 2 and nothing goes to ports 3 and 4. Similarly, a signal input to port 2 is connected to port 3 and so on

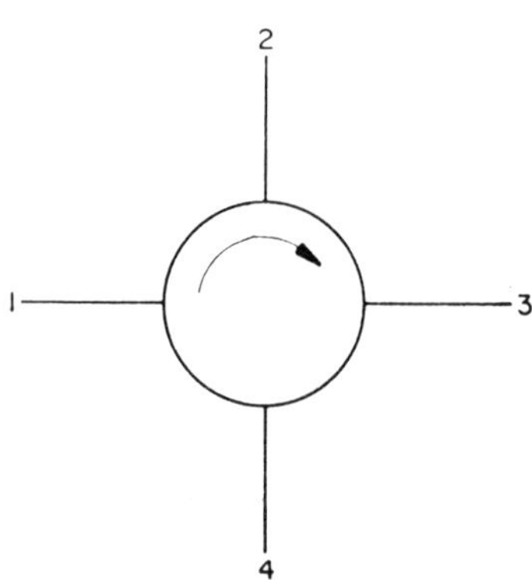

FIG. 13.1. Circulator circuit symbol.

in the direction of the arrow. A typical use of a circulator is shown in Fig. 13.2, where a three-port circulator enables a radar transmitter and receiver to use the same aerial. The circulator connects the transmitter to the aerial and then connects the return signal to the receiver. Practical radar systems are much more sophisticated than that shown in Fig. 13.2. No practical circulator is ideal and additional protection is needed to shield the receiver from transmitter signals.

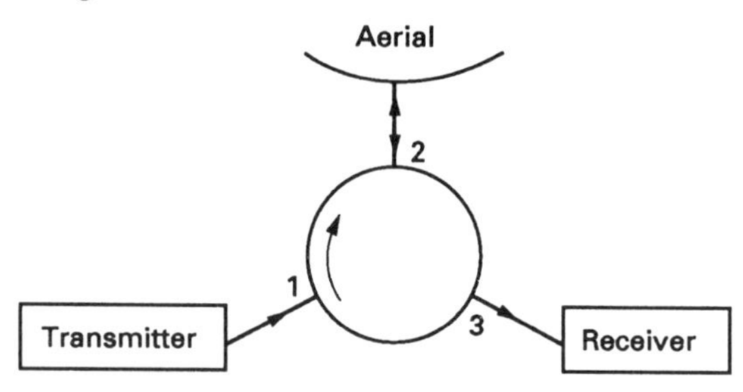

FIG. 13.2. A simplified description of a radar set using a circulator to connect both the transmitter and the receiver to one aerial.

13.2. Directional Phase Changer

A directional phase changer is a device in which the phase change for transmission in one direction is different from that for transmission in the opposite direction. The nonreciprocal ferrite phase changer shown in Fig. 8.11(a) is a directional phase changer. The magnetic field causing the directional phase change can be provided either by a permanent magnet giving a constant directional phase change or it can be provided by some kind of electromagnet giving a variable directional phase change. If the differential phase change is 180°, the device is called a *gyrator*. Another method of providing the variable phase change is to make use of the low frequency hysteresis loop of the ferrite material. Most microwave ferrite materials have a square hysteresis loop. If the magnetic circuit is completed

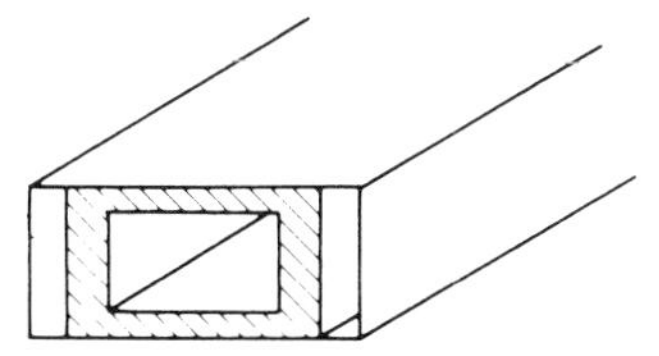

FIG. 13.3. Latching directional phase changer.

with a rectangular section ferrite core as shown in Fig. 13.3, unless deliberately demagnetized, the core will be in the state of remanent magnetization. The vertical arms of the ferrite core act as a differential phase changer similar to that shown in Fig. 8.11 (a) while the horizontal arms have little effect on the microwave field. If a magnetizing wire is threaded through the core along the centre of the waveguide, a single pulse of current is sufficient to magnetize the ferrite to its remanent condition which continues to provide a given amount of differential phase change until an opposite current pulse reverses the remanent magnetization and the direction of the differential phase change. Such a device is called a *latching phase changer*.

Ferrite spanning the cross-section of the waveguide is difficult to cool so that thin strips of ferrite adjacent to the waveguide walls, as shown in Fig. 13.7, are used for high-power operation.

The nonreciprocal effects of magnetized ferrite are also used to produce directional phase change in stripline circuits. The stripline has ferrite material as its dielectric which is magnetized in the plane of the strip and the circuit is devised so as to produce circularly polarized magnetic fields in the ferrite. A meander line with close coupling between adjacent conductors is used having each section of line a quarter wavelength long. At a point midway along one of the strips, the microwave magnetic field is shown by the dotted lines in Fig. 13.4. These magnetic fields are perpendicular to one another halfway between the two lines and, as there is a quarter wavelength between them, the two fields are 90° out of phase. Therefore this resultant magnetic field is circularly polarized in a plane perpendicular to the conductors, and the hand of circular polarization is different for the two waves travelling through the meander in different directions. If the ferrite is magnetized parallel to the line of the conductors, the wave experiences a different effective permeability depending on its direction of propagation along the meander, so that the device is a directional phase changer. The device is also a variable phase changer dependent on the strength of the magnetic field.

13.3. Waveguide Circulator

The nonreciprocal properties of ferrites may be used to make a *circulator*. The general properties of a circulator are given by reference to Fig. 13.1,

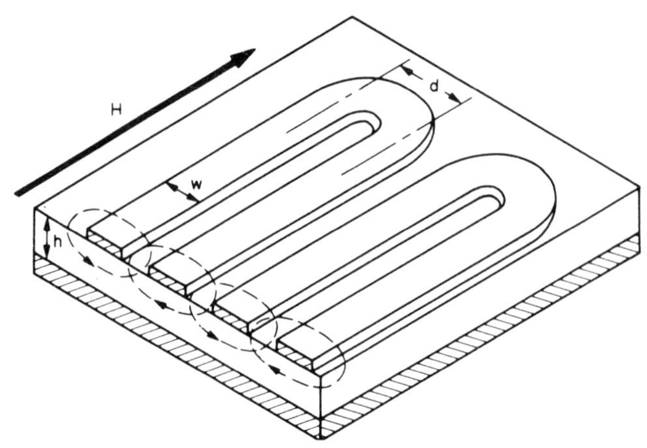

FIG. 13.4. A cross-section through a meander-line circuit used as a ferrite directional phase changer.

which shows its circuit symbol. If power is incident on port 1 it comes out of port 2 and there is no power coupled to the other ports. Similarly power incident on port 2 comes out of port 3, etc. The diagram shows a four-port circulator, but there is no restriction in principle to the number of ports in a circulator. Circulators can be made using a number of different ferrite devices suitably interconnected.

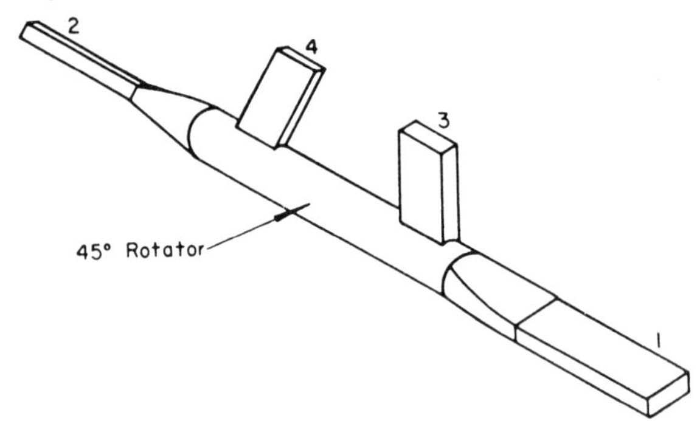

FIG. 13.5. Faraday rotation circulator.

A Faraday rotation circulator is shown in Fig. 13.5. Faraday rotation due to magnetized ferrite is described in section 8.9 and a ferrite rod rotator in section 8.11. A ferrite rod is situated on the axis of the central section of waveguide and a magnetic field is applied externally so that the plane of polarization of an incident wave is rotated 45° on transmission through this section. The magnetic field may be applied by means of a permanent magnet or by a current-carrying coil round the outside of the waveguide. The ferrite waveguide section is a 45° *rotator*.

The side arms (ports 3 and 4) shown on the diagram are so arranged that a wave polarized perpendicular to port 1 comes out of port 3 and one perpendicular to port 2 comes out of port 4. In the rotary attenuator, a mode absorber is used to absorb that linearly polarized component of the wave which cannot be accepted by the round to rectangular waveguide transformer. In this circulator, a mode duplexer in the waveguide, similar to that shown in Fig. 11.13, is used to deflect the perpendicular component into the side arm. The waveguide transformer changes the TE_{10}-mode in rectangular waveguide into the TE_{11}-mode in round waveguide. Circulation is obtained from this device as follows: a wave entering port 1 is rotated 45° and exits by port 2 with no power going to ports 3 and 4; a wave entering port 2 is rotated 45° and is in a plane perpendicular to port 1 so it exits by port 3. This method can be continued to show complete circulator action between the other ports.

A disadvantage of the Faraday rotation circulator is that it is a bulky waveguide device and is now rarely used. The ferrite rod is at the centre of the waveguide making it difficult to magnetize or to cool. However, the device is described here because it provides an easily understood description of circulator action. Another combination of components which provides circulator action is the directional phase change circulator. In this device, the ferrite is adjacent to the waveguide wall and is easily magnetized and cooled. It consists of two 3 dB directional couplers and a directional phase changer as shown in Fig. 13.6. A signal input at port 1 is split equally at the first directional coupler. The phase change due to the ferrite and to the dielectric are both 90° and the signals combine in the second directional coupler to give an output at port 2 and nothing out of port 4. A reverse wave entering at port 2 experiences equal and opposite phase changes of 90° due to the dielectric slab and the ferrite slab and the two output signals are

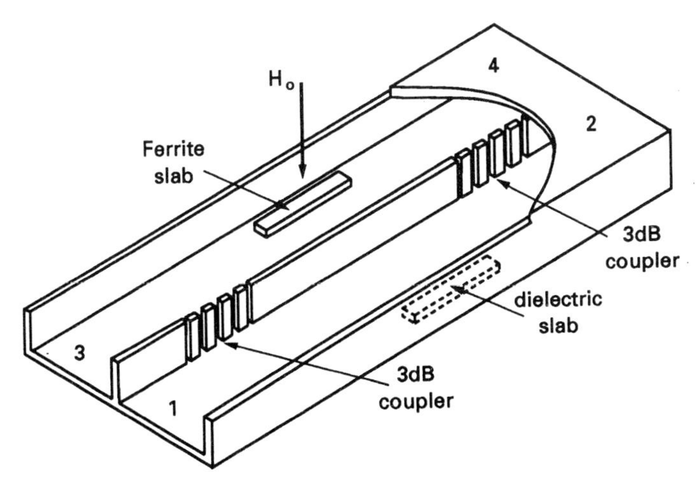

FIG 13.6. A differential phase change circulator

out of phase and combine in the directional coupler to give an output at port 3 and nothing out of port 1. Similarly a signal input to port 3 comes out of port 4 and an input to port 4 comes out of port 1. If a differential phase changer using two slabs of ferrite as shown in Fig. 8.11 (a) is used in each arm of the circulator, the device has a maximum phase change in a given length. For high-power use, the ferrite needs to be a thin strip adjacent to the waveguide wall. The differential phase change circulator most often consists of two differential phase change units using four ferrite slabs each as shown in Fig. 13.7. Such a differential phase change circulator is used because it is capable of operating in high-power microwave systems.

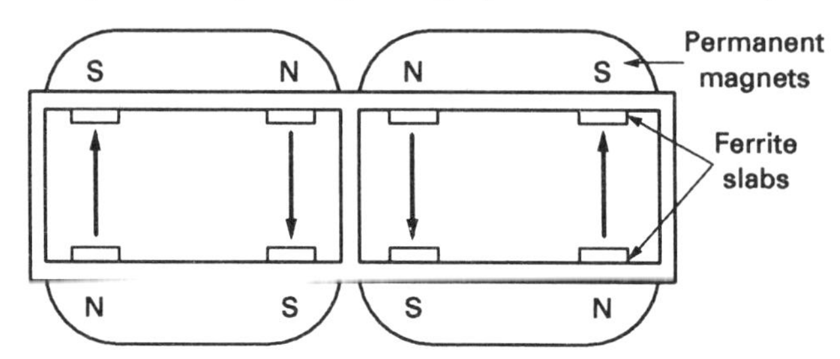

FIG. 13.7. Positions of the ferrite slabs in a high-power differential phase change circulator.

13.4. Y-junction Circulator

The Y-junction circulator is shown in Fig. 13.8. The diagram shows a particular shape of ferrite in the waveguide together with some reduction of waveguide height in the centre of the junction, but there are many other shapes of ferrite that also make satisfactory Y-junction circulators.

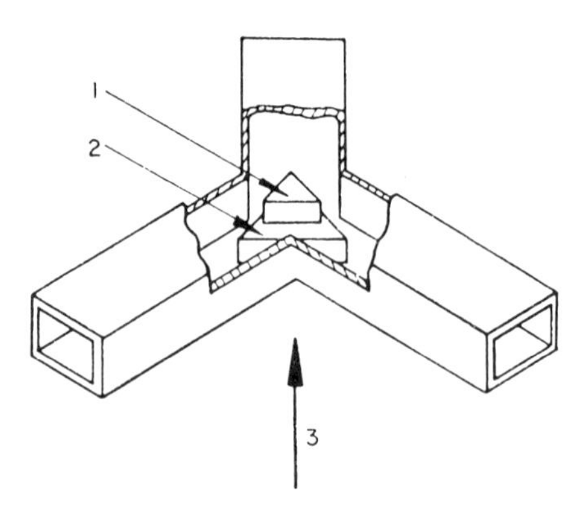

FIG. 13.8. Waveguide Y-junction circulator, partially cut away to show a particular triangular configuration of ferrite situated at the intersection of the waveguide arms. The magnet to provide the biasing magnetic field is not shown. (1) Ferrite. (2) Metal. (3) Magnetic field.

The action of this Y-junction circulator may be explained by reference to the reciprocal ferrite phase changer described in section 8.11 and shown in Fig. 8.11 (b). Assume that the situation shown in Fig. 8.11 (b) is not constrained by any waveguide walls. If the effect of the ferrite on the left-hand side is to increase the wavelength of the wave and that of the ferrite on the right-hand side to reduce the wavelength, the wavefront will turn to the right. In the Y-junction the wave will tend to travel up one arm of the junction and not up the other. It is found that at one value of biasing magnetic field, all the microwave power is coupled between two of the ports of the circulator and the other is isolated. In the reverse direction of power flow, coupling is to the other port and circulator action occurs.

An alternative description of the mode of operation of the Y-junction circulator appears particularly sensible if the nonreciprocal element is a circular rod or disc of ferrite. It is shown later in this section that perfect circulation occurs when the junction is matched. The ferrite loaded junction then behaves like a resonant cavity supporting two modes whose field patterns are contrarotating in the plane of the junction. With the ferrite unmagnetized, the resonant frequencies of the two modes are the same. When the ferrite is magnetized, however, the two contrarotating modes are similar to the circularly polarized modes of a plane wave which is propagating through an infinite ferrite medium magnetized in the direction of propagation. Such a system has already been described in sections 8.6 to 8.8. The two contrarotating magnetic fields experience different values of effective permeability in the ferrite so they rotate at different speeds. For the correct value of magnetic field, the rotating fields combine to give a standing wave pattern which presents a null in the field at the isolated port of the junction. Then all the power is transmitted to the other port of the junction.

Scattering parameters are useful in the design of symmetrical junction circulators. For a symmetrical nonreciprocal Y-junction, the scattering matrix is similar to eqn. (4.18),

$$\mathbf{S} = \begin{vmatrix} s_1 & s_2 & s_3 \\ s_3 & s_1 & s_2 \\ s_2 & s_3 & s_1 \end{vmatrix} \tag{13.1}$$

For a lossless junction, similar to eqns. (4.19) and (4.20) it can be shown that

$$|s_1|^2 + |s_2|^2 + |s_3|^2 = 1 \tag{13.2}$$

and

$$s_1 s_2^* + s_3 s_1^* + s_2 s_3^* = 0 \tag{13.3}$$

When the junction is matched, the input reflection coefficient is zero so that $s_1 = 0$. Then from eqns. (13.2) and (13.3), either $s_2 = 0$ and $s_3 = 1$, or $s_3 = 0$ and $s_2 = 1$. Therefore, if the junction is matched, the power entering port 1 must go to port 2 and power entering port 2 must go to port 3 etc. The junction must also be acting as a circulator. The scattering matrix of an ideal three-port junction circulator is

$$\mathbf{S} = \begin{vmatrix} 0 & 1 & 0 \\ 0 & 0 & 1 \\ 1 & 0 & 0 \end{vmatrix} \tag{13.4}$$

or

$$\mathbf{S} = \begin{vmatrix} 0 & 0 & 1 \\ 1 & 0 & 0 \\ 0 & 1 & 0 \end{vmatrix} \tag{13.5}$$

depending on the direction of circulation through the junction.

A Y-junction circulator is realized simply in stripline if the conductor at the junction point of a symmetrical Y is expanded to form a disc as shown in Fig. 13.9 and ferrite material magnetized perpendicular to the plane of the circuit is provided to give circulator action. The ferrite usually takes the form of a thin disc of about the same diameter as the conductor at the centre of the Y. The ferrite can be inserted into a circular hole where it replaces the dielectric material of the stripline or microstrip, or the whole device can be constructed using ferrite as the dielectric as shown in Fig. 13.9. Bridging the conductors across the joint between the ferrite and the dielectric causes difficulties with a simple ferrite disc. However, larger areas of ferrite slab are obviously more expensive and the partially magnetized ferrite adjacent to the circulator is going to have gyromagnetic properties which may be

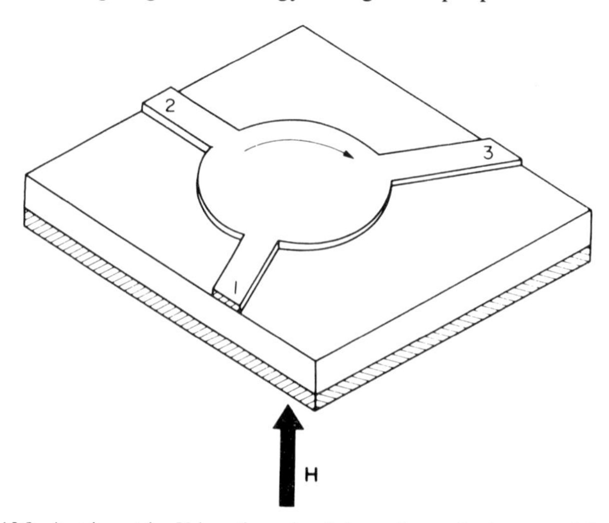

Fig. 13.9. A microstrip Y-junction circulator using a ferrite material as the dielectric.

undesirable. One solution to this has been to have a dielectric consisting of non-magnetic ferrite material into which is built a circular disc of magnetic ferrite. Because the magnetic and non-magnetic ferrite can be chosen to have the same crystal structure, they can be made into a mechanically uniform slab having no surface discontinuity between the two types of ferrite.

In the Y-junction circulator, the ferrite is magnetized perpendicular to the plane of the stripline circuit. The incident electromagnetic field sets up in the ferrite disc two contrarotating systems of electromagnetic field. For the correct value of magnetic field, the contrarotating fields combine to give a standing wave pattern in the disc so that there is an output at one arm of the junction and no output at the other arm of the junction. Therefore the device operates as a circulator. A signal into port 1 is transferred to port 2 with nothing out of port 3 and a signal into port 2 is transferred to port 3. The biasing magnetic field can be supplied by a small magnet on the opposite side of the earth conductor to the ferrite disc.

13.5. Iso-circulator

If a matched termination is attached to the third port of a three-port circulator, the assembly acts as an isolator. That is, power is transmitted without loss in the forward direction but it is absorbed in the reverse direction. The isolator is very useful because it can be used to decouple one part of a waveguide system from events further along the waveguide. For example, an isolator is used to prevent power reflected from the output of a system returning to the generator. The name iso-circulator is a contraction of isolator-circulator. In stripline or microstrip systems, it is easier to use a standard Y-junction circulator with a matched termination on one arm than to use a device specifically designed as an isolator. In high-power systems, the unwanted power can be dissipated in a specially designed termination and no power is deliberately absorbed in the ferrite device. The bandwidth of an iso-circulator is a function of the performance of the matched termination as well as that of the circulator itself.

One particular application of the iso-circulator is in the manufacture of very short waveguide isolators. Consider the waveguide Y-junction circulator shown in Fig. 13.8. For many applications, two of the output waveguides may have 30° bends so that the external ports of the device appear to be part of a T-junction. Then it is found that a T-junction as shown in Fig. 11.6(b) with suitably magnetised ferrite at the junction, also acts as a circulator. The matched termination is inserted into the side arm of the T-junction. If the absorbing material of the matched termination has a large value of effective permittivity and completely fills the side arm, the broad dimension of that side-arm waveguide is reduced, so that the iso-circulator is not much longer than the diameter of the ferrite element producing the circulation, as shown in Fig. 13.10. These are sometimes called *flange isolators* as it is often possible to keep the cross-section dimensions to that of the waveguide coupling flange and the length to between 10 and 20 mm.

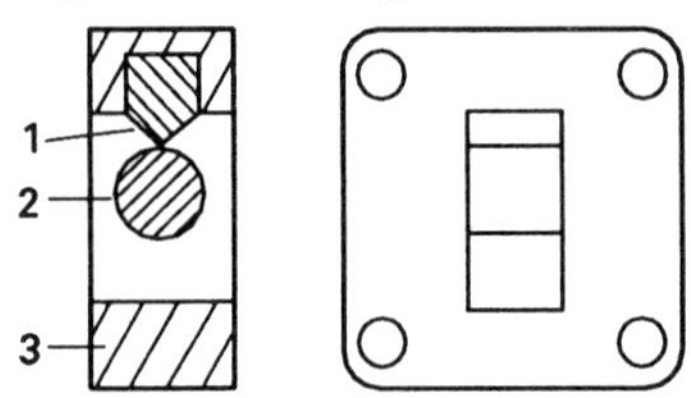

FIG. 13.10. A compact iso-circulator or flange isolator. (1) Absorber, (2) Ferrite, (3) Metal.

13.6. Resonance Isolator

An isolator can be constructed using the phenomenon of resonance absorption in ferrites. As described in section 8.11, observation of the field patterns of the TE_{10}-mode in rectangular waveguide in Fig. 8.9 shows that the magnetic field is circularly polarized in the plane of the broad face of the waveguide at a distance about a quarter of the way across the waveguide. If a slab of ferrite is placed in the waveguide at the position of circular polarization and it is magnetized perpendicular to the waveguide, as shown in Fig. 8.10, then coupling occurs between the precession in the ferrite and the magnetic field of the TE_{10}-mode. The construction of a *resonance isolator*

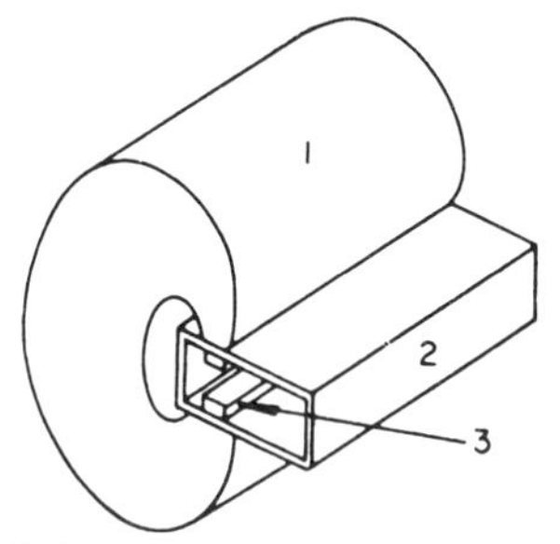

FIG. 13.11. Resonance isolator. (1) Permanent magnet. (2) Waveguide. (3) Ferrite.

is shown in Fig. 13.11. The ferrite is in small slabs adjacent to the waveguide walls so that the heat generated in the ferrite by absorption of microwave power can be easily dissipated. The magnetic fields required for resonance are high, but these can easily be supplied by permanent magnets at frequencies up to 15 GHz. The resonance isolator absorbs little power in the forward direction because the hand of rotation of the magnetic field is the opposite to that required to couple to the precession in the ferrite. A wave travelling in the reverse direction, however, has a magnetic field rotating in the same sense as the precession in the ferrite and power is coupled from the electromagnetic wave to the resonance in the ferrite and the wave is absorbed.

13.7. Field-displacement Isolator

The effective permeability of suitably magnetized ferrite material can become zero, as shown in Fig. 8.5. If the ferrite slab in rectangular waveguide, as shown in Fig. 8.10, is magnetized to that position of zero effective permeability, then the microwave field tends to be rejected from the ferrite for one direction of propagation. However, for propagation in the opposite direction, the field is concentrated in the ferrite. For the zero permeability condition, the field distribution in ferrite loaded rectangular waveguide is shown in Fig. 13.12. If an attenuating vane is placed adjacent to the ferrite

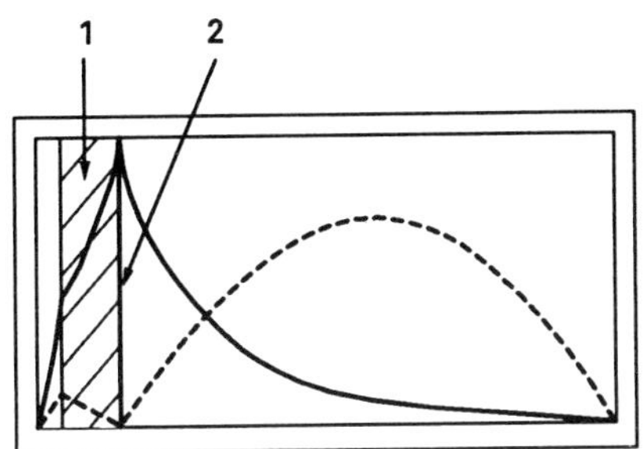

FIG. 13.12. The microwave electric field strength in a field-displacement isolator. (1) ferrite, (2) attenuating film. Forward low-loss wave – – – – Reverse attenuated wave ————

slab at the position of minimum microwave field, a field-displacement isolator is produced. The forward wave is rejected from the ferrite, the attenuating vane is at the position of minimum field and the wave is unattenuated. The backward wave, however, is concentrated in the ferrite slab with a peak of the microwave field at the position of the attenuating vane and it is attenuated. Such a device has two advantages over the resonance isolator. It is easier to provide broadband isolation because, as seen from Fig. 8.5, the permeability is changing slowly with frequency around the zero permeability condition whereas resonance is a direct function of frequency. Secondly the field displacement isolator requires a much smaller magnetic field than a resonance isolator for the same frequency of operation. However, the rectangular waveguide field-displacement isolator does need a slab of ferrite displaced from the waveguide walls so it is unsuitable for high-power applications.

The field displacement effect may also be used in stripline and is the basis of a very wide band isolator. The use of magnetized ferrite material as the dielectric in stripline has little effect on the propagation conditions of the wave unless coupled lines are involved or excessively wide stripline conductors are used. The wide conductor supports an edge-guided mode if the ferrite is magnetized perpendicular to the plane of the stripline. The fields are strong on one side of the strip conductor and they are weak on the other. A wave propagating in the reverse direction has the strong fields on the opposite side of the strip due to the gyromagnetic properties of the ferrite material.

Such an edge-guided mode can be used to construct a field-displacement isolator. If a resistive film is placed down one edge of the strip conductor, as shown in Fig. 13.13, it absorbs the microwave field for one direction of propagation and has little effect on the wave propagating in the opposite direction. The edge-guided effect occurs over a wide frequency range so that edge-guided-mode devices have a useful performance for a frequency band greater than an octave.

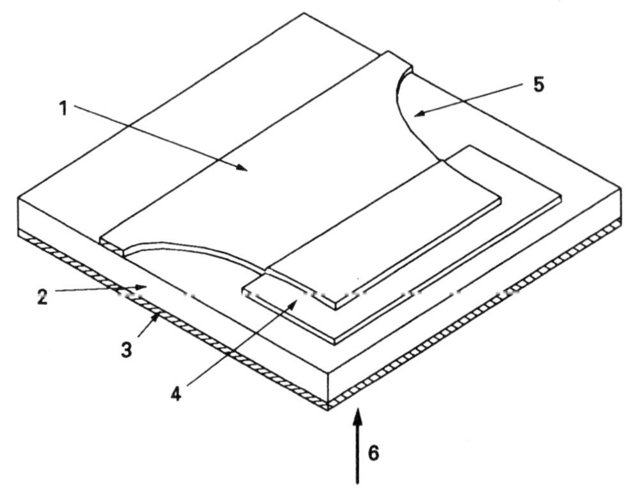

FIG. 13.13. Edge guided mode isolator. (1) Conductor. (2) Ferrite substrate. (3) Ground plane. (4) Resistive film. (5) Taper to wide microstrip. (6) Magnetizing field.

13.8. Variable Attenuator

If the 45° rotator in the circulator shown in Fig. 13.5 is replaced by a variable rotation element, the amount of power coupled from port 1 to port 2 depends on the angle of rotation which is governed by the strength of the static magnetic field. Any linearly polarized wave can be resolved into perpendicular components so that the power that is not coupled into port 2 is coupled to port 4. Such a device can be used as a variable attenuator or as a switch. It is usually designed so that ports 1 and 2 are parallel and there is no attenuation when there is no rotation. Ports 3 and 4 are fitted with matched terminations so that any unwanted power is absorbed. Alternatively the power may be absorbed in mode absorbers, which makes a device similar to the rotary attenuator shown in Fig. 12.7 where the rotating section of waveguide is replaced by a Faraday rotator.

A longitudinally magnetised ferrite rotator element also provides variable phase change to a circularly polarized wave in circular waveguide. The amount of phase change depends on the hand of circular polarization so that the rotator is a nonreciprocal variable phase changer. Construction of the variable phase changer is similar to that of the rotary phase changer shown in Fig. 12.3 except that the rotating half-wave plate is replaced by a ferrite rotator. However, variable phase change is provided more conveniently by using a variable magnetic field on either of the phase changers shown in Fig. 8.11.

13.9. Ferrite Resonant Cavity

Yttrium iron garnet, YIG, is ferrite which is a useful material for many nonreciprocal microwave devices. However, it also has the property that a very small sphere of the material behaves like a high Q resonant cavity. High purity single crystal YIG is needed to make the resonant spheres which are then ground to shape with an exceptionally fine surface finish. The quality of the surface finish determines the microwave Q-factor of the ferrite sphere. A typical sphere has a surface finish of about 0.1 μm and is about 1 mm diameter. The sphere is biased with a magnetic field which determines its resonant frequency according to the equation given in section 8.5

$$2\pi f_0 = \omega_0 = \gamma H_0 \tag{13.6}$$

The small size of the sphere makes it compatible with stripline and miniature circuits used at microwave frequencies. It can be used as a tuned filter or to control the frequency of oscillation of an oscillator. If the input and output microwave fields to the sphere are perpendicular, there is no coupling between the input and output circuits unless the sphere is in resonance. Therefore the sphere can be used to provide coupling between two stripline conductors which are perpendicular to one another. Such a circuit with coaxial transmission lines is shown in Fig. 13.14.

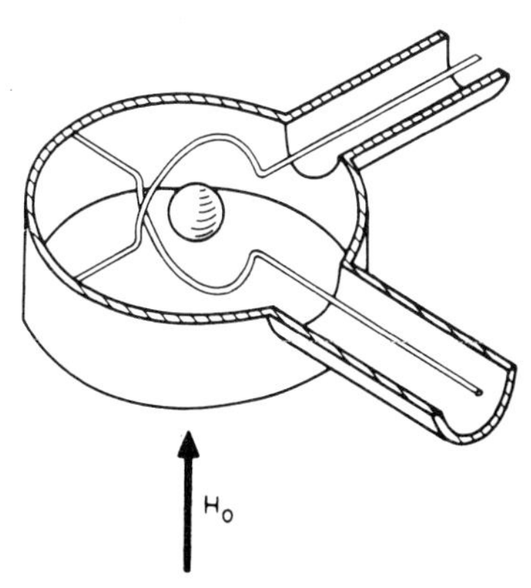

FIG. 13.14. A ferrite resonant sphere as a frequency selective coupling between two coaxial transmission lines.

13.10. Summary

13.2. In a ***directional phase changer***, the phase change for transmission in one direction is different from that for transmission in the opposite direction. It is realized using the nonreciprocal properties of transversely magnetized ferrite slabs in rectangular waveguide.

If the ferrite is in the form of a rectangular tube in the waveguide, the ferrite is magnetized by its remanent magnetization forming a ***latching phase changer***.

A meander line conductor configuration in stripline on a ferrite dielectric material magnetized in the line of the conductors acts as a ***directional phase changer***.

13.3. The ***circulator*** is a nonreciprocal device. Power incident on port 1 exits from port 2 and no power is coupled to any other ports. Power incident on port 2 exits from port 3 and nothing is coupled to any other ports and so on. A Faraday rotation circulator and a directional phase change circulator are described in section 13.3.

13.4. The ***Y-junction circulator*** in waveguide uses magnetized ferrite material at the centre of the Y-junction to provide circulator action. In stripline, it uses ferrite material instead of the dielectric at the point of the *Y*, magnetized perpendicular to the plane of the circuit.

13.5. The ***isolator*** is a device which permits free flow of microwave energy in the forward direction but which absorbs the energy flowing in the reverse direction. Isolators may be constructed by adding a matched termination to one port of a three-port circulator, so forming an ***iso-circulator***. One very compact form of the iso-circulator in waveguide is called a ***flange isolator***.

13.6. A ***resonance isolator*** makes use of the phenomenon of resonance absorption in ferrites.

13.7. A suitably magnetized ferrite slab in rectangular waveguide rejects the microwave field for one direction of propagation. An absorbing vane is placed at the position of minimum electromagnetic field to make a ***field-displacement isolator***.

The ***edge-guided-mode isolator*** uses a wide conductor on a ferrite dielectric material stripline magnetized perpendicular to the plane of the conductor to construct a field-displacement isolator. The magnetized ferrite displaces the microwave field to one side of the strip and a resistive film is used to absorb the field for one direction of propagation.

13.8. An electrically controlled ***variable attenuator*** can be made using Faraday rotation in ferrite. A portion of the incident wave is absorbed, the portion being dependent on the angle through which the wave is rotated on passing through the rotator.

Similarly, an electrically controlled ***variable phase changer*** may be constructed using Faraday rotation. The angle of rotation is added to the phase change of a circularly polarized wave in circular waveguide.

13.9. The ***ferrite cavity*** consists of a minute sphere of YIG material which acts as a high Q resonant cavity resonating at a frequency determined by the biasing magnetic field.

14

Measurements

14.1. Microwave Measurements

Engineering science is essentially experimental and the theoretical analysis outlined in this book can be verified by experiment. This chapter contains an outline of the experimental technique required to make certain measurements on microwave components and devices. Such measurements are involved in the design of microwave components that can be part of a microwave system such as a radar set or a microwave communication link. A description of these systems is considered to be beyond the scope of this book, but some microwave systems used to make measurements are described in this chapter.

The most comprehensive and sophisticated microwave measurement equipment is the microwave network analyser. The relative amplitude and phase of the signal reflected from and transmitted through the device under test is measured, giving the complex *s*-parameters of the device. The network analyser is particularly used to characterize microwave circuit components such as transistors and amplifiers. It usually operates over a very wide band so that its measurement ports have to be in a wide-band transmission line, usually a $50\,\Omega$ coaxial line. To measure the performance of stripline, microstrip or waveguide components, the measurement ports must be connected through suitable transformers. The microwave network analyser is usually coupled to a computer so that the analyser can be calibrated with any necessary transformers in place. The calibration is then used by the computer to eliminate any errors in the analyser or due to the transformers. The instrument is described in section 14.9. However, a number of simpler laboratory measurements using waveguide are described first. These measurement systems give some indication as to how waveguide components and devices may be inter-connected to provide useful manipulation of the electromagnetic wave. Only the simpler systems are described and the reader might expect to find most of the apparatus available in a teaching laboratory. Detailed experimental instructions are not given, but there is sufficient information to enable the reader to use any apparatus available to him unaided.

All the diagrams in this chapter make use of standard circuit symbols which are each identified by name the first time that they are used. However, they are also listed in Appendix 3. Sections 14.2 to 14.6 detail techniques that are used to make measurements at one *fixed frequency*, while sections 14.7 to 14.9 detail techniques used to provide a *swept frequency* display of the property to be measured.

14.2. Waveguide Test Bench

Before discussing any particular microwave measurements, we consider the basic waveguide equipment used for fixed frequency measurements. All the components have been described in Chapters 10 to 13. This chapter discusses their use. The basic waveguide test bench is shown in Fig. 14.1.

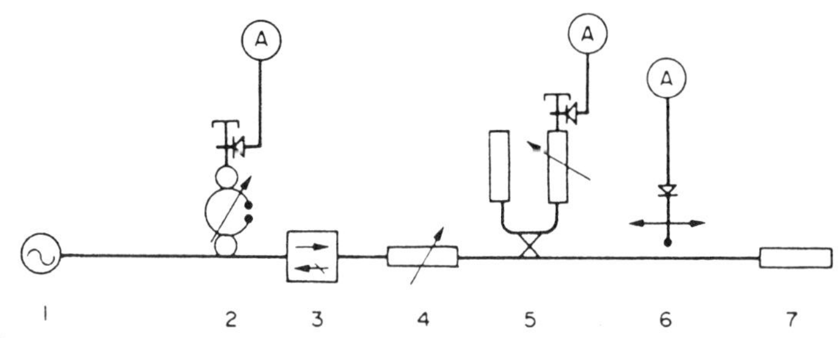

FIG. 14.1. Basic waveguide test bench used for impedance measurement. (1) Gunn diode oscillator. (2) Frequency measurement and monitor by means of an indicating wavemeter. (3) Isolator. (4) Power level setting by means of a variable attenuator. (5) Power level monitor consisting of a directional coupler, a variable attenuator and a crystal receiver. (6) VSWR measurement by means of a standing wave detector. (7) The unknown which consists of a matched termination.

The Gunn diode oscillator is driven by a stabilized power supply which is not shown in the diagram. The indicating transmission wavemeter could be replaced by an absorption wavemeter, but a crystal receiver must then be connected temporarily to the end of the bench to detect the dip in output on tune. Alternatively a directional coupler may be used with the absorption wavementer followed by a crystal receiver connected to its side arm. The isolator is used to prevent any mismatch at the end of the line from affecting the frequency or power output of the oscillator. The isolation must be at least 12 dB.

The variable attenuator is used to set the power level in the rest of the line. It is followed by a power level monitor consisting of a directional coupler, to sample a small proportion of the power in the main line, and a variable attenuator followed by a crystal receiver. The attenuator is used to vary the sensitivity of the monitor circuit. It may be omitted without materially affecting the operation of the circuit, because some sensitivity control can be attached to the meter connected to the crystal receiver. All this first part of the waveguide test bench is standard to any fixed frequency test system, and is sometimes all contained as part of a sophisticated signal source.

14.3. Voltage Standing Wave Ratio (VSWR)

In order to perform impedance matching, it is necessary to know the reflection coefficient, in amplitude and phase, of the device causing the mismatch. The reflection coefficient is completely specified by the microwave network analyser, but it is also possible to obtain the information from VSWR measurements in a transmission line or waveguide. The reflection coefficient and VSWR are related as given in eqns. (1.36) and (1.37). The standing wave detector is the item concerned with measurements of standing wave ratio and is the next item shown in the test bench of Fig. 14.1. The detector crystal output is indicated on a meter and the ratio of maximum to minimum output is the standing wave ratio. If the detector crystal in the standing wave detector is square law, as it usually is, the VSWR is the square root of the ratio of the indicated outputs. A number of standing wave indicators have meters whose calibration is in VSWR allowing for the square law of the crystal. It is usually necessary to alter the sensitivity of the indicator so that the maximum gives a meter reading of one and then the minimum gives the correct reading on the VSWR scale.

If an impedance plot is required for use with a Smith chart, the position of the minimum of the standing wave pattern needs to be measured. Most standing wave detectors are provided with a scale and pointer to measure the position of the probe. The scale gives the distance from the output flange face of the standing wave detector. If the device whose impedance is under investigation is connected directly to the standing wave detector, the impedance plot gives the effective impedance at the entry to the unknown device. If the effective impedance at any other point is required, allowance can be made for this in the calculation of the impedance plot.

14.4. Attenuation

The attenuation of a passive device is given by s_{21} and s_{12} which are measured by the microwave network analyser. Alternatively attenuation may be measured by comparing the performance of two attenuators. A fundamental method of calibrating precision attenuators consists of a comparison of attenuation at microwave frequencies with attenuation at lower frequencies, often provided by a piston attenuator. This involves complicated apparatus which most microwave engineers are not required to use. Measurement of attenuation and calibration of attenuators from one another are essentially the same process.

The technique of attenuation or insertion loss measurement by substitution is described (see Fig. 14.2). As well as the basic measurement bench already

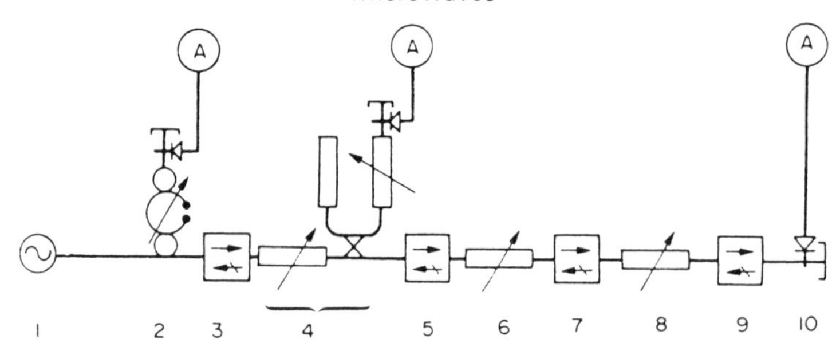

FIG. 14.2. Waveguide test bench used for attenuation and insertion loss measurements. (1) Gunn diode oscillator. (2) Frequency measurement and monitor. (3) Isolator. (4) Power level setting and monitor. (5) Isolator. (6) Precision variable attenuator. (7) Isolator. (8) Unknown. (9) Isolator. (10) Crystal receiver.

shown in Fig. 14.1, there is a precision calibrated variable attenuator, the unknown and a crystal receiver all separated by isolators. The isolators are not always necessary. Isolator (5) is only necessary if the precision attenuator reflects an appreciable amount of power which is unlikely with most waveguide attenuators. Isolator (7) is necessary if the unknown device reflects an appreciable amount of power. It is not necessary if the unknown is another attenuator that is being calibrated or if it is some device in common use that has been designed to be a good match. If the unknown is a bad match, the isolator is necessary because any large standing wave pattern in the precision attenuator will give a non-uniform distribution of electric field through the attenuator which might alter its calibration. Isolator (9) is used because a crystal receiver is not a very good match. It removes the likelihood of a standing wave generated by the crystal receiver altering the conditions in either an unknown attenuator or the calibrated attenuator.

The method of measurement is to set some level on the output of the crystal detector when the unknown is in place and the calibrated attenuator set to zero. The unknown is removed from the circuit and the output rises as some attenuation has been removed from the circuit. The calibrated attenuator is altered to introduce some more attenuation into the circuit to bring the output of the crystal receiver to its original value. Then the attenuation reading of the calibrated attenuator is the insertion loss of the unknown device.

If the unknown is another variable attenuator, to be calibrated as shown in Fig. 14.2, the unknown attenuator is not removed from the circuit. One attenuator is set to a maximum and the other to zero. While attenuation is removed from one attenuator it is introduced by the other so that the output remains constant. Equivalent readings of the two attenuators are obtained giving a calibration of the unknown in terms of the precision calibrated attenuator readings.

If the unknown device is some process or condition under an external control, the method of measurement is similar to that used for the calibration of an attenuator. As the control is varied, the attenuation through the device varies and the setting of the variable attenuator is altered to keep the indicated output constant. If the unknown process causes an appreciable amount of reflected power, the attenuation or power loss consists of both reflected power and absorbed power. It may be necessary to monitor the VSWR as well as the attenuation so as to be able to separate the two sources of power loss. If this is required, a standing wave detector is inserted into the waveguide bench between items (7) and (8) on Fig. 14.2.

Since all these attenuation measurements involve bringing the output of the waveguide bench to the same position on the output crystal receiver current meter, the measurements are independent of the detector crystal calibration. Sometimes small attenuations may be measured without using a calibrated attenuator by measuring the change in detected output current level, assuming a square law relationship between the detected current output and the microwave electric field, so that output current is proportional to output power.

14.5. Power

At d.c. and low frequencies it is convenient to measure potential difference and current. At high frequencies it becomes difficult to measure these quantities and in waveguide they have little meaning. At d.c. and low frequencies, it is usual to determine power as the product of potential difference and current but at microwave frequencies it becomes easier to measure power directly. Power is measured in a bolometer mount for low powers and in a water calorimeter for high powers. The bolometer mount is added onto the end of the waveguide system in which it is required to measure the power level; for example, it would be added in place of the unknown in the basic waveguide test bench shown in Fig. 14.1. If the bolometer mount is not a good match and some of the microwave power is being reflected from the mount rather than being absorbed into the power measurement element, the bolometer mount may be preceded by a stub tuner which is adjusted to give unity VSWR at the standing wave detector. Then it can be assumed that all the incident power is absorbed in the bolometer element. The bolometer element is connected to some form of wheatstone bridge.

The method of measurement is that the bridge is balanced with no power going into the bolometer mount. The balance may be controlled by the heating effect of the direct current flowing through the bolometer element. The bolometer mount is connected to the waveguide system in which it is required to measure the power, and the bridge becomes unbalanced. A direct reading bridge has the unbalance current in the galvanometer calibrated directly in power. Alternatively, the direct current through the bolometer element is reduced to bring the bridge back to balance and the microwave power is the same as the d.c. power change in the bolometer element.

To measure high powers, a water calorimeter is connected as a matched termination to a waveguide system. The rate of flow of water through the calorimeter and its temperature rise measure the microwave power being absorbed in the calorimeter. The water calorimeter is not sufficiently sensitive for use in low-power measurements.

14.6. Phase

The measurement of phase at microwave frequencies is the same as the measurement of electrical length. Normally interest is only centred on change of phase or electrical length or on the difference in electrical length between some device and the same physical length of standard waveguide. Phase angle and electrical length are directly related. One wavelength is the same as a phase change of 2π radians or 360°. Change of electrical length may be measured by two methods, by the change of phase of a standing wave pattern through the device or by using a microwave bridge sensitive to both attenuation and phase.

The standing wave pattern method uses the basic test bench shown in Fig. 14.1 with the unknown device followed by a short circuit. The position of the minimum of the standing wave pattern is noted. It has already been shown in the consideration of microwave resonators in section 3.11 that the minima of the standing wave pattern occur at an integral number of half wavelengths in front of a short circuit terminating any transmission line. The voltage standing wave pattern is shown in Fig. 3.14. Since the position of the minimum gives the electrical length of the waveguide between that position and the short circuit, any change of electrical length can be measured. Measurement of phase is often made by a substitution method where the electrical length of the device to be measured is compared with the electrical length of the same physical length of standard waveguide.

The phase delay due to a passive device is given by the phase angle of s_{21} or s_{12} as measured by the microwave network analyser. Alternatively a more general investigation into the phase properties of a device can be performed by means of the phase-sensitive bridge. The phase of the device under test is compared with a phase reference supplied by a reference signal. The bridge is balanced using a phase invariant variable attenuator and an attenuation invariant calibrated phase changer. In waveguide, these devices are most often the rotary attenuator and the rotary phase changer.

14.7. Swept Frequency Techniques

The Gunn diode oscillator may be tuned over a wide frequency range by a resonant circuit incorporating a varactor diode or by a resonant ferrite sphere. It is tuned by varying the bias voltage on the varactor diode or the biasing magnetic field on the ferrite sphere. If a Gunn diode is mounted in a waveguide cavity, it may be tuned mechanically through a frequency range of an octave or more. A transistor oscillator also may be tuned by a varactor diode or by a circuit containing a resonant ferrite sphere. The tuning range of a waveguide system is limited by the frequency range of the waveguide but there is no such frequency limitation on microstrip or transmission line systems.

Many microwave network analyser systems operating over a very wide band are not true swept frequency measurements. Calibration of the device is performed at a number of spot frequencies and subsequent measurements are performed at those frequencies so that the inherent errors in the network analyser may be eliminated from the measurements. In order to provide repeatability of measurement calibration, the spot frequencies are synthesized by multiplication from a master oscillator at a lower frequency. Even when the spot frequencies are so close together that any display of results provides a continuous line, it is still possible to miss unacceptable microwave performance at some intermediate frequency. Therefore true swept frequency microwave measurements are often necessary.

If a swept frequency oscillator is connected into a microwave test system giving a direct reading of the quantity to be measured, quick and easy measurements may be made over any band of frequencies and only the worst condition needs accurate measurement. The measurements already described in this chapter can only be made at a fixed frequency. Sometimes waveguide devices are not well behaved and then swept frequency measurements are necessary to give an indication of the performance of the device. Most waveguide measurements involve the comparison between some output power level and the input power to the device. The power output of the wideband swept frequency oscillators varies with frequency so that for swept frequency measurements, it is either necessary to calibrate the output, as with the microwave network analyser, or it is necessary to level the output of the oscillator.

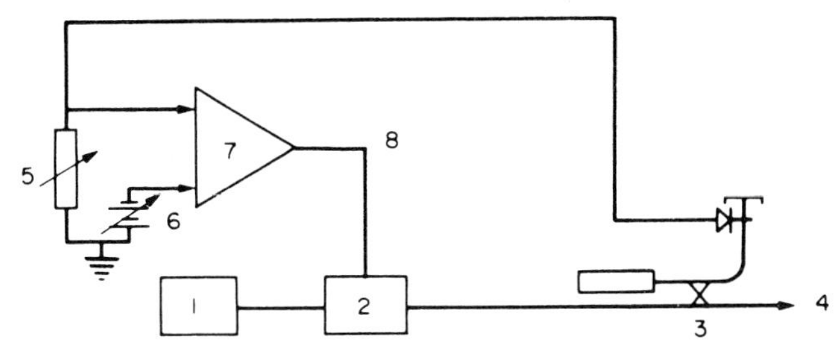

FIG. 14.3. Basic levelling circuit of a swept frequency signal generator. (1) Oscillator. (2) PIN diode attenuator. (3) Directional coupler and crystal receiver. (4) Levelled r.f. output. (5) Automatic level control gain. (6) Power level reference voltage. (7) Levelling amplifier. (8) D.C. control signal.

In most swept frequency measurement systems, the power output of the microwave oscillator is levelled using a PIN diode attenuator. A typical system is shown in Fig. 14.3. In many swept frequency signal generators, the complete levelling system shown in Fig. 14.3 is included in the generator so that a levelled output is delivered to the output port. If the output from the signal generator is a coaxial connector and measurements are to be made in waveguide, it is better if the directional coupler and level detector are in the size of waveguide that is to be used for measurements. The waveguide-to-coaxial transition is then situated between the PIN diode attenuator and the directional coupler, items (2) and (3) of Fig. 14.3. The efficiency of the levelling system is dependent on the performance of the directional coupler and crystal receiver. Multi-hole directional couplers and crystal receivers specially designed for this purpose have a good performance over the whole waveguide bandwidth.

14.8. Swept Frequency Measurements

A system for the swept frequency measurement of attenuation is shown in Fig. 14.4 which uses a levelled oscillator output and an oscilloscope display. If the two directional couplers in the system are the same and the two crystal receivers have the same microwave performance, then any errors in the levelling are reproduced at the output, so compensating for these errors. If the detector crystal has a square law characteristic, the attenuation can be read directly from the oscilloscope trace. Alternatively the oscilloscope may be replaced by a pen recorder and the system may be calibrated by first

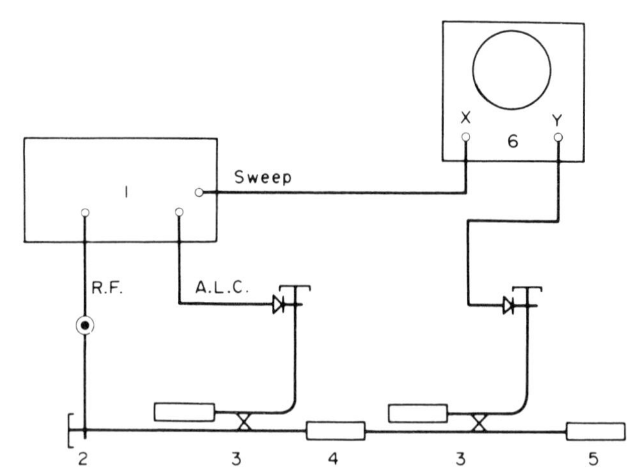

FIG. 14.4. Waveguide system for swept frequency measurements of attenuation. (1) Sweep generator. (2) Waveguide to coaxial transformer. (3) Directional coupler and crystal receiver. (4) Unknown. (5) Matched termination. (6) Cathode ray oscilloscope.

measuring a number of devices with known attenuation.

The sophisticated microwave network analyser gives a display of impedance on a cathode ray tube having a Smith chart graticule. A simpler system gives a swept frequency display of reflection coefficient, using the same apparatus as the attenuation measurement system shown in Fig. 14.4. The system for measurement of reflection coefficient is shown in Fig. 14.5. The first directional coupler is used to level the power in the waveguide system and the second measures the power reflected from the device under test. Such a system relies on the high directivity of multi-hole directional couplers. Their directivity is often greater than 40 dB. The error due to a directivity of 40 dB in a 10 dB coupler is equivalent to a VSWR of 1.005. If the coupling factor of the two couplers varies similarly with frequency, the other errors cancel as do those due to the crystal receivers. If the output detector crystal has a square law response, the output display can be calibrated in reflection coefficient, with a square law calibration.

14.9. Microwave Network Analyser

A network analyser measures both the real and imaginary parts of all four s-parameters of a two-port device. The signal which is transmitted through or reflected from the device under test is compared in amplitude and phase with a reference signal. The apparatus which is used for making the amplitude and phase comparison is not described here, but in the manufacturers' literature. A microwave system for applying the signals to the device under test is shown in Fig. 14.6. A signal of a known frequency is supplied by the oscillator and a portion is taken by the directional coupler (no. 4) to supply the reference signal to the phase sensitive bridge. Then switch no. 2 directs

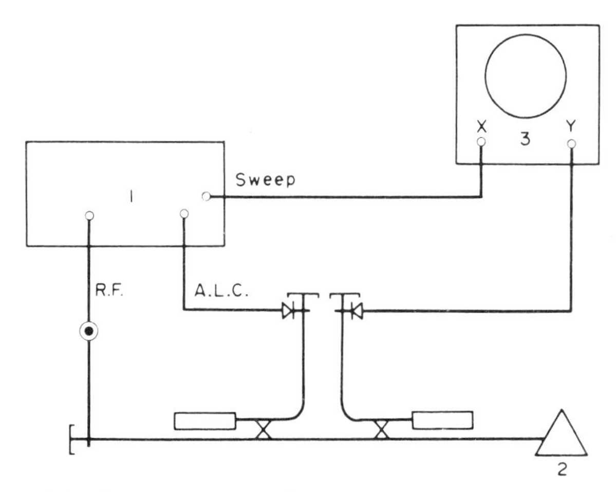

Fig. 14.5. Waveguide reflectometer with oscilloscope display. (1) Sweep generator. (2) Unknown. (3) Cathode ray oscilloscope.

the signal to one of the two ports of the unknown. The two directional couplers (no. 9) direct the output to the phase sensitive bridge. With switch no. 7 in the position shown in Fig. 14.6, the bridge is measuring the reflection coefficient from the unknown in the same way as the reflectometer shown in Fig. 14.5. With the switch in the opposite position, the bridge is measuring the transmission coefficient through the unknown in the same way as the attenuation measurement system shown in Fig. 14.4. Switch no. 2 is then moved to make the same measurements at the alternative port of the device under test.

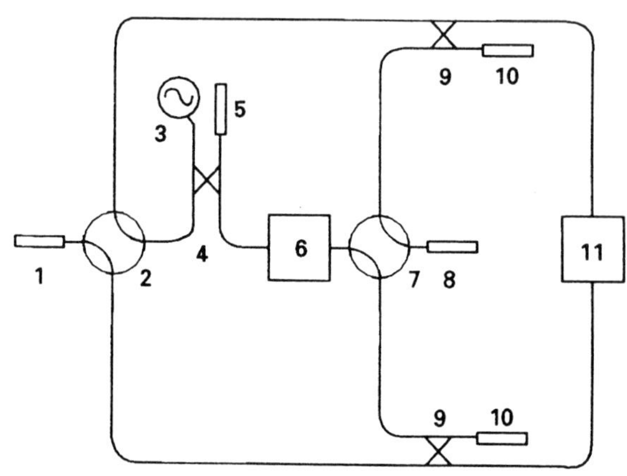

FIG. 14.6. Microwave network analyser system. (1) Matched termination, (2) Switch, (3) Oscillator, (4) Directional coupler, (5) Matched termination, (6) Phase sensitive detector, (7) Switch, (8) Matched termination, (9) High directivity directional coupler, (10) Matched termination, (11) Device under test (unknown).

A swept frequency or stepped frequency oscillator may be used to give the variation of each of the *s*-parameters over a band of frequencies. If the *s*-parameters are plotted on a polar diagram giving amplitude and phase, a Smith chart overlay may be used to read off the input and output impedance of the device under test. In the automatic network analyser, the analyser is coupled to a computer system to perform analysis of the results and to organize the switching necessary to measure the different *s*-parameters. Before making measurements, a number of known test devices are measured. The computer uses these results to eliminate many of the inherent errors in the analyser and to correct subsequent measurements. The computer corrected results may subsequently be processed to present the results in any convenient format.

14.10. Summary

14.3. The standing wave detector is used to measure the ***voltage standing wave ratio***. The output from the detector plots the standing wave pattern in the waveguide. Waveguide wavelength may be measured because the wavelength of the standing wave pattern is half the wavelength of the travelling wave. If the standing wave detector is fitted with a square law crystal, the VSWR is the square root of the ratio of the maximum to minimum outputs.

14.4. ***Attenuation*** is measured by a substitution method. The detected output at the end of the waveguide test bench is maintained constant and the setting of a calibrated attenuator is altered to compensate for change in attenuation due to the removal of the device under test.

14.5. ***Power*** is measured using a bolometer mount connected to some form of wheatstone bridge calibrated in microwave power.

14.6. ***Phase*** may be measured using a phase-sensitive bridge. Alternatively, a simple phase measurement is made, by using a standing wave detector to measure the distance between a minimum of the standing wave pattern and a short circuit situated on the other side of the device.

14.7. A voltage tunable oscillator may be used to provide a swept frequency source for swept frequency measurements.

The output from the ***oscillator*** is ***levelled*** so that it is at a constant power level as it sweeps over the band.

14.8. ***Attenuation*** may be measured by displaying the output from the waveguide bench on an oscilloscope to give an output of power level against frequency.

Reflection coefficient is measured by using a high directivity directional coupler to sample the reflected portion of the wave in the waveguide. The reflected power is displayed on an oscilloscope against frequency.

14.9. The ***microwave network analyser*** measures both the real and imaginary parts of all four s-parameters of a two-port device. Coupled to a computer, it gives error corrected measurements at a number of spot frequencies and becomes the ***automatic network analyser***.

Bibliography

THIS is a list of books for further reading on the different topics indicated.

Microwaves in general and particularly Chapters 1–7

Collin, R. E. *Foundations for Microwave Engineering.* McGraw-Hill, 1966.
Rizzi, P. A. *Microwave Engineering.* Prentice-Hall, 1988.
Sander, K. F. *Microwave Components and Systems.* Addison-Wesley, 1987.
Ramo, S., Whinnery, J. R. and Van Duzer, T. *Fields and Waves in Communication Electronics.* 2nd ed. Wiley, 1984.

Mathematics—vector analysis

Any mathematical textbook which includes the subject. A specialist textbook is:
Spiegel, M. R. *Theory and Problems of Vector Analysis.* S.I. ed. McGraw-Hill, 1974.

Mathematics—Bessel functions

There are a number of textbooks about Bessel functions. This book is not a textbook but tabulates values of many different functions and lists their properties:
Abramovitz, M. and Stegun, I. A. *Handbook of Mathematical Functions.* Dover, 1965.

Electromagnetic waves, Chapter 2

Hammond, P. *Electromagnetism for Engineers.* 3rd ed. Pergamon, 1986.
Hammond, P. *Applied Electromagnetism.* Pergamon, 1971.
Hayt, W. H. *Engineering Electromagnetics*, 5th ed. McGraw-Hill, 1989.
Liao, S. W. *Engineering Applications of Electromagnetic Theory.* West, 1988.
Stratton, J. A. *Electromagnetic Theory.* McGraw-Hill, 1941.

Microwave circuits. Chapter 4

Gonzalez, G. *Microwave Transistor Amplifiers.* Prentice-Hall, 1984.
Ha, T. T. *Solid-state Microwave Amplifier Design.* Wiley, 1981.
Hoffmann, R. K. *Handbook of Microwave Integrated Circuits.* Artech, 1987.
Vendelin, G. D. *Design of Amplifiers and Oscillators by the s-parameter Method.* Wiley, 1982.

Waveguide theory. Chapters 5 and 6

Marcuvitz, N. *Waveguide Handbook.* McGraw-Hill, 1951. (Reprinted, P. Peregrinus, 1986)

Ferrite media. Chapter 8

Baden Fuller, A. J. *Ferrites at Microwave Frequencies.* P. Peregrinus, 1987.
Clarricoats, P. J. B. *Microwave Ferrites.* Chapman & Hall, 1961.
Helszajn, J. *Principles of Microwave Ferrite Engineering.* Wiley, 1969.

Plasma and oscillators and amplifiers. Chapters 9 and 10

Beck, A. H. W. *Space Charge Waves.* Pergamon, 1958.
Howes, M. J. and Morgan, D. V. *Microwave Devices.* Wiley, 1976.

Components, devices and measurements. Chapters 11–14

Adam, S. F. *Microwave Theory and Applications.* Prentice-Hall, 1969.
Cross, A. W. *Experimental Microwaves.* 3rd ed. Marconi Instruments Ltd., 1977.
Saad, T. S. *Microwave Engineers Handbook.* (2 vol.) Airtech, 1971.

Worked Solutions to Selected Problems

(Nos. 1.8; 2.9; 3.9; 4.5; 5.7; 5.8; 6.5; 6.10; 7.6; 7.7; 8.6; 9.7)

PROBLEM 1.8. A swept frequency measurement (see section 14.8) gives a plot of reflected power ratio (in dB) against frequency. The table below gives some readings from such a measurement of the effect of a line terminated in an inductor whose series resistance is 50 Ω at all frequencies. Find the value of its inductance when the characteristic impedance of the line is 50 Ω.

f GHz	Reflected power dB
0.60	9.5
1.00	6.0
1.30	4.4
1.55	3.5
1.90	2.5
2.30	1.9

Answer. As the real part of the impedance is 50 Ω, the plot of the impedance on the Smith chart (see Fig. 1.5) is on the locus of $Z = 1 + jX$. The intersection of this locus and the VSWR locus appropriate to any measurement will give the reactive impedance at the measurement frequency and hence the value of the inductance may be found. The calculation is given in the table below. The relationship between the VSWR, S, and the normalized reactance, X, is found graphically from the Smith chart as shown in Fig. 15.1. The other relationships used are:

$$\text{Reflected power} = -20\log_{10}\rho \text{ dB}$$

$$S = \frac{\rho + 1}{\rho - 1}$$

$$\omega L = Z_0 X$$

f GHz	Reflected power, dB	ρ	S	X	L nH
0.60	9.5	0.336	2.0	0.70	9.28
1.00	6.0	0.502	3.0	1.15	8.90
1.30	4.4	0.602	4.0	1.50	9.18
1.55	3.5	0.667	5.0	1.80	9.23
1.90	2.5	0.752	7.0	2.2	9.21
2.30	1.9	0.800	9.0	2.6	9.00
				Mean	9.1

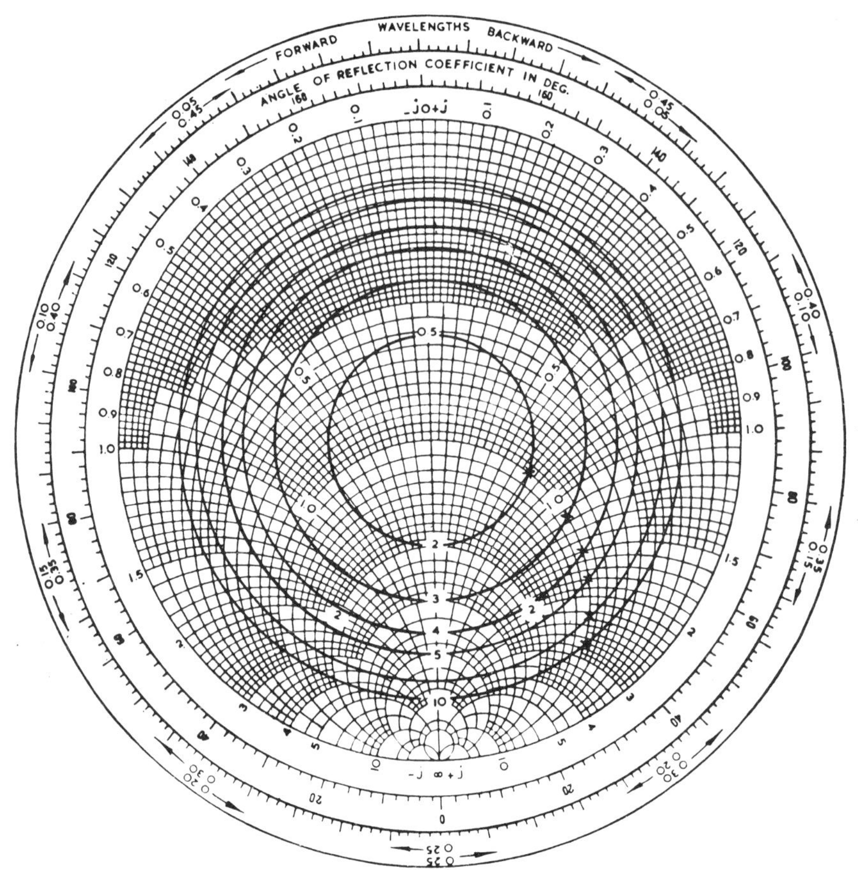

FIG. 15.1. Illustrating the solution to problem 1.8 and showing constant VSWR loci on the impedance diagram.

PROBLEM 2.9. The boundary condition at a conducting surface is that the tangential electric field components of an electromagnetic wave are zero at the plane of the surface. A plane wave is normally incident onto a plane conducting sheet. Find an expression for the field components of the reflected wave and the position of the first minimum of the standing wave pattern.

Answer. Consider the reflection of a normally incident plane wave. In the plane of the conducting sheet, the electric field strength is zero. Hence there must be generated another wave whose electric field is equal and opposite to that of the incident wave in the plane of the conducting sheet. The magnetic field, however, is unaffected by the conductor and hence a new wave is generated whose magnetic field is in the same direction but whose electric field is the negative of that of the incident wave. A diagram of the field vector relationships does show that the new wave is of the same magnitude but travelling in the reverse direction to the incident wave. If the incident wave components are E_x and H_y where

$$E_x = \eta H_y \tag{2.28}$$

and z is the direction of propagation, the reflected wave components will be $-E_x$ and H_y and negative z is the direction of propagation. As the waves are of equal magnitude, the VSWR is infinite and the distance to the first minimum of the standing wave pattern is half a wavelength in front of the conducting sheet.

PROBLEM 3.9. A low-loss transmission line (where $\alpha = 0$) has short circuits applied to it at distances along it of 25 cm and 40 cm. By using the expression for the voltage on a transmission line, eqn. (1.17), find the lowest frequency at which an electromagnetic wave can exist on the transmission line between the short circuits. Are there other frequencies also at which these fields can exist? If so, derive an expression for their frequency.

Answer. When $\alpha = 0$, eqn. (1.17) becomes

$$V = A\exp j(\omega t - \beta z) + B\exp j(\omega t + \beta z) \tag{15.1}$$

At the short circuit, $V = 0$ at $z = z_1$, hence

$$A\exp j(\omega t - \beta z_1) + B\exp j(\omega t + \beta z_1) = 0 \tag{15.2}$$

Then

$$\frac{A}{B} = -\frac{\exp j(\omega t + \beta z_1)}{\exp j(\omega t - \beta z_1)} = -\exp j(2\beta z_1) \tag{15.3}$$

For eqn. (15.3) to be valid, $\exp j(2\beta z_1) = 1$ and $A = -B$. This is true if $2\beta z_1 = 2n\pi$, where n is zero or an integer. As the position of the zero of z is arbitrary, it will be made coincident with one of the short circuits and then z_1 is the distance between them. Hence substituting from eqn. (1.19),

$$z_1 = \frac{n\pi}{\beta} = \tfrac{1}{2}n\lambda \tag{15.4}$$

The distance between the short circuits is 0.15 m and this must be a multiple number of half-wavelengths. $n = 0$ is appropriate to d.c. (The device is a resonator with a number of discrete resonant frequencies.)

The frequencies of operation are given by

$$f = \frac{c}{\lambda} = \frac{nc}{2z_1} = \frac{n \times 3 \times 10^8}{2 \times 15} = n \times 10^9\,\text{Hz}$$

Hence: The first answer is: 1 GHz.
The second answer is: Yes—integer multiples of 1 GHz.

PROBLEM 4.5(a). A transistor has the following s-parameters at 4.0 GHz. $s_{11} = 0.96\angle -40°$, $s_{12} = 0.063\angle 65°$, $s_{21} = 2.54\angle 150°$ and $s_{22} = 0.65\angle -20°$, measured relative to 50 Ω. It is connected into the circuit shown in Fig. 4.10. Using the Smith chart, find the dimensions l_1 and l_2 at 4.0 GHz of the input matching circuit. The transmission lines all have 50 Ω characteristic impedance.

(b) An input matching circuit for the transistor at 4.0 GHz could consist of an inductor in series with the input and another inductor in shunt to earth at the opposite end of the series inductor. Using the Smith chart, find the values of the two inductors. It is all required to be matched to 50 Ω lines.

Answer. The question relates to the reflection coefficient due to the input impedance of the transistor. The reflection coefficient is the same as s_{11}. In order to plot the input impedance on the Smith chart, we use the amplitude and phase of the reflection coefficient. The radial distance on the Smith chart is the magnitude of the reflection coefficient with a value of unity at the outside of the chart. The input impedance is given by the point A on the Smith chart with the radial distance suitably scaled, as shown in Fig. 15.2, and the phase angle of $-40°$ taken off the outside of the chart. Alternatively, the input impedance may be found by first calculating the VSWR. The VSWR is given by eqn. (1.36)

$$S = \frac{1.96}{0.04} = 49$$

A value of 49 is difficult to locate with any accuracy on the real axis of the Smith chart, but its reciprocal is easy to plot. $1/S = 0.02$, and this gives the radius of the VSWR circle on Fig. 15.2. The angle of $-40°$ is given by the outside of the chart.

(a) Consider the matching circuit of Fig. 4.10 which has been drawn again as Fig. 15.3(a). l_1 is a shunt stub so it is necessary to use the Smith chart as an admittance diagram. The point B is diametrically opposite the point A on the Smith chart so that B represents the admittance value of the input impedance. We then move backward along the line away from the termination

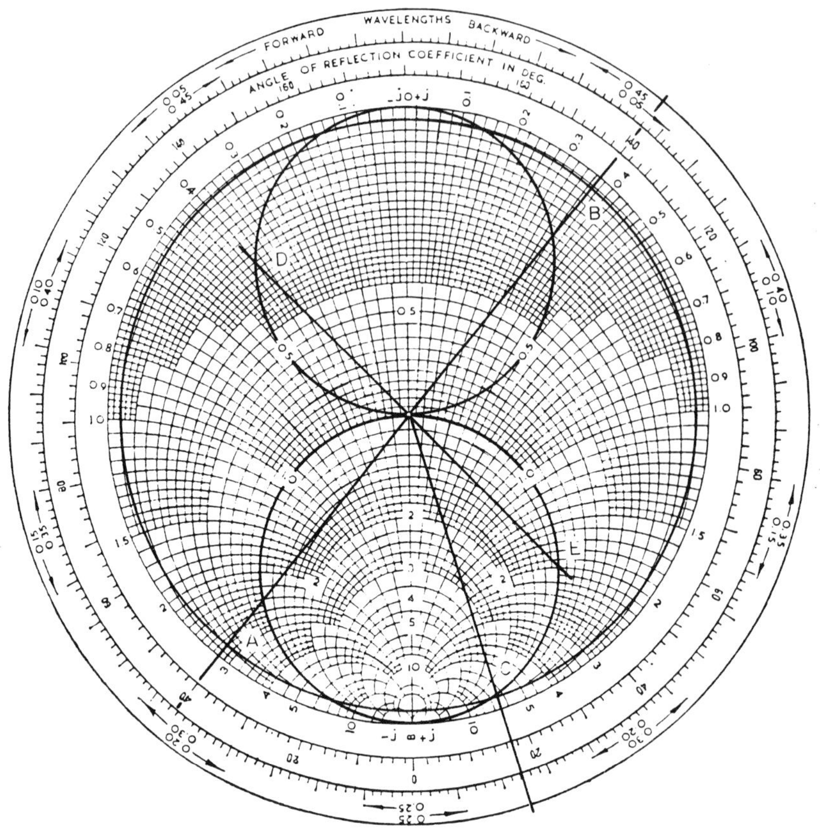

FIG. 15.2. Illustrating the solution to problem 4.5.

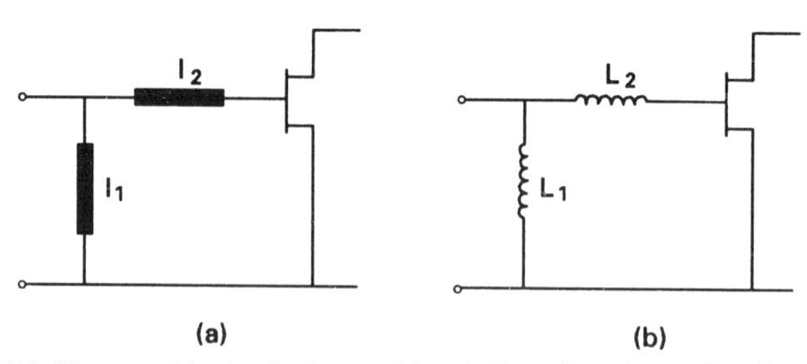

FIG. 15.3. Two possible circuits for matching the input impedance of a microwave transistor.

(the transistor) to the $G = 1.0$ circle, to the point C. In moving to the point C we move a distance l_2,

$$l_2 = (0.226 - 0.056)\lambda = 0.170\lambda$$

At C the admittance is $1.0 + j6.5$, so that the shorted stub has to present an admittance of $-j6.5$. The short circuit is given by an infinite conductance, the point at the bottom of the diagram, so that the length l_1 is given by

$$l_1 = (0.250 - 0.225)\lambda = 0.025\lambda$$

(b) The matching circuit is shown in Fig. 15.3(b). This time it is necessary to use the Smith chart as an impedance chart to determine L_2 and to use the chart as an admittance chart to determine L_1. On the admittance chart it is still necessary to move backward from the termination to the $G = 1.0$ circle, but this time we move along the R = constant line on the impedance chart. This operation is most easily performed by transposing the $G = 1.0$ circle onto the impedance chart. This has been drawn on Fig. 15.2. Then we move from the point A along the $R = 0.20$ line to the point D. A positive reactance is added in series with the terminating impedance to make this move. The reactance is given by

$$X = (2.8 - 0.4)Z_0 = 2.4Z_0$$

Therefore,

$$L_2 = \frac{2.4 \times 50}{2\pi \times 4.0 \times 10^9} = 4.77\,\text{nH}$$

The admittance at point D is given by point E which has the value

$$Y = (1.0 + j2.0)Y_0$$

Therefore, a susceptance of $-j2.0Y_0$ needs to be added by the inductance L_1. Therefore

$$L_1 = \frac{0.5 \times 50}{2\pi \times 4.0 \times 10^9} = 1.00\,\text{nH}$$

PROBLEM 5.7. (a) Sketch graphs of the amplitude of the different components of the field of the TE_{10}-mode against position inside the waveguide.

(b) Sketch graphs of the amplitude of the components of the wall current of the TE_{10}-mode against position in the waveguide.

Answer. (a) The field components of the TE_{10}-mode are given by eqn. (5.37).

The distribution of the components E_y and H_x are the same. They are in time antiphase with one another and 90° out of phase with H_z so that H_x leads H_z by 90° and E_y lags H_z by 90°. Each component is constant with variation in the y-direction so that the variation in the x- and z-directions only is needed. These are plotted for any instant of time in Fig. 15.4.

(b) In section 5.10 it states that: in the broad wall J_x is proportional to H_z and J_z to H_x, and in the narrow wall J_y is proportional to H_z and J_z is proportional to H_y which is zero. Hence the current components are proportional to the magnetic fields.

In the narrow wall, the only current flows perpendicular to the direction of propagation and its only variation is the normal sinusoidal variation in the direction of propagation.

In the broad wall, the currents are proportional to the magnetic fields whose variations in the plane of the wall are given in Fig. 15.4.

PROBLEM 5.8. Which of the sections of slotted waveguide shown in Fig. 5.6 radiate for the dominant mode, on the principle that slots which do not cut lines of current flow do not radiate? In particular, which section preferentially accepts the TE_{01}-mode without radiating, whilst radiating for other modes?

Answer. The slot in the waveguide section shown in Fig. 5.6(a) cuts the current component J_z of the dominant mode and it will radiate.

The slot in the waveguide section shown in Fig. 5.6(b) cuts the current component J_x and it will radiate.

The slot in the waveguide section shown in Fig. 5.6(c) also cuts the current component J_x, but the slot is on the centre line of the waveguide where $J_x = 0$ so that the slot will not radiate.

The slot in the waveguide sections shown in Fig. 5.6(d) and (e) both cut the current component J_y and they will both radiate.

The slot in the waveguide section shown in Fig. 5.6(f) is parallel to the current in the narrow wall and does not cut any current, so that it does not radiate.

The TE_{01}-mode is the mode occupying a perpendicular orientation in the waveguide compared with the TE_{10}-mode. Hence it can be seen that the waveguide slotted sections which will not radiate for this mode are given in Fig. 5.6(a) and (e). However, Figs. 5.2 and 5.3 show that all other possible modes have current components which cut the slot shown in Fig. 5.6(a).

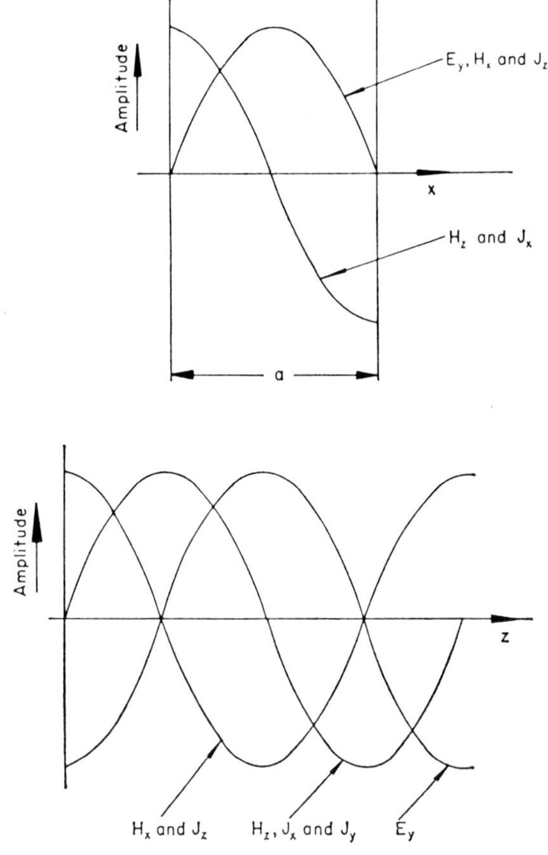

FIG. 15.4. Field component amplitudes for the TE_{10}-mode in rectangular waveguide.

(The TE_{01}-mode of propagation is used because it provides low-loss propagation in oversize waveguide, and slots similar to that shown in Fig. 5.6 (a) can be used to filter out *other* unwanted modes.)

PROBLEM 6.5. A circular polarizer can be made from material with directional dielectric properties. In the plane of the material, for an electromagnetic plane wave with its electric field parallel to the plane of the material, the permittivity of the material appears to be $\varepsilon_0\varepsilon_r$. Perpendicular to the plane of the material, for an electromagnetic plane wave with its electric field perpendicular to the plane of the material, the permittivity appears to be ε_0. Write down an expression for the field components of the plane wave which entered the material as a linearly polarized plane wave with its electric field at an angle of 45° to the plane of the material. What length of material is required to give an output which is a circularly polarized plane wave? What is the output if twice this length of material is used? In the latter condition, what is the effect of altering the angle between the plane of polarization of the incident wave and the plane of the material?

Answer. Let the incident plane wave have an electric field strength, $E_0 \exp j(\omega t - \beta z)$. It may be resolved into two components parallel and perpendicular to the dielectric material where θ is the angle between the electric field and the plane of the dielectric material.

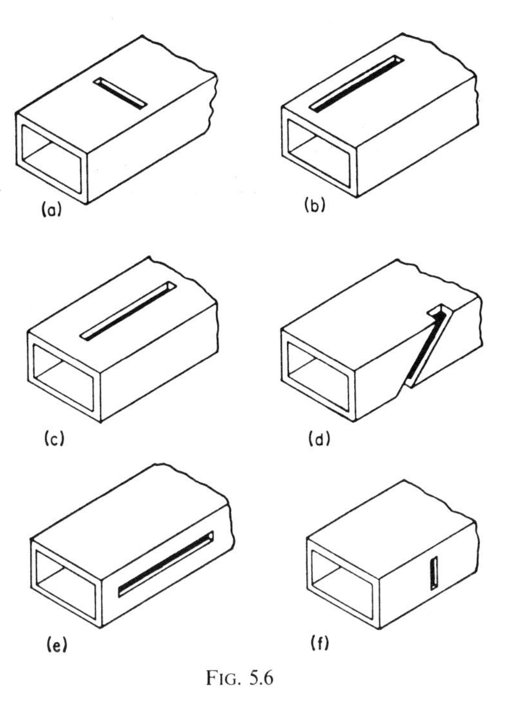

FIG. 5.6

Then

$$\left.\begin{aligned} E_{\parallel} &= E_0 \sin\theta \exp j(\omega t - \beta_{\parallel} z) \\ E_{\perp} &= E_0 \cos\theta \exp j(\omega t - \beta_{\perp} z) \end{aligned}\right\} \tag{15.5}$$

where

$$\beta_{\parallel} = \omega\sqrt{(\mu_0\varepsilon_0\varepsilon_r)} = \frac{\omega}{c}\sqrt{\varepsilon_r}$$

and

$$\beta_{\perp} = \omega\sqrt{(\mu_0\varepsilon_0)} = \frac{\omega}{c}$$

Then at a distance z

$$\left.\begin{aligned} E_{\parallel} &= E_0 \sin\theta \exp j\omega\left(t - \frac{\sqrt{\varepsilon_r}}{c} z\right) \\ E_{\perp} &= E_0 \cos\theta \exp j\omega\left(t - \frac{z}{c}\right) \end{aligned}\right\} \tag{15.6}$$

If $[\sqrt{(\varepsilon_r)} - 1]\omega z/c = \frac{1}{2}\pi$, the two components of the wave are 90° out of phase and the sum effect is a combination of a linearly polarized wave and a circularly polarized wave which is called an elliptically polarized wave. If the angle $\theta = 45°$, the two components of the wave are equal and the resultant is circular polarization. Hence the length of the section of dielectric is

$$z = \frac{\pi c}{2[\sqrt{(\varepsilon_r)} - 1]\omega} \tag{15.7}$$

If twice this length of material is used, the two components of the wave are 180° out of phase or they could be considered to be in phase but with the relative direction of the parallel component reversed. The resultant is a linearly polarized plane wave whose electric field lies in a plane at an angle 2θ to that of the incident plane wave. If $\theta = 45^\circ$, the output plane wave is linearly polarized in a direction perpendicular to that of the input. Altering the angle of the dielectric alters the angle of the plane of polarization of the output plane wave relative to that of the input. (If such a system is mounted in waveguide, the circular polarizer is called a *half-wave plate* and the longer section is called a *full-wave plate* as described in section 6.12.)

PROBLEM 6.10. It is desired to design a cylindrical cavity to be resonant to two frequencies, one twice the other. Suggest approximate values for d/l and fd and identify the modes for a system to satisfy this requirement.

Answer. In order to answer this problem it is necessary to refer to the mode chart (Fig. 6.16). For one cavity to be resonant at two different frequencies, d and l will be specified and f_0 will be different. Hence the condition is given by any vertical ordinate, constant $(d/l)^2$, on the diagram which cuts two mode lines, one at twice the vertical distance of the other. The other constraint is that it is undesirable to operate any cavity close to a condition when two different modes may be supported simultaneously. Applying these considerations and studying the diagram shows that one possible answer lies in the region of $(d/l)^2 \approx 0.7$ and the resonant modes are TE_{111} and TE_{311}, $(fd)^2 \approx 4.2$ and 16.8×10^{16}.

PROBLEM 7.6. Calculate the VSWR in the air space for a plane wave in air normally incident onto the plane surface of a medium of conductivity σ.

Answer. In order to simplify the algebra, it is necessary to assume that the conducting medium is a medium of high conductivity where

$$\sigma \gg \omega\varepsilon$$

Then the fields in the material are

$$E_x = (1+j)\sqrt{\left(\frac{\omega\mu_0}{2\sigma}\right)}H_0 \exp(j\omega t - \alpha z - j\alpha z) \tag{7.14}$$

$$H_y = H_0 \exp(j\omega t - \alpha z - j\alpha z) \tag{15.8}$$

Consider a situation similar to that posed in the example in section 2.13. Medium 1 is the air space with the same properties as free space and medium 2 is the conducting medium. Use the notation for the forward and reverse waves in medium 1 given there and the transmitted wave in medium 2 is given by eqns. (7.14) and (15.8).

The boundary relationships are given by eqns. (2.45) and (2.47), hence eqn. (2.50) applies,

$$E_f + E_r = E_x \tag{2.50}$$

and also

$$H_f + H_r = H_y + J_p\delta \tag{15.9}$$

If the material is a good conductor, the term $J_p\delta$ cannot be neglected; however, there is complete reflection of the wave at the surface and the VSWR $= \infty$.

Otherwise δ can be made so small that even the contribution of the current to the boundary condition may be neglected and then eqn. (2.49) also applies:

$$H_f + H_r = H_y \tag{2.49}$$

Then from eqn. (2.50 a)

$$\eta H_f - \eta H_r = (1+j)\sqrt{\left(\frac{\omega\mu_0}{2\sigma}\right)}H_y \tag{15.10}$$

Solving eqns. (2.45 a) and (15.10) gives

$$2H_f = \left[1 + \frac{(1+j)}{\eta}\sqrt{\left(\frac{\omega\mu_0}{2\sigma}\right)}\right]H_y$$

$$2H_r = \left[1 - \frac{(1+j)}{\eta}\sqrt{\left(\frac{\omega\mu_0}{2\sigma}\right)}\right]H_y$$

Let

$$\frac{1}{\eta}\sqrt{\left(\frac{\omega\mu_0}{2\sigma}\right)} = \sqrt{\left(\frac{\omega\varepsilon_0}{2\sigma}\right)} = \frac{1}{g}$$

Then, similar to eqn. (2.53)

$$\frac{E_f}{E_r} = \frac{g + 1 + j}{g - 1 - j}$$

But according to the initial assumption

$$g \gg 1$$

which gives the relationship

$$E_r \approx E_f$$

and all the incident microwave power is reflected giving

$$S = \infty$$

PROBLEM 7.7. Discuss in terms of skin depth and calculate approximate sizes, guessing values for material parameters where appropriate, for the following:

laminated transformer cores for use at mains supply frequency;
laminated transformer cores for use at high frequencies;
copper-plated steel wire for use at high frequencies;
microstrip;
thin wall waveguide;
copper-plated waveguide.

Answer.

Laminated transformer cores for use at mains supply frequency:
Iron for a mains frequency transformer core might have the properties: $\mu_r = 2 \times 10^3$, $\sigma = 8 \times 10^6$ S/m. The mains supply frequency is taken to be 50 Hz. The skin depth is given by eqn. (7.19).

$$z_0 = \sqrt{\left(\frac{2}{\omega\mu\sigma}\right)} \tag{7.19}$$

Substituting values into the equation gives $z_0 = 0.564$ mm. If the transformer core is made of thin laminations of iron, insulated from one another and not thicker than 0.5 mm, there will not be any appreciable loss of magnetic field due to eddy currents.

Laminated transformer cores for use at high frequencies:
At 10 kHz the transformer core is made of special high permeability alloy. Typical properties are: $\mu_r = 5 \times 10^4$, $\sigma = 1 \times 10^6$ S/m. Substituting these values into eqn. (7.19) gives $z_0 = 22.5\ \mu$m. This shows that for the high-frequency transformer core, the laminations have to be about a tenth of the thickness of those of the mains frequency transformer.

Copper-plated steel wire for use at high frequencies:
It might appear that, because of the skin effect, a cheap conductor could be made from copper-plated steel wire. If so, it is necessary to ensure a sufficient thickness of copper coating on the steel wire so that there is negligible current flowing in the steel. The copper plating needs to be greater than the skin depth in thickness. As the current is entirely confined to the copper, the skin depth in a copper wire is needed. Using the properties of copper given, the skin depth is:

$$\begin{aligned} z_0 = {} & 0.71 \text{ mm at } 10\text{ kHz},\\ & 0.225 \text{ mm at } 100\text{ kHz},\\ & 71\ \mu\text{m at } 1\text{ MHz},\\ & 22.5\ \mu\text{m at } 10\text{ MHz}. \end{aligned}$$

These results show that such plating might be suitable at frequencies above 10 MHz, i.e. at the higher radio frequencies.

Microstrip:
In the microwave frequency range, the skin depth in copper varies from 2.25 μm at 1 GHz to 0.225 μm at 100 GHz. Therefore, the conductors in microstrip line need to be only the thickness of copper foil or normal electroplating.

Thin wall waveguide:

As seen above, as far as microwave fields are concerned, the waveguide wall needs to be only the thickness of normal electroplating. Hence there are no electrical disadvantages in thin wall waveguide.

Copper-plated waveguide:

From the principles of skin depth, there are no reasons why a copper-plated skin on the inside of a waveguide should not behave like copper waveguide. (However, there are other considerations which result in the attenuation due to copper-plated waveguide being higher than that of solid copper waveguide.)

PROBLEM 8.6. A plane wave in free space is normally incident onto the plane face of a semi-infinite ferrite medium magnetized normally to the plane face. Find an expression for the VSWR of the standing wave in the free space.

Answer. The solution to this problem is similar to the example given in section 2.13. The conditions in the ferrite material are given by eqns. (8.20) to (8.23). Using the notation of section 8.6, these equations become

$$E_x^+ = jE_y^+ \tag{8.20}$$

$$E_x^- = -jE_y^- \tag{8.21}$$

$$\eta^+ H_y^+ = E_x^+ \tag{8.22a}$$

$$\eta^- H_y^- = E_x^- \tag{8.22b}$$

$$\eta^+ H_x^+ = -E_y^+ \tag{8.23a}$$

$$\eta^- H_x^- = -E_y^- \tag{8.23b}$$

The two waves defined by eqns. (8.20) to (8.23) are the waves transmitted through the boundary into the ferrite. Let the incident and reflected waves in free space be the same as those defined in section 2.13, except that due to the rotational effect of the ferrite there is a reflected wave component perpendicular to the incident wave. Let the perpendicular component of the reflected wave have the fields E_s and H_s. If we define $E_x^+ = E_1$ and $E_x^- = E_2$ then substitution into eqns. (8.20) to (8.23) gives

$$E_x^+ = E_1$$

$$H_y^+ = \frac{E_1}{\eta^+}$$

$$E_y^+ = -jE_1$$

$$H_x^+ = \frac{jE_1}{\eta^+}$$

$$E_x^- = E_2$$

$$H_y^- = \frac{E_2}{\eta^-}$$

$$E_y^- = jE_2$$

$$H_x^- = -\frac{jE_2}{\eta^-}$$

Summation of all the field components of the different waves at each side of the boundary give for the electric components:

x-direction: $$E_1 + E_2 = E_f + E_r \tag{15.11}$$

y-direction: $$E_1 + E_2 = E_s \tag{15.12}$$

and for the magnetic field components:

x-direction: $$\frac{E_1}{\eta^-} - \frac{E_2}{\eta^-} = -\frac{E_s}{\eta} \tag{15.13}$$

y-direction:

$$\frac{E_1}{\eta^+} + \frac{E_2}{\eta^-} = \frac{E_f}{\eta} - \frac{E_r}{\eta} \tag{15.14}$$

From eqns. (15.12) and (15.13)

$$-E_1 - E_2 = \frac{\eta}{\eta^+}E_1 - \frac{\eta}{\eta^-}E_2$$

Therefore

$$\frac{E_2}{E_1} = \frac{[(\eta/\eta^+) - 1]}{[(\eta/\eta^-) - 1]} \tag{15.15}$$

Equation (15.11) gives

$$E_{max} = E_f + E_r = E_1 + E_2$$

and eqn. (15.14) gives

$$E_{min} = E_f - E_r = \frac{\eta}{\eta^+}E_1 + \frac{\eta}{\eta^-}E_2$$

Then

$$S = \frac{E_{max}}{E_{min}} = \frac{E_1 + E_2}{(\eta/\eta^+)E_1 + (\eta/\eta^-)E_2}$$

and substitution from eqn. (15.15) and simplifying gives

$$S = \frac{\eta(\eta^+ + \eta^-) - 2\eta^+\eta^-}{2\eta^2 - \eta(\eta^+ + \eta^-)}$$

The perpendicular component of the reflected wave does not contribute to the VSWR in the free space provided that the standing wave detector is directional and that it is arranged to detect the maximum amplitude of the incident wave. The meanings of all the symbols that have not been explained here are given in Chapters 2 and 8.

PROBLEM 9.7. Discuss whether a uniform plasma can be considered to be exactly similar to a uniform dielectric material of relative permittivity less than one. In particular consider whether the group velocity is the same as the phase velocity. Plot the relationship between plane-wave wavelength and frequency and compare it with the relationship between waveguide wavelength and frequency for air-filled waveguide.

Answer. The propagation constant in a uniform lossless plasma is given by eqn. (9.17)

$$\beta = \frac{\omega}{c}\sqrt{\left[1 - \left(\frac{\omega_p}{\omega}\right)^2\right]} \tag{9.17}$$

Substitution for β and ω in terms of wavelength in eqn. (9.17) and defining a plasma wavelength λ_p to be the characteristic wavelength corresponding to the plasma frequency gives

$$\frac{1}{\lambda_p^2} + \frac{1}{\lambda^2} = \frac{1}{\lambda_0^2} \tag{15.16}$$

It is seen that eqn. (15.16) is of the same form as eqn. (5.16) where the propagating wavelength through the plasma is equivalent to waveguide wavelength and the plasma wavelength is equivalent to waveguide cut-off wavelength. There is a correspondence between electromagnetic propagation along waveguide and through plasma. Plasma exhibits the properties of cut off, etc., and hence there is a phase velocity greater than the speed of light and a group velocity less than the speed of light.

Substitution from eqn. (9.17) into eqns. (3.5) and (3.9) gives

$$v_p = \frac{c}{\sqrt{[1 - (\omega_p/\omega)^2]}}$$

$$v_g = c\sqrt{[1 - (\omega_p/\omega)^2]}$$

The graphs for the answer are plotted by substituting numbers into eqn. (15.16).

APPENDIX 1

Physical Constants

Velocity of light	$c = 2.998 \times 10^{8}$ m/s
Charge of the electron	$e = 1.602 \times 10^{-19}$ C
Mass of the electron	$m = 9.109 \times 10^{-31}$ kg
Planck's constant	$h = 6.626 \times 10^{-34}$ J s
Boltzmann's constant	$k = 1.380 \times 10^{-23}$ J/K
Permeability constant	$\mu_0 = 4\pi \times 10^{-7}$ H/m
Permittivity constant	$\varepsilon_0 = \dfrac{1}{c^2\mu_0} \approx \dfrac{1}{36\pi \times 10^9}$ F/m

APPENDIX 2

Notation

A	arbitrary constant; voltage attenuation
$\boldsymbol{A}$	arbitrary vector
a	input wave amplitude; broad dimension of rectangular waveguide; inner radius of round waveguide; inner radius of outer of coaxial line
$\boldsymbol{a}$	area
B	arbitrary constant
$\boldsymbol{B}$	arbitrary vector; magnetic flux density
b	susceptance; output wave amplitude; narrow dimension of rectangular waveguide; outer radius of central conductor of coaxial line
C	arbitrary constant; capacitance; capacitance of a transmission line
$\boldsymbol{C}$	arbitrary vector
c	speed of light; length of rectangular waveguide cavity
D	arbitrary constant
$\boldsymbol{D}$	electric flux density
d	differential coefficient; diameter
$\boldsymbol{E}$	electric field
e	base of napierian logarithms; electronic charge
$\boldsymbol{F}$	arbitrary vector
f	frequency; arbitrary function
f_0	resonant frequency
G	leakage conductance of a transmission line; gain
g	constant for a conducting medium
$\boldsymbol{H}$	magnetic field
H_0	static magnetic field
h	Planck's constant
I	current
$\boldsymbol{J}$	current density; angular momentum
J_n	Bessel function of the first kind, order n
j	$\sqrt{(-1)}$
K	arbitrary constant
k	wave number; stability factor of an amplifier; Boltzmann's constant
L	inductance; series inductance of a transmission line
l	a length; an integer—cavity mode number
$\boldsymbol{M}$	magnetization
M_0	static magnetization
m	an integer—waveguide mode number; constant for a conducting medium; mass of an electron
N	an integer—the number
n	an integer—waveguide mode number
P	power
Q	Q-factor of resonator; charge
R	resistance; series resistance of a transmission line
R_s	equivalent surface resistance
r	radius; radial coordinate in polar coordinate systems

S	voltage standing wave ratio
$\mathbf{S}$	the scattering matrix; Poynting vector
s	a scattering parameter
T	absolute temperature
t	time; transverse; tangential
$\mathbf{u}$	the unit vector
V	potential difference
$\mathbf{v}$	velocity
v_p	phase velocity
v_g	group velocity
v	volume
W	energy density; energy states in matter
X	reactance
x	a dimensional coordinate in the rectangular coordinate system
Y	admittance
$\mathbf{Y}$	arbitrary vector; the admittance matrix
Y_n	Bessel function of the second kind, order n
y	an admittance parameter; a dimensional coordinate in the rectangular coordinate system
Z	impedance
Z_0	characteristic impedance—wave impedance
Z_t	terminating impedance
$\mathbf{Z}$	the impedance matrix
z	an impedance parameter; a dimensional coordinate in the rectangular coordinate system—the direction of propagation; the longitudinal dimension in the cylindrical polar coordinate system—the direction of propagation
z_0	skin depth
α	attenuation constant
β	phase constant
γ	propagation constant; gyromagnetic ratio
δ	loss angle; length; differential operator for a small quantity
$\boldsymbol{\varepsilon}$	permittivity tensor
ε	permittivity
ε_0	permittivity constant
ε_t	diagonal component of the permittivity tensor
ε_z	z-component of the permittivity tensor
ζ	rotation
η	impedance of free space
η_t	cross-diagonal component of the permittivity tensor
θ	angle; angular coordinate in the polar coordinate system
κ	cross-diagonal component of the permeability tensor
λ	wavelength
λ_0	characteristic wavelength
λ_c	cut-off wavelength
λ_g	waveguide wavelength
$\boldsymbol{\mu}$	permeability tensor
μ	permeability; diagonal component of the permeability tensor
μ_0	permeability constant
ν	effective collision frequency of the electrons
ρ	reflection coefficient; charge density; resistivity; surface resistivity
σ	conductivity
τ	time delay
ϕ	angular coordinate in spherical polar coordinate system
χ	magnetic susceptance
ψ	rotation
ω	angular frequency
ω_0	resonant frequency

ω_p	plasma frequency
ω_g	gyrofrequency
Δ	determinant of the scattering matrix
∇	differential coefficient
∂	partial differential coefficient

APPENDIX 3

Circuit Symbols

Coaxial line probe coupled to waveguide

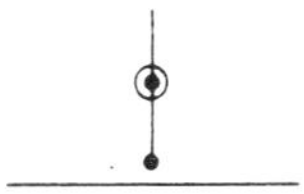

Indicating wavemeter

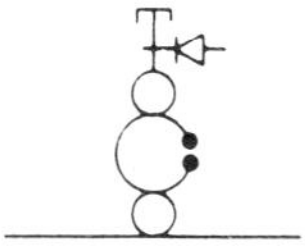

Directional coupler

Matched termination

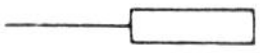

Short-circuit termination

Crystal receiver

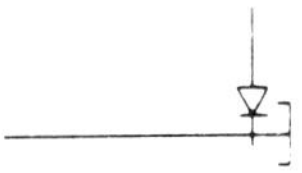

Standing wave detector

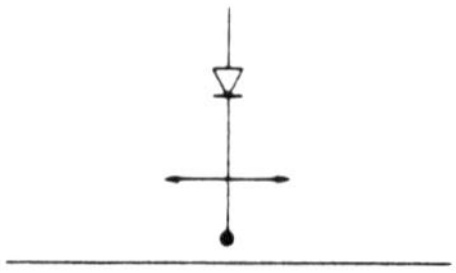

Slotted line probe

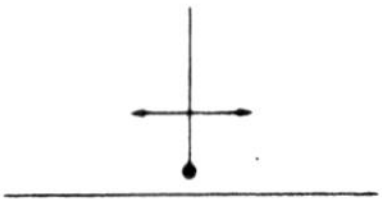

Attenuator

Phase changer

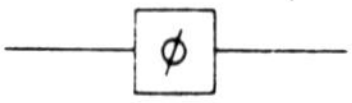

Isolator

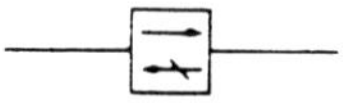

Circulator

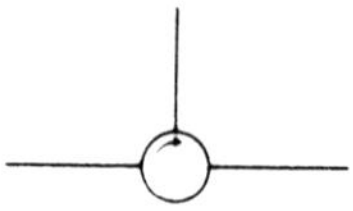

Index